16.25

Seepage
and
Groundwater Flow

WILEY SERIES IN GEOTECHNICAL ENGINEERING

Consulting Editors
T. W. Lambe
R. V. Whitman

Department of Civil Engineering
Massachusetts Institute of Technology

Applications of the Theory of Plasticity in Soil Mechanics

J. Salençon

Ecole Polytechnique
Ecole Nationale des Ponts et Chaussées, Paris

Seepage and Groundwater Flow

K. R. Rushton
S. C. Redshaw

University of Birmingham

Seepage and Groundwater Flow

Numerical Analysis by Analog and Digital Methods

K. R. Rushton
Reader in Civil Engineering
University of Birmingham

S. C. Redshaw
Emeritus Professor in Civil Engineering
University of Birmingham

A Wiley–Interscience Publication

JOHN WILEY & SONS
Chichester · New York · Brisbane · Toronto

Library of Congress Cataloging in Publication Data:
Rushton, K. R.
Seepage and Groundwater Flow.

(Wiley series in geotechnical engineering)
'A Wiley–Interscience publication.'
Bibliography: p.
Includes index.

1. Seepage—Mathematical models. 2. Seepage—Data processing. 3. Groundwater flow—Mathematical models. 4. Groundwater flow—Data processing. I. Redshaw, S. C., joint author. II. Title. III. Series.

TC176.R87   551.4'9   78-23359

ISBN 0 471 99754 4

Set and printed in Great Britain by
Page Bros (Norwich) Ltd, Mile Cross Lane, Norwich

Preface

Seepage and groundwater flow draw together several different disciplines. A number of excellent books describe the basic theory of seepage and groundwater flow, dealing both with the properties of the medium and analytical and graphical methods of solution. Nevertheless, in our opinion, there is a lack of information enabling investigators to obtain solutions to problems of practical interest.

It is with this in mind that our main objective is to demonstrate that, by the use of simple analog and digital computer methods, solutions may be obtained to problems that would otherwise be intractable by formal analytical and graphical means. In spite of the physical complexity of porous flow, the number of basic governing equations which have to be solved are surprisingly few. We have endeavoured to show how these well-known governing equations, which occur in many other branches of mathematical physics, can be solved by simple methods which, with minor modifications, can be applied to almost any problem in seepage and groundwater flow. The methods we describe permit solutions to be obtained for both two- and three-dimensional problems in steady-state and time-variant situations.

This work is the outcome of many years experience in the utilization of analog and digital methods in the solution of a variety of problems in elasticity and fluid motion. The need for a working description of the methods has become apparent whilst teaching postgraduate and post-experience courses and advising industry. The types of problem which we describe are typical of those to which our attention has been drawn during collaborative studies with the water industry.

The book is divided into four parts. The first part considers the fundamental equations for seepage and groundwater flow and the basis of the finite-difference approximations. Seepage is the subject of the second part; both steady and time-variant seepage are considered with special emphasis on the free water surface. The third part is concerned with the analysis of regional groundwater flow; a variety of numerical techniques using either analog or digital computers are presented with both external and internal boundaries considered in detail. In the final part a numerical method of analysing pumping tests is described; this technique is used to interpret pumping test results which cannot be analysed by conventional methods.

We wish to acknowledge the valuable help of Mrs. L. M. Tomlinson, Mr. P. M. Thompson, and Mr. K. S. Rathod in the preparation of the practical examples and computer programs. Also we wish to thank members of the secretarial staff and in particular Mrs. U. M. Brown for the great care taken in preparing the manuscript. We are indebted to many of our colleagues for their advice and encouragement. The support of a Leverhulme Trust Fund Emeritus Fellowship for one of the authors is gratefully acknowledged.

Birmingham
August 1978.

K. R. RUSHTON
S. C. REDSHAW

Contents

PART II. SEEPAGE

PART III. ANALYSIS OF REGIONAL GROUNDWATER FLOW

Symbols

Most of the commonly used symbols are listed below; others are defined where they appear in the text. Any consistent set of units may be used. Dimensions are indicated in square brackets.

a	$\log_e(r)$
A	area $[L^2]$
A, B, C, D, E, F	coefficients of finite-difference equations $[T^{-1}]$
b	breadth of aquifer $[L]$
BH	groundwater potential of previous time step $[L]$
c	coefficient of $\partial h/\partial t$ $[T/L^2]$
C	capacitance, well loss constant $[T^2/L^5]$
d	day $[T]$
D	depth of aquifer $[L]$
E_p, E_a	potential and actual evaporation $[L/T]$
F	well factor
$F(1), F(2), F(3)$	scaling factors between physical and electrical quantities
g	acceleration due to gravity $[L^2/T]$
h	groundwater potential $[L]$
H	groundwater potential on free surface $[L]$
$H(\ )$	equivalent horizontal hydraulic resistance $[T/L^2]$
$HU(\ ), HL(\ )$	upper and lower horizontal hydraulic resistance $[T/L^2]$
i	hydraulic gradient
I	electrical current, nodal position
J	nodal position
k	permeability or hydraulic conductivity $[L/T]$
k_s	modified permeability close to singularity $[L/T]$
k'	permeability of leaky stratum $[L/T]$
l	length of river represented by a nodal point $[L]$
L	suffix indicating lower part of aquifer
m	thickness of aquifer $[L]$
m'	thickness of leaky stratum $[L]$
M	iteration counter
n	normal to boundary, power
N	number of nodal points
p	porosity (Part II); period of wave $[T]$; penetration of well $[L]$

P precipitation [L/T]

q inflow rate per unit area [L/T]

Q flow rate, discharge rate from abstraction well [L^3/T]

r radial coordinate [L]

r_r radius of river [L]

r_w radius of well [L]

R electrical resistance

R_L electrical resistance representing leakage

R_w well resistance

RC   root constant [L]

RECH   recharge [L/T]

RO   runoff [L/T]

RS total inflow to aquifer [L/T]

s drawdown [L]

s_f free-surface drawdown [L]

s_w drawdown in well [L]

S storage coefficient

S_c confined storage coefficient

S_s specific storage [L^{-1}]

S_y specific yield

SMD   soil-moisture deficit [L]

t time [T]

t_e electrical time [T]

T transmissivity [L^2/T]

$TX(\ )TY(\ )$ transmissivities in x- and y-directions [L^2/T]

u pumping test parameter, $r^2S/4Tt$

U   upper part of aquifer

v, V velocity [L/T]

v_x, v_y, v_z velocities in x-, y-, and z-directions [L/T]

V electrical potential, volume [L^3]

x, X coordinate [L]

y, Y coordinate [L]

z, Z coordinate [L]

α, β angle below horizontal of free-water surface in x- and y-directions

δ increment

θ angle

ρ density

ψ stream function [L^2/T]

ϕ velocity potential [L^2/T]

ω over-relaxation factor

Δ increment

∇ Laplace operator

PART I

General Considerations

CHAPTER 1

Introduction

1.1 Practical Implications of Seepage and Groundwater Flow

The underground flow of water has significant practical consequences and refers to a wide range of problems. It is usual to divide underground flow problems into two categories, seepage and groundwater flow. Seepage problems are generally classified as the percolation of water through dams, river banks or into excavations. The dimensions of the areas through which seepage takes place will usually be measured in tens or hundreds of square metres. On the other hand, groundwater flow problems are concerned with the natural process of the infiltration and subsequent flow of water through pervious strata until discharge occurs at springs, rivers or abstraction wells. A regional groundwater flow study will often cover a region of tens or hundreds of square kilometres. Though these two types of problem have different practical implications, mathematical formulation and numerical techniques for their solution are similar.

Controlling *seepage* is of great importance. One of the objectives is to prevent, or at least reduce to a small value, the magnitude of the seepage flow. For example, losses due to flow through dams must be minimized and seepage flow into excavations and basements which are below the water table must be controlled.

The control of seepage flow is not the only reason for analysing seepage problems. Other consequences such as excessive saturation, seepage forces and uplift pressures can lead to failures. Indeed, the large deleterious effects seem out of all proportion to the slow movement of water through pores in the soil.

Frequently, seepage forces are only considered when they reach their final steady-state value. However, whenever seepage paths are disturbed the transition from the original to the new conditions can take a significant time. In certain instances critical conditions can occur during this change and it is therefore important to investigate time-variant seepage behaviour.

Groundwater flow is of considerable significance since aquifers are an important source of water. Because of the low velocities of the water, misuse of the aquifer can have serious long-term effects. If the consequences of abstraction from an aquifer are examined, the water levels in the vicinity of the wells and the quality of the water may be acceptable yet there may be detrimental effects elsewhere in the aquifer. For example, poor-quality water may be drawn slowly into the aquifer, advancing a few hundred metres each year. Then, when this poor-quality water reaches a distance of about 1 km from the well, it will be drawn in much more rapidly. Once contamination has occurred it is difficult to force the poor-quality water out of the aquifer.

Understanding the flow regime within an aquifer is therefore of great importance. Furthermore it is advantageous to be able to *predict* the effect of changes of parameters, such as the abstraction rate, on groundwater levels within the aquifer and on the base flow to rivers. Any technique of analysis must therefore be capable of helping to identify the aquifer flow mechanisms by comparing with historic data and then predicting the future aquifer behaviour. The effect of *time* in the study of regional groundwater flow cannot be over-stressed.

In regional groundwater flow studies it is important to examine the flow mechanism close to the abstraction wells. Therefore techniques are needed to study the detailed flow pattern close to the well. This is usually achieved using conventional methods of pumping-test analysis based on curve-matching techniques. Due to the difficulty of obtaining analytical solutions, these methods often do not allow a fully detailed representation of the many factors which have a substantial influence on the flow to a well. Therefore, in the analysis of the drawdown due to abstraction wells, a flexible method which can represent all the different features is essential.

1.2 Range of Techniques

A wide range of techniques are available for the study of seepage and groundwater flow. Each of the techniques is suited to a particular class of problems and may well lead to inaccurate results when applied in other situations. It is not the purpose of this book to present a critical or exhaustive review of the different techniques; reference should be made to the literature, quoted below, which indicates the type of problem to which each method is best suited.

Analytical

Certain analytical solutions are available for seepage and groundwater flow problems. In a few cases analytical expressions can be obtained by direct integration of the appropriate differential equation. However, for most situations alternative mathematical techniques are used. Often the solution is in terms of some form of infinite integral or series. A range of problems can also be analysed by conformal transformation techniques. Many analytical techniques are described by Polubarinova-Kochina (1962), Harr (1962), De Wiest (1965), and Carslaw and Jaeger (1959).

Though a considerable number of analytical solutions have been derived, there are many practical situations for which analytical solutions cannot be obtained. This arises because seepage and groundwater problems have a number of non-linear features which are not amenable to inclusion in analytical solutions, such as variable permeability, moving boundaries, or long-term time-dependent effects.

Flow nets

Much of the understanding of seepage problems has been gained through the techniques of constructing flow nets. A thorough description of the methods of flow-net construction is given by Cedergren (1977) and his book also shows how flow nets can be used to investigate a wide range of seepage problems.

There are, unfortunately, many limitations in flow-net construction techniques. Significant changes in permeability present difficulties in the construction of flow nets, and moving boundaries cause even greater complications, but perhaps the greatest limitation is that it is not possible to obtain direct estimates of the quantity of fluid flowing.

Physical models

Sand-tank models provide a useful means of examining the flow of water in soils (Prickett, 1975). The actual physical shape of the medium is modelled and the boundary conditions are simulated as heads of water or as drains. Measurement can either be made by means of piezometers or by using dye to trace the streamlines. Though a sand tank does give some indication of the flow pattern, the technique suffers from a number of disadvantages, including the difficulty of representing the correct permeability, capillary effects in the unsaturated region, and poor representation of time-variant effects.

An alternative method which overcomes some of these difficulties is the viscous-flow model. The basis of the method is that the equations for viscous flow between parallel plates and the equations of groundwater flow are identical. Both horizontal and vertical flow models can be used (Bear, 1972; Prickett, 1975). Studies have included seepage through dams, seawater intrusion, regional groundwater flow, and multiple-layered aquifers. Steady and non-steady effects can be included. Groundwater potentials are measured by means of piezometers and streamlines can be identified by injecting dyes. Major disadvantages of this technique are the complicated construction and operation procedures. For example, the viscosity of the fluid varies significantly with temperature and therefore a careful control of the temperature is essential. Despite these limitations, the viscous-flow model has been useful in many practical investigations.

Mathematical models

Much of the development in recent years in the analysis of seepage and groundwater flow has been by means of mathematical models; an excellent review of the methods is given by Prickett (1975). In a mathematical model the equations describing the flow through an aquifer are solved with the appropriate initial and boundary conditions. Mathematical models are usually based on either *analog* or *digital* computers.

There are three main types of *analog model*. The first uses *conducting paper* (Stallman, 1959). A two-dimensional section of the aquifer is represented by a scaled model cut from paper on which there is a conducting layer. The equation for the flow of electricity through the conducting medium is identical to the equation for the flow of water through the soil. Limitations of the conducting-paper model are that variable permeability is difficult to represent and that it is not convenient to represent time-variant behaviour. Variable conductivity and lack of isotropy in the paper also cause difficulties. The *electrolytic tank* is similar to the conducting-paper technique but with the advantage that three-dimensional behaviour can be represented (Prickett, 1975; De Brine, 1970). However, it has similar limitations to the conducting-paper method but with the added disadvantage that a more difficult alternating current measuring system is required.

Resistance and *resistance–capacitance* electrical networks represent the third type of analog model (Karplus, 1958; Rushton and Bannister, 1970; Prickett, 1975). With these analog computers the equation of the electrical model is identical to a finite-difference form of the appropriate differential equation. Wide variations in permeability and storage coefficients can be included and the model can be time-dependent. For relatively straightforward problems the network and associated input and measuring equipment are very simple, although for complex regional groundwater flow problems sophisticated ancillary equipment is required.

Digital mathematical models are based on a wide variety of techniques. In each case they are essentially alternative methods of solving simultaneous equations which represent the flow process within the aquifer. The earliest methods were based on finite-difference approximations and solutions to seepage problems were obtained by the hand calculations of the relaxation method (Allen, 1954). In recent years a wide range of finite-difference techniques suited to the digital computer have been devised. The mathematical basis of many of these techniques is described by Remson *et al.* (1971) and detailed explanations of certain techniques including program listings are given by Prickett and Lonnquist (1971), Thomas (1973), and Trescott *et al.* (1976). A comparison of various techniques can be found in Kitching *et al.* (1975). Variable parameters and time-dependent effects can be included without difficulty.

As an alternative to the finite-difference approach, the flow mechanism can be described by a finite-element formulation. This is based on a variational approach and makes extensive use of matrix techniques. Descriptions of the technique can be found in Zienkiewicz (1977) and Pinder and Gray (1977). A useful comparison between finite-difference and finite-element approaches has been made by Davis (1975).

A third digital computer technique has recently been introduced, called the boundary integral equation method. The essential feature of the method is that solutions are obtained in terms of the functions at certain positions on the boundary of the field (Liggett, 1977).

In each of the mathematical model techniques, approximations are introduced by describing the continuous functions of seepage and groundwater potential in terms of values on some form of discrete grid. Furthermore, approximations are introduced in the digital approach by dividing the time into discrete steps. With care, the errors due to these approximations can be made sufficiently small to be neglected.

1.3 Reasons for Choosing Finite Differences

Throughout this book solutions to practical problems will be obtained using finite-difference techniques which have been chosen for the following reasons.

Finite-difference techniques are fundamentally very simple. Indeed, they are used as the means of first introducing calculus to students, since infinitesimal calculus is the limit of finite-difference approximations as the interval between discrete points becomes vanishingly small. Because of the basic simplicity of the finite-difference approach, the approximation leads directly to simultaneous equations without the need to introduce further mathematical manipulations.

Secondly, the equations resulting from the finite-difference approximations can be solved using either analog or digital computers. The resistance or resistance–capacitance network can be arranged to be directly analogous to the finite-difference equations and therefore solutions for the groundwater potential or the quantity flowing can be obtained directly. Digital computer solutions to the finite-difference equations can also be obtained by a variety of methods without the use of matrices.

From a number of post-experience courses and numerous contacts with industry, it has become apparent that those involved in seepage and groundwater investigations, whether their training is in geology, engineering or mathematics, are able to understand and use finite-difference techniques. One important reason is that with both analog and digital computer techniques, the geometry of the field is maintained in the solution. This is particularly important in digital computer solutions where the regional parameters are usually stored in two-dimensional arrays which correspond directly to the two-dimensional grid. One consequence is that solutions can often be obtained on digital minicomputers.

A wide variety of techniques are available for formulating and solving the finite-difference equations on digital computers (Remson et al., 1971). In this book attention will be confined to the simpler techniques which, though they may not be the most efficient, take only a few lines of computing coding and are therefore simple to understand. Particular emphasis will be placed on the *point successive over-relaxation* technique.

Techniques such as successive over-relaxation are iterative techniques which converge to the correct solution of the finite-difference equations. Such iterative techniques are convenient for seepage and groundwater flow analyses since they allow non-linear effects to be included in the solutions. For example, if the storage coefficient is a function of the saturated depth, the actual value of the storage oefficient will depend on the unknown groundwater potential. Within the iterative calculation improved values of the storage coefficient can be included as the calculation converges to the correct values of the groundwater potentials.

Another advantage of the finite-difference techniques is that conditions on external or internal boundaries can be changed directly. Therefore, if there is uncertainty about the nature of a boundary condition, two successive runs can be made with alternative conditions on the boundary. Another instance is the representation of an intermittent river. If the groundwater potentials in an aquifer are such that they cannot maintain the flow in the river, then the conditions must change. This can be represented in the electrical analog solution by a switch and in the digital computer program by a conditional statement.

1.4 Accuracy

Care must be taken to identify all the sources of inaccuracy in assessing the reliability of a mathematical model solution of seepage or groundwater flow.

It is usual to suggest that the major source of inaccuracy in the modelling of seepage and groundwater flow problems is the lack of precise information about the permeability distribution. Thus Cedergren (1977) states that 'the answers are no better than the input'. This statement suggests that an uncertainty in the value of the permeability leads to similar uncertainty about the predicted groundwater flows. This is not necessarily correct. Indeed in a number of studies it has been found that tenfold changes in the permeability of the confined portion of an aquifer leads to only minor differences in the groundwater potentials and the flow. On the other hand, in the unconfined region of an aquifer small changes in the permeability could cause significant variations in the flow. The only means of determining the dependence of the flow on the permeability, or other parameters, is to carry out a *sensitivity analysis*. This requires that separate solutions be obtained with a range of possible values of the parameters. From these different solutions it becomes apparent whether further fieldwork is required.

Since many of the important parameters in seepage and groundwater problems are open to considerable uncertainty, it is often suggested that approximate methods of analysis are adequate. Consequently the problem is modified to conform to available analytical solutions. This can lead to results that are seriously in error. For instance, the flow pattern for a problem with variable permeability may be totally different from the flow patterns when the average of the permeabilities is taken. Alternatively, the lack of precise input data is used as the reason for using inexact numerical techniques. However, such techniques may lead to results which are seriously in error, particularly if the techniques do not satisfy the conditions of conservation of mass. One important feature of the finite-difference approximations which are used in the following chapters is that conservation of mass is automatically satisfied.

Potential sources of error in numerical solutions

Because of the large number of simultaneous equations resulting from the finite-difference approximations, it can be difficult to obtain solutions. With the electrical analog method, electrical circuits are used which have identical equations to those of the finite-difference equations. Solutions can then be obtained automatically. However, tolerances of the components and inadequate experimental techniques can lead to errors but these errors can be made small enough to be neglected if adequate precautions are taken.

When digital computers are used to solve the finite-difference equations, the library routines for the solution of simultaneous equations are not usually sufficiently flexible or economical in computing time. Therefore various forms of iterative techniques are introduced. Certain of the standard approaches have proved to be unreliable for seepage and groundwater problems but this is not surprising when it is recognized that very rapid changes occur. For example, an abstraction well is a point sink which suddenly changes from zero to a large flow; such conditions are a vigorous test for any numerical method. Consequently, all numerical solutions must be examined carefully to ensure that adequate solutions to the finite-difference equations are obtained.

Even if satisfactory solutions to the finite-difference equations are obtained, the finite-difference formulation itself may not be an adequate representation of the flow patterns. Inadequate representations are likely to occur when rapid changes in the velocity of flow occur. For instance, velocities are high close to sheet piles or when a drain is adjacent to an impermeable boundary, and the flow patterns cannot be represented adequately by the standard finite-difference approximations. Sudden changes in the flow pattern are called *singularities*; special finite-difference approximations are then required. In the vicinity of singularities, the velocities may be so high that the flow ceases to be laminar and Darcy's law will no longer apply. Correction for non-laminar flow close to singularities can be made but for most regions in which seepage and groundwater flow occur, Darcy's law is valid.

1.5 Scope of the Book

The primary aim of this book is to explain how practical seepage and groundwater flow problems can be analysed using numerical techniques on analog and digital computers. There are already many texts which describe the fundamentals of seepage and groundwater flow. This book will not reproduce the material; instead the reader is referred to Muskat (1937), Todd (1959), Harr (1962), Polubarinova-Kochina (1962), De Wiest (1965), Aravin and Numerov (1965), Childs (1969), Walton (1970), Domenico (1972), Bear (1972), UNESCO (1975), Cedergren (1977), and Bouwer (1978).

Material concerning the fundamental differential equations, boundary conditions, and finite-difference techniques is selected to provide only what is needed for later sections. For a fuller discussion of the fundamental equations, reference should be made to the texts listed above and to the book edited by De

Wiest (1969). Detailed discussions of the finite-difference method can be found in Forsythe and Wasow (1960) and Smith (1965); the material presented on finite differences in Chapter 3 of this book has been chosen to give the reader sufficient information and experience for the solution of practical problems.

Throughout the book techniques are introduced by means of examples. It is the authors' experience that the greatest hindrance to the successful use of a technique is the lack of detailed instructions. Therefore details are frequently given of steps in the technique, and the final numerical results are also quoted. In other examples computer programs, the input data, and the results are presented. It is recognized that each worker has his own distinctive approach to a problem and it is not suggested that the techniques presented in this book are the best or the only ones available. However, the material should form a basis from which to start an analysis and the reader should then feel free to adapt and modify the techniques and programs if desired.

In preparing the examples for this book, many of the problems have been simplified. For instance, straight-line boundaries are often introduced for clarity of the presentation. Modifications to include irregular boundaries are straightforward. In addition, the examples tend to use insufficient mesh intervals; this prevents important results from being swamped by a mass of detail. Nevertheless, recommendations are made of the number of mesh intervals required for practical problems.

Seepage and groundwater flow draw together a number of different disciplines. Consequently the fundamental objective of this work is to provide information and techniques that can be used by all those involved in the study of seepage and groundwater flow. Most of the material and many of the examples have been developed whilst carrying out practical studies in association with industrial organizations. Valuable experience has also been gained by presenting the material to postgraduate and post-experience courses.

A number of important topics on which work is still proceeding are not included in this book. These include flow in the unsaturated zone, the dispersion of contaminants in aquifers, and the management of aquifer resources. Further developments are also occurring in the study of certain topics that are considered in this book. For example, attempts are being made to understand the flow mechanisms in fissured aquifers. Another topic which is being re-examined is the interaction between rivers and aquifers.

The book is divided into four parts and five appendices. The first part considers the fundamental equations and the basis of finite-difference approximations. Seepage is the subject of the second part; both steady and unsteady seepage are considered with special emphasis on the free water surface. The third part is concerned with the analysis of regional groundwater flow; a variety of numerical techniques using either analog or digital computers are presented with both external and internal boundaries considered in detail. In the final part a numerical method of analysing pumping tests is described; this technique can be used to interpret pumping test results which cannot be analysed by conventional methods.

CHAPTER 2

Review of Fundamental Equations

2.1 Introduction

Groundwater flow and seepage are concerned with the flow of fluids through porous media. The topic may be broken down into several subdivisions according to the dimensional character of the flow, its time dependency, the boundaries of the flow region or domain, and the properties of the medium and of the fluid.

All groundwater flow is to a certain extent three-dimensional in space and is also a function of time. This means that the velocity vector at any point has components along three mutually perpendicular axes, x, y, z, of a cartesian coordinate system, thus the velocity is a function of x, y, z, and t.

The difficulty in solving groundwater problems depends on the degree to which the flow is three-dimensional. It is virtually impossible to solve a three-dimensional groundwater problem analytically unless a condition of axial symmetry exists. Although a problem may sometimes be simplified by reducing it dimensionally, errors may be introduced which are difficult to assess.

Groundwater flow may be evaluated quantitatively when the velocity, pressure, density, temperature, and viscosity of water flowing through the ground are known. These characteristics of water are the unknown variables of the problem and may vary from place to place, and in time. If the unknown or dependent variables are assumed to be functions of the space variables x, y, z alone then the flow is defined as being *steady*. However, if the unknowns are also functions of time, then the flow is defined as being *unsteady* or *time-dependent*.

Groundwater flow is *confined* when all the boundaries or bounding surfaces of the space through which the water percolates are fixed in space for different states of flow. Groundwater flow is *unconfined* when it possesses a free surface, the position of which varies with the state of flow.

In studying the flow through porous media, the problem must be specified in two parts. Initially, the type of flow through the media must be defined; and secondly, the conditions at the boundaries of the media must be stated.

This chapter reviews the fundamental equations of seepage and groundwater flow with particular emphasis on the equations used in numerical solutions. For full derivations and discussions of the assumptions and approximations, the reader is referred to the texts of Scott (1963), Harr (1962), De Wiest (1965, 1969), Bear (1972), McWhorter and Sunada (1977), and Bouwer (1978).

11

2.2 Darcy's Law

The flow through the porous media is slow frictional flow governed by Darcy's law (Darcy, 1856; De Wiest, 1965) which may be stated in the form

$$v = ki \qquad (2.1)$$

where

v = seepage velocity [L/T]

k = permeability of the medium (or hydraulic conductivity) [L/T]

i = hydraulic gradient

all in a specified direction. The direction in which the hydraulic gradient is measured must be one of the principal directions of the permeability tensor. The generalized form of Darcy's law is discussed by Bear (1972, Chapter 5).

Throughout this book the factor k in Darcy's law will be called the *permeability* since this is commonly used both in the context of seepage and groundwater flow. Strictly this coefficient should be called the *hydraulic conductivity* since it is a function of the intrinsic permeability of the medium and the dynamic viscosity and density of the water.

It is important to note that the Darcy or seepage velocity is an artificial velocity since it equals the total quantity flowing divided by the cross-sectional area. The average velocity v_w at which the water moves can therefore be calculated as the Darcy velocity divided by the porosity, $\bar{p}$ (Bouwer and Rice, 1968).

$$v_w = v/\bar{p} \qquad (2.2)$$

The hydraulic gradient results from a difference in groundwater potential across an element of the medium. If Δh represents the total potential loss of the fluid over a distance Δs then, in the limit,

$$i = -\frac{dh}{ds}$$

the sign convention denoting a negative gradient. The term h, the groundwater potential, equals the sum of the kinetic energy, the pressure and the potential energy heads

$$h = \frac{v^2}{2g} + \frac{p}{\rho g} + z \qquad (2.3)$$

where ρg is the specific weight of the liquid. Note that the term h is sometimes called the *groundwater head* or the *piezometric head*. It may easily be verified that the first term, which represents the kinetic energy, is in effect usually negligible in seepage studies (Cedergren, 1977, pp. 87–9). Therefore Eq. 2.3 reduces to

$$h = \frac{p}{\rho g} + z \qquad (2.4)$$

However, high velocities do occur in some situations; for example, adjacent to the ends of cut-off walls and in the vicinity of wells. Modifications can be made to the permeability to allow for high velocities (Volker 1975); see also Section 12.2.

The components of the seepage velocity, v, are

$$v_x = -k_x \frac{\partial h}{\partial x}$$

$$v_y = -k_y \frac{\partial h}{\partial y} \tag{2.5}$$

$$v_z = -k_z \frac{\partial h}{\partial z}$$

where k_x, k_y, and k_z are the component permeabilities. These equations only hold when the x, y, and z axes coincide with principal permeability axes (Bear, 1972, pp. 143–8).

For solving problems of groundwater flow, Darcy's law alone is not sufficient. It gives only three relations between four unknown quantities, the three components of the velocity vector and the groundwater potential. An additional equation may be obtained by realizing that the flow has to satisfy the principle of the conservation of mass. Whatever the pattern of flow, no mass can be gained or lost. This leads to an equation of continuity. Initially the equation of continuity for steady-state flow will be discussed.

2.3 Steady-state Flow

Consider the steady-state flow through a small elementary volume of space (Fig. 2.1). By assuming that the fluid filters in such a manner that the porous ground is incompressible, the principle of the conservation of mass stipulates that the sum of the mass flow entering the three faces is equal to the sum of the mass flow leaving by the opposite faces. Since the derivation will be restricted to water having a constant density, the conservation principle can be enforced by ensuring that the total volume rate of flow entering the three faces is equal to the volume rate of flow leaving the opposite three faces.

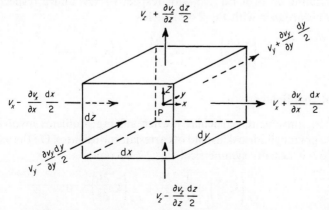

Figure 2.1. Flow for an elementary volume of fluid

At the central point $P(x = y = z = 0)$ of an element (see Fig. 2.1), the velocities are v_x, v_y, and v_z. Consequently on the plane $x = -0.5\,dx$, the velocity in the x-direction is

$$v_x - \frac{\partial v_x}{\partial x}\frac{dx}{2}$$

The total volume rate of flow of water entering the element across this plane is given by the velocity times the cross-sectional area,

$$\left(v_x - \frac{\partial v_x}{\partial x}\frac{dx}{2}\right) dy\,dz$$

Similarly, the volume rate of flow of water leaving the element across the plane $x = +0.5\,dx$ is given by

$$\left(v_x + \frac{\partial v_x}{\partial x}\frac{dx}{2}\right) dy\,dz$$

Therefore the net total volume rate of flow from the element due to flow in the x-direction is

$$\frac{\partial v_x}{\partial x}dx\,dy\,dz$$

Similar expressions hold for the volumes in the y- and z-directions, and therefore the total volume rate of flow leaving the element is

$$\left(\frac{\partial v_x}{\partial x} + \frac{\partial v_y}{\partial y} + \frac{\partial v_z}{\partial z}\right) dx\,dy\,dz \tag{2.6a}$$

According to the conservation principle this must equal zero; hence

$$\frac{\partial v_x}{\partial x} + \frac{\partial v_y}{\partial y} + \frac{\partial v_z}{\partial z} = 0 \tag{2.6b}$$

This is the *steady-state continuity equation*.

Differentiating each of Eq. 2.5 with respect to x, y, and z respectively and adding, in accordance with Eq. 2.6(b), yields

$$\frac{\partial}{\partial x}\left(k_x\frac{\partial h}{\partial x}\right) + \frac{\partial}{\partial y}\left(k_y\frac{\partial h}{\partial y}\right) + \frac{\partial}{\partial z}\left(k_z\frac{\partial h}{\partial z}\right) = 0 \tag{2.7}$$

Cylindrical coordinates

It is often convenient, particularly when solving problems involving radial flow, to use polar cylindrical instead of cartesian coordinates. On the assumption that the flow is radially symmetrical Eq. 2.7 becomes

$$\frac{\partial}{\partial r}\left(k_r\frac{\partial h}{\partial r}\right) + \frac{1}{r}k_r\frac{\partial h}{\partial r} + \frac{\partial}{\partial z}\left(k_z\frac{\partial h}{\partial z}\right) = 0 \tag{2.8}$$

where k_r and k_z are the coefficients of permeability in the radial and vertical direction and $r = (x^2 + y^2)^{\frac{1}{2}}$.

Boundary conditions

The different types of boundary conditions which can apply in steady flow can be grouped as follows:

(a) The groundwater potential takes known values on the boundary.
(b) For an impermeable boundary no water crosses the boundary. Since the flow is proportional to the potential gradient, this condition is satisfied by setting $\partial h/\partial n = 0$, where n is in the direction normal to the boundary.
(c) A similar condition applies when the magnitude of the flow crossing the boundary is known. This is represented by the normal gradient $\partial h/\partial n$ taking a specified value $\partial h/\partial n = -$(velocity normal to boundary divided by permeability normal to the boundary).
(d) On a steady free surface, the actual position of the boundary is unknown, but two conditions must be satisfied simultaneously on the boundary. The first condition is that no flow crosses the boundary, hence $\partial h/\partial n = 0$; and the second is that the pressure is atmospheric, thus from Eq. 2.4, $h = z$.

Certain of these conditions can hold on both internal and external boundaries. Conditions (a), (b), and (c) also apply for time-dependent problems; but a time-dependent free surface requires a more complex definition than (d) above; see Section 2.4.

2.4 Time-variant Problems

In many seepage and groundwater problems, the variation of the groundwater potential with time is of considerable significance. In considering time-dependent problems, it is important to consider separately the two alternative mechanisms for confined and unconfined aquifers.

Confined aquifer

For a confined aquifer (Fig. 2.2(a)) a fall in the groundwater potential (piezometric head) of Δh results in a reduction in pressure. The volume of water released *per unit volume* of aquifer due to a unit decrease in head is termed the *specific storage coefficient*, S_s.

An examination of the mass balance for a saturated element in a confined aquifer (De Wiest, 1965), indicates that both the compressibility of the water and the change in pore volume due to vertical compression of the aquifer contribute to the specific storage. Thus the specific storage is a function of the density of water, the porosity, the pore volume compressibility, and the compressibility of water. Though an expression can be written for the specific storage, its magnitude is usually determined from field tests. Usually S_s is within the range 10^{-5} to

10^{-7}/m. Note that the specific storage applies to the whole saturated volume of the aquifer.

The differential equation including the effect of the specific storage can be derived by extending the steady-state formulation of Section 2.3. Since the effects of the compressibility and density changes are included in the specific storage coefficient, it is permissible to continue to work in terms of the conservation of volume.

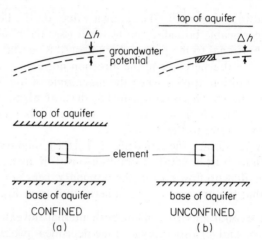

Figure 2.2. Diagrammatic representation of storage effect for confined and unconfined aquifers

Following the derivation of Eq. 2.6(a), the net volume leaving the element of the aquifer *during a time* δt due to the changing velocities equals

$$\mathrm{d}x\,\mathrm{d}y\,\mathrm{d}z\left(\frac{\partial v_x}{\partial x} + \frac{\partial v_y}{\partial y} + \frac{\partial v_z}{\partial z}\right)\delta t$$

During the time increment δt, the groundwater potential at point P increases by δh, thus the volume of water taken into storage due to this increase in groundwater potential is

$$\mathrm{d}x\,\mathrm{d}y\,\mathrm{d}z\,S_s\delta h$$

From the principle of continuity these two quantities must sum to zero, hence in the limit

$$\frac{\partial v_x}{\partial x} + \frac{\partial v_y}{\partial y} + \frac{\partial v_z}{\partial z} = -S_s\frac{\partial h}{\partial t}$$

Substituting for v_x, v_y, and v_z from Eq. 2.5 leads to the differential equation

$$\frac{\partial}{\partial x}\left(k_x\frac{\partial h}{\partial x}\right) + \frac{\partial}{\partial y}\left(k_y\frac{\partial h}{\partial y}\right) + \frac{\partial}{\partial z}\left(k_z\frac{\partial h}{\partial z}\right) = S_s\frac{\partial h}{\partial t} \tag{2.9}$$

Note that this equation holds for any element within a saturated aquifer.

Unconfined aquifer

With an unconfined aquifer (Fig. 2.2(b)) two mechanisms apply. Due to the compressibility of the aquifer and the water, the specific storage coefficient applies to all elements within the saturated portion of the aquifer. In addition, the fall of the free water surface leads to a dewatering of the aquifer. A unit fall in the free surface position results in a release of water from storage equal to S_y *per unit plan area*, where S_y is the specific yield. The specific yield is sometimes known as the effective porosity. Due to complex effects when the water table moves, such as the capillary fringe and entrapped air, it is advisable to estimate the specific yield from observations of the aquifer response. Note that the terms *free surface* or *water table* are both used to indicate the upper surface of the water at atmospheric pressure.

The release of water due to the specific yield occurs at the free surface, unlike the effect of the specific storage which is distributed uniformly throughout the saturated volume of the aquifer. In fact the release due to the specific yield may not be instantaneous, as indicated by the concept of delayed yield (Boulton, 1963).

Of the two mechanisms, only the specific storage is included in the differential equation which describes the conservation of volume, Eq. 2.9. The specific yield which results from the moving free surface is included as a boundary condition.

For a moving free surface in a three-dimensional situation, the movement of a *particle* of water on the free surface will be not only in the vertical direction; it will also have horizontal components. Thus the particle A in Fig. 2.3 will move to position A′ during a time δt. However, when setting the new free surface position on a finite-difference network, it is necessary to know where the new water surface intersects the *vertical grid* line through A. At the end of the time increment δt the free surface position vertically below A is B′. An expression for the vertical distance AB′ is given by Bear (1972); an alternative method of estimating the position by a geometric construction is described below.

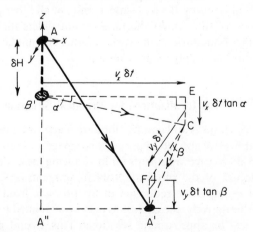

Figure 2.3. Construction to determine movement of a free surface on a vertical grid line

Assuming that the time step is sufficiently small, the vertical component of movement of particle A is

$$AA'' = \frac{-\delta t}{p} v_z \qquad (2.10)$$

The porosity p is included in the expression to convert the Darcy velocity v_z into an actual velocity (see Eq. 2.2).

From the geometry of Fig. 2.3.

$$AB' = AA'' - EC - FA' \qquad (2.11)$$

Assuming that the change of slope of the free surface is small then

$$B'E = v_x \delta t / p$$

hence

$$EC = v_x \delta t \tan \alpha / p$$

where α is the angle of the free surface below the horizontal in the x-direction. Similarly

$$FA' = v_y \delta t \tan \beta / p$$

Therefore from Eqs. 2.10 and 2.11

$$AB' = \frac{-\delta t}{p} (v_z + v_x \tan \alpha + v_y \tan \beta)$$

But the effective porosity equals the specific yield, thus

$$AB' = \frac{-\delta t}{S_y} (v_z + v_x \tan \alpha + v_y \tan \beta) \qquad (2.12)$$

This condition holds on the free surface only; at all other positions within the saturated region of the aquifer the time-variant differential equation, Eq. 2.9, applies. On other boundaries, time-variant forms of boundary conditions (a), (b), and (c) of Section 2.3 apply as appropriate.

2.5 Idealizations for Practical Situations

Different idealizations and assumptions are made when the general equations as described above are applied to particular seepage and groundwater problems. There are two alternative arguments for introducing these idealizations. The first is that the nature of the physical problem warrants a simplification in the equations, for example the reduction of the problem from three to two space dimensions. Alternatively, the equations may be simplified to make it possible to obtain analytical or approximate solutions. This second reason may lead to solutions which appear to be adequate but are, in fact, unreliable in representing the real flow mechanism.

The governing differential equations and the appropriate boundary conditions for each of the main groups of seepage and groundwater problems to be found in Parts II, III and IV of this book are *summarized* in Tables 2.1 and 2.2; these Tables are *supported* by a brief discussion of the idealizations involved. Reference should be made to the standard texts listed in Section 2.1 for more details of the reasoning behind the various approximations.

Seepage
(Part II)

Steady state

For most steady-state seepage problems, either the full three-dimensional equation, Eq. 2.7,

$$\frac{\partial}{\partial x}\left(k_x \frac{\partial h}{\partial x}\right) + \frac{\partial}{\partial y}\left(k_y \frac{\partial h}{\partial y}\right) + \frac{\partial}{\partial z}\left(k_z \frac{\partial h}{\partial z}\right) = 0$$

or the two-dimensional form

$$\frac{\partial}{\partial x}\left(k_x \frac{\partial h}{\partial x}\right) + \frac{\partial}{\partial z}\left(k_z \frac{\partial h}{\partial z}\right) = 0$$

are used. The two-dimensional equations are applicable when, for example, the flow is predominantly in the x–z plane with negligible flow in the y-direction. The appropriate boundary conditions are listed in Table 2.2.

Time-variant problems

In time-variant seepage problems, the time-dependent features usually arise from the moving free surface. This movement of the free surface yields far more water than the quantity released due to the compressibility effects. Therefore the specific storage can be neglected, and the differential equations quoted above for the steady-state problems are used.

The flow pattern is dominated by the free surface boundary condition that the movement of the water surface on a vertical grid line is given by Eq. 2.12 which becomes in terms of the free surface potential, H,

$$\delta H = \frac{+\delta t}{S_y}(v_z + v_x \tan \alpha + v_y \tan \beta)$$

where H is the groundwater potential on the free surface boundary.

Regional groundwater flow
(Part III)

The main assumption introduced in most regional groundwater flow problems is that vertical components of flow can be neglected. This is a

Table 2.1 Summary of main equations

Description	Permeability	Time-dependent	Equation	Possible boundary conditions (see Table 2.2)
Chapters 4, 5: Seepage				
Flow in a typical vertical section	variable	No	$\frac{\partial}{\partial x}\left(k_x\frac{\partial h}{\partial x}\right) + \frac{\partial}{\partial z}\left(k_z\frac{\partial h}{\partial z}\right) = 0$	1, 3, 4, 6
Flow in a typical vertical section	constant	No	$\frac{\partial^2 h}{\partial x^2} + \frac{\partial^2 h}{\partial z^2} = 0$	1, 3, 4, 6
Three-dimensional seepage	variable	No	$\frac{\partial}{\partial x}\left(k_x\frac{\partial h}{\partial x}\right) + \frac{\partial}{\partial y}\left(k_y\frac{\partial h}{\partial y}\right) + \frac{\partial}{\partial z}\left(k_z\frac{\partial h}{\partial z}\right) = 0$	1, 3, 4, 6
Three-dimensional seepage	variable	Yes	$\frac{\partial}{\partial x}\left(k_x\frac{\partial h}{\partial x}\right) + \frac{\partial}{\partial y}\left(k_y\frac{\partial h}{\partial y}\right) + \frac{\partial}{\partial z}\left(k_z\frac{\partial h}{\partial z}\right) = 0$	1, 2, 3, 4, 5, 7
Radial seepage	variable	No	$\frac{\partial}{\partial r}\left(k_r\frac{\partial h}{\partial r}\right) + \frac{1}{r}k_r\frac{\partial h}{\partial r} + \frac{\partial}{\partial z}\left(k_z\frac{\partial h}{\partial z}\right) = 0$	1, 3, 4, 6
Chapters 6 to 10: Regional groundwater flow				
One-dimensional	variable	No	$\frac{\partial}{\partial x}\left(T_x\frac{\partial h}{\partial x}\right) = -q(x)$	1, 3, 4
One-dimensional	variable	Yes	$\frac{\partial}{\partial x}\left(T_x\frac{\partial h}{\partial x}\right) = S\frac{\partial h}{\partial t} - q(x,t)$	1, 2, 3, 4, 5
Two-dimensional	variable	Yes	$\frac{\partial}{\partial x}\left(T_x\frac{\partial h}{\partial x}\right) + \frac{\partial}{\partial y}\left(T_y\frac{\partial h}{\partial y}\right) = S\frac{\partial h}{\partial t} - q(x,y,t)$	1, 2, 3, 4, 5
Chapters 11, 12: Radial flow to a well				
Vertical flow negligible	variable	Yes	$\frac{\partial}{\partial r}\left(mk_r\frac{\partial s}{\partial r}\right) + \frac{m}{r}k_r\frac{\partial s}{\partial r} = S\frac{\partial s}{\partial t} + q(r,t)$	1, 2, 3, 4, 5
Vertical flow included	variable	Yes	$\frac{\partial}{\partial r}\left(k_r\frac{\partial s}{\partial r}\right) + \frac{1}{r}k_r\frac{\partial s}{\partial r} + \frac{\partial}{\partial z}\left(k_z\frac{\partial s}{\partial z}\right) = S_s\frac{\partial s}{\partial t}$	1, 2, 3, 4, 5, 8

reasonable assumption when it is recognized that the distances in the x- and y-directions are far greater than the aquifer thickness, consequently the hydraulic gradient is generally less than 0.01.

Thus the general three-dimensional equation for compressible flow, Eq. 2.9, is integrated in the vertical direction and becomes

$$\frac{\partial}{\partial x}\left(mk_x\frac{\partial h}{\partial x}\right) + \frac{\partial}{\partial y}\left(mk_y\frac{\partial h}{\partial y}\right) = S_s m\frac{\partial h}{\partial t} \qquad (2.13)$$

where m is the aquifer thickness. The term h, though still a function of x and y, is now an average groundwater potential

$$\bar{h} = \frac{1}{m}\int_0^m h(x,y,z,t)\,dz \qquad (2.14)$$

The bar signifies that this is an average potential. For the remainder of the book this will be suppressed. However, it must be recognized that whenever a two-dimensional equation in x and y, or a one-dimensional equation in x, is used, then the groundwater potential is an average. This can lead to complications when considering recharge and unconfined storage.

Table 2.2 Summary of boundary conditions

No.	Type	Function of time	Equations
1	Fixed	No	$h = f(x,y,z)$
2	Fixed	Yes	$h = f(x,y,z,t)$
3	Zero flow	No/Yes	$\partial h/\partial n = 0$
4	Known flow	No	$\partial h/\partial n = f(x,y,z)$
5	Known flow	Yes	$\partial h/\partial n = f(x,y,z,t)$
6	Steady free surface	No	position unknown $h = z, \partial h/\partial n = 0$
7	Unsteady free surface	Yes	initial free surface defined $h(x,y,z,0) = H(x,y,0)$ then movement on vertical mesh line given by $$\delta H = \frac{\delta t}{S_y}(v_z + v_x \tan\alpha + v_y \tan\beta)$$ where α and β are inclinations of water surface in the x- and y-directions
8	Unsteady free surface	Yes	inflow = recharge + $S_y\dfrac{\partial s_f}{\partial t}$

Notes. 1. Conditions are listed as functions of the three space variables (x,y,z) but they may be functions of only two or one cartesian variable or of the radial variables (r,z). Certain conditions can also be written in terms of the drawdowns.
2. n denotes the normal to the boundary.

It is convenient to define the *transmissivities* as

$$T_x = mk_x \quad \text{and} \quad T_y = mk_y \qquad (2.15)$$

22

and the *confined storage coefficient*

$$S_c = mS_s \qquad (2.16)$$

Thus the differential equation is written as

$$\frac{\partial}{\partial x}\left(T_x\frac{\partial h}{\partial x}\right) + \frac{\partial}{\partial y}\left(T_y\frac{\partial h}{\partial y}\right) = S_c\frac{\partial h}{\partial t} \qquad (2.17)$$

For a confined aquifer, the saturated aquifer thickness m remains constant, but for an unconfined aquifer the saturated thickness is a function of h. However, in many practical situations the saturated thickness and therefore the transmissivity are assumed to remain constant, either because the change in saturated thickness is small or alternatively because the manner in which the permeability changes with depth is unknown. In the absence of sufficiently precise information, the only practical assumption for many aquifers is that the transmissivity is a constant.

In many groundwater flow situations, water enters the aquifer from above (and perhaps below). Figure 2.4 illustrates an inflow to a confined aquifer through a semi-pervious stratum where the inflow is $q(x, y, t)$, the units of q are flow rate per unit plan area, [L/T]. An approximation that is usually introduced is that this inflow is immediately distributed throughout the full depth of the aquifer. This is equivalent to considering an element of sides dx and dy but extending for the full saturated depth of the aquifer. As a consequence the differential equation becomes

$$\frac{\partial}{\partial x}\left(T_x\frac{\partial h}{\partial x}\right) + \frac{\partial}{\partial y}\left(T_y\frac{\partial h}{\partial y}\right) = S_c\frac{\partial h}{\partial t} - q(x, y, t) \qquad (2.18)$$

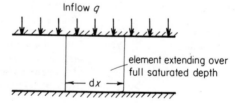

Figure 2.4. Inflow due to leakage in a confined aquifer

With an unconfined aquifer, the inflow $q(x, y, t)$ arises from recharge which may result from precipitation or perhaps the presence of a stream. In addition, a falling water table (decreasing groundwater potential) results in water released from storage. This is equivalent to an additional recharge q to the upper surface.

As was shown in Section 2.4 the influence of the moving water table is represented as a boundary condition, Eq. 2.12. However, by introducing certain assumptions, this can be expressed as an equivalent recharge, q_s.

If the water table slope is small then $\tan\alpha$ and $\tan\beta$ in Eq. 2.12 tend to zero. Writing AB′ as δh_f, where h_f is the free surface elevation,

$$\delta h_f = \frac{\delta t}{S_y}v_z$$

In the limit this reduces to

$$v_z = +S_y \frac{\partial h_f}{\partial t}$$

This Darcy velocity v_z is in effect a recharge to the aquifer at the free surface; a negative sign is introduced since q is positive downwards,

$$q_s = -v_z$$

Consequently the effective recharge due to the specific yield can be expressed as an inflow

$$q_a = -S_y \frac{\partial h_f}{\partial t} \qquad (2.19)$$

This expression could be deduced from the definition of specific yield, but the derivation given above does indicate the nature of the approximations.

Since the water table is approximately horizontal, it is also reasonable to assume that the free surface potential h_f is approximately equal to the average potential h. Hence the inflow can be written as

$$q_s = -S_y \frac{\partial h}{\partial t}$$

Figure 2.5 illustrates the situation showing that inflow to the aquifer can arise due to recharge and the moving water surface. Further, if it is assumed that this inflow is immediately distributed throughout the depth of the aquifer, then the differential equation for an unconfined aquifer becomes

$$\frac{\partial}{\partial x}\left(T_x \frac{\partial h}{\partial x}\right) + \frac{\partial}{\partial y}\left(T_y \frac{\partial h}{\partial y}\right) = S_c \frac{\partial h}{\partial t} + S_y \frac{\partial h}{\partial t} - q(x, y, t)$$

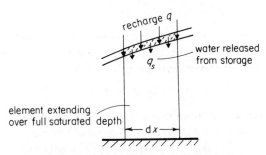

Figure 2.5. Inflow due to leakage and moving water surface in an unconfined aquifer

In practice the confined storage coefficient is very much smaller than the specific yield; hence the differential equation for an unconfined aquifer is usually written as

$$\frac{\partial}{\partial x}\left(T_x \frac{\partial h}{\partial x}\right) + \frac{\partial}{\partial y}\left(T_y \frac{\partial h}{\partial y}\right) = S_y \frac{\partial h}{\partial t} - q(x, y, t) \qquad (2.20)$$

If the flow is predominantly in one direction, then a one-dimensional idealization is used and the basic equations for confined or unconfined aquifers can be written, with the suffices for the storage coefficients suppressed, as

$$\frac{\partial}{\partial x}\left(T_x \frac{\partial h}{\partial x}\right) = S\frac{\partial h}{\partial t} - q(x,t) \tag{2.21}$$

By considering a unit width of aquifer, the inflow $q(x,t)$ retains the dimensions of flow per unit area, $[L/T]$.

Methods of representing the situation when vertical components of flow become significant are discussed in Section 9.2.

Pumping-test analysis
(Part IV)

When analysing pumping tests using a numerical technique, the flow is assumed to be radial. Consequently at any radius r the groundwater potentials and the permeabilities are taken to have the same values. Depending on the nature of the physical problem, two alternative idealizations may be adopted.

Vertical flow neglected

If the vertical components of flow can be neglected, then the same idealizations as for regional groundwater flow can be adopted.

On substituting $r^2 = x^2 + y^2$ and writing mk instead of T, Eqs. 2.18 or 2.20 become

$$\frac{\partial}{\partial r}\left(mk_r \frac{\partial h}{\partial r}\right) + \frac{m}{r}k_r\frac{\partial h}{\partial r} = S\frac{\partial h}{\partial t} - q(r,t) \tag{2.22}$$

where S may be the confined storage coefficient or the specific yield.

Rather than working in terms of the groundwater potential h, it is usually convenient to consider the drawdown s,

$$s = H - h \tag{2.23}$$

where H is the elevation of the rest water level. Thus Eq. 2.22 becomes

$$\frac{\partial}{\partial r}\left(mk_r \frac{\partial s}{\partial r}\right) + \frac{m}{r}k_r\frac{\partial s}{\partial r} = S\frac{\partial s}{\partial t} + q(r,t) \tag{2.24}$$

Vertical flow included

For unconfined aquifers where the vertical component of flow is more likely to be significant, the differential equation based on Eqs. 2.8 and 2.9 becomes

$$\frac{\partial}{\partial r}\left(k_r \frac{\partial s}{\partial r}\right) + \frac{1}{r}k_r\frac{\partial s}{\partial r} + k_z\frac{\partial^2 s}{\partial z^2} = S_s\frac{\partial s}{\partial t} \tag{2.25}$$

It is demonstrated in Chapter 11 that an approximation can be introduced which allows the vertical flow component $\partial^2 s/\partial z^2$ to be included by working in terms of the average drawdowns and the free surface drawdown s_f.

According to Eq. 2.19 the movement of the water table can be represented as an inflow

$$q = -S_y \frac{\partial h_f}{\partial t} = S_y \frac{\partial s_f}{\partial t}$$

Therefore the condition at the free surface together with the recharge or leakage can be represented as an inflow at the water table boundary.

CHAPTER 3

Finite Differences: Relevant Aspects

3.1 Introduction

The equations describing most practical seepage and groundwater problems cannot be solved by analytical means unless major simplifying assumptions are introduced. However, solutions to these equations can often be obtained by numerical means. One of the earliest numerical techniques to be used successfully was the finite-difference approximation (Richardson, 1911). Though other approaches, such as finite element (Zienkiewicz, 1977) and boundary element (Liggett, 1977) have been introduced, the simplicity and flexibility of the finite-difference technique means that it can be used with confidence by those who are not specialists in numerical analysis. A further feature of the finite-difference technique is that non-linearities arising from changes in parameter values, such as the change between confined and unconfined states, can be included without difficulty.

In the following discussion, attention will be restricted to the finite-difference approximations for elliptic and parabolic equations. Elliptic equations include all the steady-state equations which describe seepage and groundwater flow, whilst parabolic equations are those that are first-order time-dependent, containing terms of the form $\partial h/\partial t$. Both cartesian and radial coordinate systems will be considered. Only the simplest finite-difference approximations will be discussed; higher-order corrections can be introduced, leading to greater accuracy with the same number of mesh points, but they have the disadvantage of additional complexities on the boundaries. The chapter concludes with a detailed description of a resistance network solution.

Throughout this chapter, the main purpose is to provide background information for the later chapters. This has significantly effected the selection of the material presented. For a fuller treatment of the solution of differential equations using the finite-difference method, the reader is referred to Smith (1965) and Forsythe and Wasow (1960).

3.2 One-dimensional Cartesian Examples

The first finite-difference approximation to be considered is that for one-dimensional cartesian problems. Flow in an aquifer of constant transmissivity arising from recharge will be used as an example. The appropriate equation derived from Eq. 2.21 is

$$T\frac{d^2h}{dx^2} = -q \tag{3.1}$$

where h is the groundwater potential, T is the uniform transmissivity, and q is the inflow per unit area. Two methods of deducing the finite-difference approximation will be considered, by straight lines or by Taylor's series.

Approximating the function by straight lines

A representative variation in groundwater head over a section of the aquifer is plotted in Fig. 3.1(a). Rather than describing the head variation by a continuous function, the heads are defined at certain discrete points, positioned at intervals of Δx (Fig. 3.1(b)). The unknown potentials are written as h_{-1}, h_0, h_1, etc.

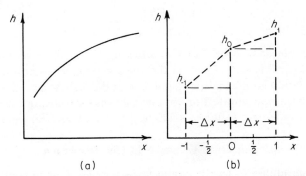

(a)   (b)

Figure 3.1. Smooth continuous function and finite-difference approximation

Approximate values of the slopes at intermediate points can be calculated as

$$\left(\frac{dh}{dx}\right)_{+\frac{1}{2}} = \frac{h_1 - h_0}{\Delta x}, \quad \left(\frac{dh}{dx}\right)_{-\frac{1}{2}} = \frac{h_0 - h_{-1}}{\Delta x}$$

where the suffices $+\frac{1}{2}$ and $-\frac{1}{2}$ signify distances $\frac{1}{2}\Delta x$ on either sides of node 0. The second differential can be written as

$$\left(\frac{d^2h}{dx^2}\right)_0 = \frac{d}{dx}\left(\frac{dh}{dx}\right) = \left[\left(\frac{dh}{dx}\right)_{+\frac{1}{2}} - \left(\frac{dh}{dx}\right)_{-\frac{1}{2}}\right]\Big/ \Delta x = \frac{(h_{-1} - 2h_0 + h_1)}{\Delta x^2} \tag{3.2}$$

Therefore the differential equation (Eq. 3.1) can be written as an algebraic equation

$$h_{-1} - 2h_0 + h_1 = -\Delta x^2 q_0/T \tag{3.3}$$

Approximation using Taylor's series

The approximation introduced by the finite-difference approach can be identified by considering the Taylor's series

$$h_1 = h_0 + \frac{\Delta x}{1!}\left(\frac{dh}{dx}\right)_0 + \frac{\Delta x^2}{2!}\left(\frac{d^2h}{dx^2}\right)_0 + \frac{\Delta x^3}{3!}\left(\frac{d^3h}{dx^3}\right)_0 + \frac{\Delta x^4}{4!}\left(\frac{d^4h}{dx^4}\right)_0 + \cdots \quad (3.4)$$

$$h_{-1} = h_0 - \frac{\Delta x}{1!}\left(\frac{dh}{dx}\right)_0 + \frac{\Delta x^2}{2!}\left(\frac{d^2h}{dx^2}\right)_0 - \frac{\Delta x^3}{3!}\left(\frac{d^3h}{dx^3}\right)_0 + \frac{\Delta x^4}{4!}\left(\frac{d^4h}{dx^4}\right)_0 - \cdots \quad (3.5)$$

Summing,

$$h_{-1} - 2h_0 + h_1 = \Delta x^2\left(\frac{d^2h}{dx^2}\right)_0 + \frac{\Delta x^4}{12}\left(\frac{d^4h}{dx^4}\right)_0 + \cdots$$

Therefore

$$\left(\frac{d^2h}{dx^2}\right)_0 = \frac{1}{\Delta x^2}(h_{-1} - 2h_0 + h_1) - \frac{\Delta x^2}{12}\left(\frac{d^4h}{dx^4}\right)_0 + \cdots \quad (3.6)$$

The terms

$$-\frac{\Delta x^2}{12}\left(\frac{d^4h}{dx^4}\right)_0 + \cdots \quad (3.7)$$

are called the *truncation error*. If two solutions are obtained, the second having a mesh interval equal to half of that for the first solution, the truncation error for the second solution is only one quarter of that for the first. As a result, the second solution will be more accurate. For certain problems the truncation error is zero; then the finite-difference solution is identical to the theoretical solution.

3.3 Numerical Examples In One Dimension

Consider an aquifer of length L, and transmissivity T, as illustrated in Fig. 3.2(a). In particular, the boundary conditions should be noted. On the left-hand boundary no flow is allowed to enter or leave; on the right-hand side of the aquifer the potential is held constant at H. A steady recharge of q per unit area enters the aquifer.

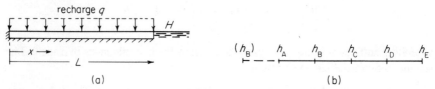

Figure 3.2. Simple example

An *analytical solution* to this problem can be obtained. The differential equation

$$T\frac{d^2h}{dx^2} = -q \quad (3.1)$$

when integrated twice becomes

$$h = -0.5qx^2/T + Ax + B$$

The two boundary conditions are that:

(a)   at $x = 0$, $dh/dx = 0$, thus $A = 0$;
(b)   at $x = L$, $h = H$, thus $B = H + 0.5qL^2/T$.

Therefore the exact solution is

$$h = 0.5q(L^2 - x^2)/T + H \qquad (3.8)$$

For the *finite-difference solution* the aquifer is divided into four equal intervals of length $0.25L$ (Fig. 3.2(b)). The unknown heads are h_A, h_B, h_C, h_D, and h_E. Now considering the boundary conditions:

(a) at $x = 0$, $dh/dx = 0$. This condition can be enforced by introducing a fictitious node at a distance $0.25L$ to the left of node A. By setting the potential at the fictitious node equal to h_B, the condition $(dh/dx)_A = 0$ is automatically satisfied.

(b) at $x = L$ the potential is fixed at H and therefore $h_E = H$.

The differential equation (3.1) can be expressed as

$$h_{-1} - 2h_0 + h_1 = -\Delta x^2 q/T \qquad (3.9)$$

This equation is written in turn for nodes A, B, C, and D, with the boundary conditions substituted where appropriate.

$$h_B - 2h_A + h_B = -\Delta x^2 q_A/T$$
$$h_A - 2h_B + h_C = -\Delta x^2 q_B/T$$
$$h_B - 2h_C + h_D = -\Delta x^2 q_C/T \qquad (3.10)$$
$$h_C - 2h_D + H = -\Delta x^2 q_D/T$$

Since the recharge is uniform, $q_A = q_B = q_C = q_D = q$. There are four equations with four unknowns which can be solved to give

$$h_A = 0.5qL^2/T + H$$
$$h_B = 0.46875qL^2/T + H$$
$$h_C = 0.375qL^2/T + H$$
$$h_D = 0.21875qL^2/T + H$$

These results are identical to the exact solutions. This occurs because the truncation error

$$\frac{\Delta x^2}{12} \frac{d^4h}{dx^4} + \cdots$$

when evaluated from the analytical expression of Eq. 3.8 is found to equal zero.

For a *second example* consider the same aquifer, but with a parabolic distribution of recharge (Fig. 3.3). The analytical solution, obtained by integration, is

$$h = q_m(L^2 - 2x^3/L + x^4/L^2)/3T + H \qquad (3.11)$$

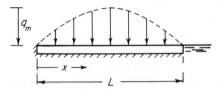

Figure 3.3. Example with parabolic
distribution of recharge

Finite-difference equations are identical to Eq. 3.10 but the parabolic distribution of recharge is represented by

$$q_A = 0, \qquad q_B = q_D = 0.75q_m, \qquad q_C = q_m$$

The finite-difference results are recorded in the third column of Table 3.1. These numerical results are slightly lower than the theoretical.

Table 3.1 Results for parabolic recharge

Node	x	Numerical	Theoretical
A	0	$H + 0.313qL^2/T$	$H + 0.333qL^2/T$
B	0.25L	$H + 0.313qL^2/T$	$H + 0.324qL^2/T$
C	0.5L	$H + 0.266qL^2/T$	$H + 0.271qL^2/T$
D	0.75L	$H + 0.156qL^2/T$	$H + 0.158qL^2/T$

These discrepancies arise from the truncation error which, in this case, is equivalent to a uniform reduction in recharge over the whole aquifer. The reduction in recharge, q_r, deduced from Eqs. 3.1, 3.6, and 3.11, is

$$q_r = T\frac{\Delta x^2}{12}\frac{d^4h}{dx^4} = T\frac{L^2}{12 \times 16}\frac{8q_m}{TL^2} = \frac{q_m}{24} \qquad (3.12)$$

With smaller mesh intervals, the truncation error becomes smaller and the numerical values become closer to the theoretical values. As a general guide, eight mesh intervals are usually adequate for this type of one-dimensional problem provided that there are no sudden changes in the inflows or outflows.

3.4 Finite-Difference Approximations in Two or Three Dimensions

Finite-difference approximations in two and three dimensions can be derived directly using Taylor's series. Taking the general case of the three-dimensional cartesian network shown in Fig. 3.4(a),

$$\left(\frac{\partial^2 h}{\partial x^2} + \frac{\partial^2 h}{\partial y^2} + \frac{\partial^2 h}{\partial z^2}\right)_0 = \frac{1}{\Delta x^2}(h_1 - 2h_0 + h_3) + \frac{1}{\Delta y^2}(h_2 - 2h_0 + h_4)$$

$$+ \frac{1}{\Delta z^2}(h_5 - 2h_0 + h_6) \qquad (3.13)$$

Nodal points are not always uniformly spaced; with non-uniform spacing special finite-difference approximations are required; see later in this section and Section 4.6. Various approximations are possible; one that has proved to be useful is described below. The discussion will be restricted to two dimensions, but three dimensions can be included in a similar manner. The aim is to devise a suitable finite-difference approximation to the expression

$$\frac{\partial^2 h}{\partial x^2} + \frac{\partial^2 h}{\partial y^2} = \frac{-q}{T} \qquad (3.14)$$

using the graded mesh of Fig. 3.4(b).

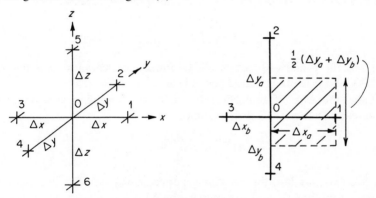

Figure 3.4. (a) Three-dimensional grid; (b) Determination of coefficients of finite-difference equations

The slopes midway between nodes 3 to 0 and 0 to 1 are respectively $(h_0 - h_3)/\Delta x_b$ and $(h_1 - h_0)/\Delta x_a$. It follows, therefore, that

$$\left(\frac{\partial^2 h}{\partial x^2}\right)_0 = \frac{2}{\Delta x_a + \Delta x_b}\left[\frac{h_1 - h_0}{\Delta x_a} - \frac{h_0 - h_3}{\Delta x_b}\right]$$

and $\qquad\qquad\qquad\qquad\qquad\qquad\qquad\qquad\qquad\qquad\qquad\qquad (3.15)$

$$\left(\frac{\partial^2 h}{\partial y^2}\right)_0 = \frac{2}{\Delta y_a + \Delta y_b}\left[\frac{h_2 - h_0}{\Delta y_a} - \frac{h_0 - h_4}{\Delta y_b}\right]$$

Multiplying these terms by $0.25\,(\Delta x_a + \Delta x_b)(\Delta y_a + \Delta y_b)$ and substituting in Eq. 3.14,

$$\frac{\Delta y_a + \Delta y_b}{2}\left[\frac{h_1 - h_0}{\Delta x_a} + \frac{h_3 - h_0}{\Delta x_b}\right] + \frac{\Delta x_a + \Delta x_b}{2}\left[\frac{h_2 - h_0}{\Delta y_a} + \frac{h_4 - h_0}{\Delta y_b}\right]$$

$$= \frac{-0.25q(\Delta x_a + \Delta x_b)(\Delta y_a + \Delta y_b)}{T} \qquad (3.16)$$

These expressions have a physical meaning. Consider the shaded area of Fig. 3.4(b); for nodes 0–1, the expression

$$(h_1 - h_0) \times \text{cross-sectional area/distance between nodes}$$

where the cross-sectional area equals $0.5(\Delta y_a + \Delta y_b)$ and the distance between the nodes equals Δx_a, is equivalent to the first term of the finite-difference equation, Eq. 3.16. Therefore the coefficients for graded nets in one-, two- or three-dimensional problems can be deduced from the cross-sectional area and distance between nodes. Finite-difference equations based on graded meshes are used for several examples; see Sections 4.10 and 5.4.

3.5 Finite Differences in Radial Flow

The differential equation for steady radially symmetrical flow in an aquifer when vertical flow components are neglected (see Eq. 2.22) is

$$\frac{d}{dr}\left(mk_r \frac{dh}{dr}\right) + \frac{m}{r} k_r \frac{dh}{dr} = -q \tag{3.17}$$

where r is the radial ordinate, k_r is the radial permeability, m is the aquifer thickness, and q is the recharge per unit area.

Initially the case of uniform transmissivity, $T = mk_r = \text{constant}$, will be considered. Equation 3.17 then becomes

$$T\left(\frac{d^2h}{dr^2} + \frac{1}{r}\frac{dh}{dr}\right) = -q \tag{3.18}$$

There are two standard finite-difference approaches used for this equation.

In one, a mesh with constant increments of radius is used, as shown in Fig. 3.5(a). Equation 3.18 can then be written as

$$T\left[\frac{h_{-1} - 2h_0 + h_1}{\Delta r^2} + \frac{1}{r_0}\left(\frac{h_1 - h_{-1}}{2\Delta r}\right)\right] = -q \tag{3.19}$$

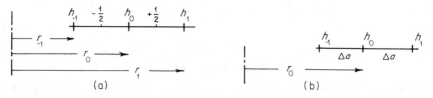

Figure 3.5. Radial finite-difference mesh: (a) regular mesh intervals; (b) logarithmic increase in mesh interval

In certain situations, this is a perfectly satisfactory finite-difference approximation.

As an alternative, the radial ordinate is divided into a discrete mesh which increases logarithmically by an amount Δa. This is achieved by setting

$$a = \log_e (r) \qquad (3.20)$$

Since

$$\frac{\partial a}{\partial r} = \frac{1}{r}$$

then

$$\frac{\partial h}{\partial r} = \frac{1}{r} \frac{\partial h}{\partial a}$$

and

$$\frac{\partial^2 h}{\partial r^2} = \frac{\partial}{\partial r} \left(\frac{1}{r} \frac{\partial h}{\partial a} \right) = -\frac{1}{r^2} \frac{\partial h}{\partial a} + \frac{1}{r^2} \frac{\partial^2 h}{\partial a^2}$$

When substituted in Eq. 3.18, this leads to the equation

$$T \frac{d^2 h}{da^2} = -qr^2 \qquad (3.21)$$

For the regular logarithmic mesh of Fig. 3.5(b), Eq. 3.21 in finite-difference form becomes

$$T \left(\frac{h_{-1} - 2h_0 + h_1}{\Delta a^2} \right) = -qr_0^2 \qquad (3.22)$$

This equation has a smaller truncation error than Eq. 3.19 and is usually more suitable for analysing radial flow towards a well.

When the transmissivity varies due to a changing saturated depth or changing permeability, or alternatively when the mesh spacing varies, then the finite-difference form of Eq. 3.19 becomes

$$\frac{2}{\Delta r_a + \Delta r_b} \left[m_a k_a \frac{h_1 - h_0}{\Delta r_a} - m_b k_b \frac{h_0 - h_{-1}}{\Delta r_b} \right] + \frac{m_0}{r_0} k_0 \frac{h_1 - h_{-1}}{\Delta r_a + \Delta r_b} = -q_0 \qquad (3.23)$$

where the intervals are as defined in Fig. 3.6.

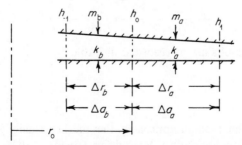

Figure 3.6. Radial finite-difference mesh, variable parameters

If the governing equation in terms of a is used, then for variable parameters, Eq. 3.22 becomes

$$\frac{2}{\Delta a_a + \Delta a_b}\left[\frac{m_a k_a(h_1 - h_0)}{\Delta a_a} - \frac{m_b k_b(h_0 - h_{-1})}{\Delta a_b}\right] = -qr_0^2 \qquad (3.24)$$

Example

As a simple example of the technique for solving radial-flow problems, a particular case of flow in a confined aquifer is considered. Radial flow occurs from an outer boundary, $r_{max} = 10$ m with a piezometric head $h_{max} = 5$ m to a well of radius $r_w = 1$ m at which the head is $h_w = 0$ m. The transmissivity has a constant value of $T = 400$ m^2/d.

This is a standard problem with a solution

$$h - h_w = (Q/2\pi T)\log_e (r/r_w) \qquad (3.25)$$

where $Q = 2\pi T(h_{max} - h_w)\log_e (r_w/r_{max})$

Finite-difference solutions can be obtained using the two alternative approaches. In each case, the area is divided into only four mesh intervals. This mesh is too coarse for most practical problems but it is useful in comparing the two different techniques.

Method 1

If a regular mesh spacing is used with the interval from 1.0 to 10.0 divided into four equal intervals, then $\Delta r = 2.25$. For convenience, the nodes at radii of 3.25, 5.5, and 7.75 will be denoted by the letters A, B, and C. The finite-difference equations derived from Eq. 3.19 are

$$\frac{0 - 2h_A + h_B}{5.0625} + \frac{1}{3.25}\cdot\frac{h_B - 0}{4.5} = 0$$

$$\frac{h_A - 2h_B + h_C}{5.0625} + \frac{1}{5.5}\cdot\frac{h_C - h_A}{4.5} = 0$$

$$\frac{h_B - 2h_C + 5.0}{5.0625} + \frac{1}{7.75}\cdot\frac{5.0 - h_B}{4.5} = 0$$

On solving, the results are as recorded in Table 3.2.

Method 2

When using the second approach, the mesh spacing is chosen so that $\Delta a = $ constant. Now $\log_e (1.0) = 0.0$ and $\log_e (10.0) = 2.3026$, so

$$\Delta a = 0.25 \times 2.3026 = 0.57565$$

Using this increment, the third column of Table 3.3 can be constructed; the variations in radius can then be deduced. At each node the finite-difference equation takes the form

$$h_{-1} - 2h_0 + h_1 = 0$$

Table 3.2 Results for radial flow, method 1

Node	Radius	Potential	Theoretical potential	Difference
—	1.0	0	0	0
A	3.25	2.444	2.600	−0.156
B	5.5	3.631	3.702	−0.071
C	7.75	4.415	4.446	−0.031
—	10.0	5.0	5.000	0

In this case, the solution is particularly simple with the piezometric head increasing linearly with $\log(r)$. The resultant values are, in fact, identical to the theoretical expression.

Table 3.3 Results for radial flow, method 2

Node	Radius	a	Potential	Theoretical potential
—	1.0	0	0	0
A	1.782	0.57565	1.25	1.25
B	3.162	1.1513	2.5	2.5
C	5.624	1.72695	3.75	3.75
—	10.0	2.3026	5.0	5.0

The main purpose of the above example is to show how the calculations are carried out. However, the results also demonstrate that, for the case of radial flow towards a well, the second method with a logarithmic increase in radius is to be preferred. Nevertheless, there are situations, particularly when the flow is not directed towards the origin, when the first method has certain advantages.

3.6 Non-standard Finite-difference Approximations

It has been shown by Rushton (1964a) that finite-difference solutions do not always converge to the exact solution as the number of mesh subdivisions is increased. If the variation of the function throughout the field follows a roughly parabolic form, then the standard-difference approximation is usually adequate. However, if the flow is concentrated towards a particular region the variation may not approximate to a parabolic form and as a result the standard finite-difference approximation may not provide an adequate representation. Even when a very small mesh interval is used, the finite-difference solution may be in error by 10 or 20 per cent.

Typical situations where the standard cartesian approximation is inadequate include flow towards a well where the potential varies logarithmically with the radius from the well, and the flow near to the end of a drain or sheet pile.

Considering the situation in a region surrounding the junction of an impermeable base and a drain, the potential distribution is of the form (Lamb, 1932)

$$h - h_0 = C_1 r^{1/2} \sin(\theta/2) + C_2 r^{3/2} \sin(3\theta/2) + \cdots \tag{3.26}$$

In this expression, r is the radial distance from the end of the drain, h_0 is the potential at the junction of the drain and impermeable boundary, and θ is an angle as defined in Fig. 3.7. The constants C_1, C_2, etc., depend on the quantity flowing and the nature of the outer boundaries. There are two possible approaches when incorporating this type of singularity into a standard finite-difference solution.

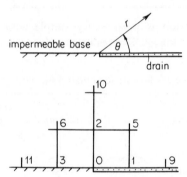

Figure 3.7. Singularity at junction of impermeable base and drain

Alternative equation method

In this first method, the first two terms of Eq. 3.26 are written for nodes 2, 3, 10, and 11 of the square grid (Fig. 3.7). The four equations are

$$h_2 - h_0 = C_1(\Delta x)^{1/2} \sin(\pi/4) + C_2(\Delta x)^{3/2} \sin(3\pi/4)$$

$$h_3 - h_0 = C_1(\Delta x)^{1/2} \sin(\pi/2) + C_2(\Delta x)^{3/2} \sin(3\pi/2)$$

$$h_{10} - h_0 = C_1(2\Delta x)^{1/2} \sin(\pi/4) + C_2(2\Delta x)^{3/2} \sin(3\pi/4) \tag{3.27}$$

$$h_{11} - h_0 = C_1(2\Delta x)^{1/2} \sin(\pi/2) + C_2(2\Delta x)^{3/2} \sin(3\pi/2)$$

It should also be noted that since the drain passes through nodes 0, 1, and 9 that $h_0 = h_1 = h_9$.

Eliminating C_1 and C_2 leads to the equations

$$h_2 = 0.5303h_{10} + 0.125h_{11} + 0.3447h_0$$

$$h_3 = 0.25h_{10} + 0.5303h_{11} + 0.2197h_0 \tag{3.28}$$

These special singularity equations are introduced to calculate the potentials at nodes 2 and 3. They can be included with the standard finite-difference equations which apply at the other nodes and therefore become part of the overall set of simultaneous equations. The importance of this form of correction is demonstrated by Rushton (1964a).

Modified permeability

The approach adopted in the second method is to introduce an effective permeability into the standard finite-difference equation which will represent the modified flow pattern in the vicinity of the singularity.

The governing differential equation is

$$mk\left(\frac{\partial^2 h}{\partial x^2} + \frac{\partial^2 h}{\partial y^2}\right) = 0 \tag{3.29}$$

where m is the saturated depth of the aquifer. When written for a mesh with $\Delta x = \Delta y$ and centred at node 2 (Fig. 3.7), Eq. 3.29 becomes

$$m[k(h_5 - h_2) + k(h_{10} - h_2) + k(h_6 - h_2) + k(h_0 - h_2)] = 0$$

In order to represent the modified flow, the permeability between nodes 2 and 0 is modified to k_s, and the equation becomes

$$m[k(h_5 - h_2) + k(h_{10} - h_2) + k(h_6 - h_2) + k_s(h_0 - h_2)] = 0 \tag{3.30}$$

Taking only the first term of Eq. 3.26,

$$h_5 - h_0 = C_1(2)^{1/4}\Delta x^{1/2} \sin(\pi/8)$$

$$h_{10} - h_0 = C_1(2)^{1/2}\Delta x^{1/2} \sin(\pi/4)$$

$$h_6 - h_0 = C_1(2)^{1/4}\Delta x^{1/2} \sin(3\pi/8) \tag{3.31}$$

$$h_2 - h_0 = C_1\Delta x^{1/2} \sin(\pi/4)$$

Substituting these expressions into Eq. 3.30 leads to

$$k_s = 0.6115k \tag{3.32}$$

Using this modified permeability, an improved representation of the flow in the vicinity of the end of the drain is achieved. Examples are discussed by Herbert (1968a) and further details of the singularity corrections are given in Sections 4.10 and 9.2.

3.7 Curved Boundaries

In certain problems the boundaries do not coincide with the mesh points of cartesian networks and therefore some modifications must be made to represent these curved boundaries. This can be achieved by devising special finite-difference equations at the boundaries. The coefficients of the finite-difference equations are calculated using the length and cross-sectional area as described by Eq. 3.16.

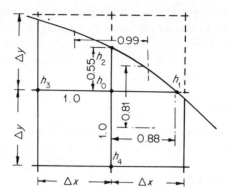

Figure 3.8 Representation of curved boundaries

Therefore, for the boundary of Fig. 3.8, the finite-difference form of the governing equation

$$T\left(\frac{\partial^2 h}{\partial x^2} + \frac{\partial^2 h}{\partial y^2}\right) = -q$$

becomes

$$\frac{0.81\Delta y}{0.88\Delta x}(h_1 - h_0) + \frac{0.99\Delta x}{0.55\Delta y}(h_2 - h_0) + \frac{\Delta y}{\Delta x}(h_3 - h_0) + \frac{\Delta x}{\Delta y}(h_4 - h_0)$$

$$= \frac{-(1.88\Delta x)(1.55\Delta y)q}{4T} \tag{3.33}$$

This form of boundary condition can be used when the conditions on the boundary are known values of the function or known values of the slope.

MacNeal (1953) has proposed an alternative approach in which the region is covered by a random array of nodal points, chosen to give a good representation of the boundary shape. However, as was shown by Rushton (1964b), a square array of mesh points should be used whenever possible with irregular mesh lengths on the boundary, since this leads to smaller truncation errors.

As a demonstration of the adequacy of the technique of using a square mesh for problems in which the boundaries are curved, the radial problem of flow towards a well previously considered in Section 3.5 will be studied using a cartesian network. On the outer boundary of radius 10 units, a condition of constant inflow is enforced. At the inner boundary, $r = 5$ units, the piezometric head is held fixed at zero. The coefficients of the finite-difference equations are recorded in Fig. 3.9 with the values close to the boundaries calculated as described above. The inflow condition on the outer boundary is enforced by injecting a flow proportional to the length of the boundary modelled by that node; values of the inflow are recorded in Fig. 3.9. The total inflow on this 45° sector equals 0.25π. On the inner boundary nodes, the head is held fixed at zero. Numerical results are quoted above the line in Fig. 3.10.

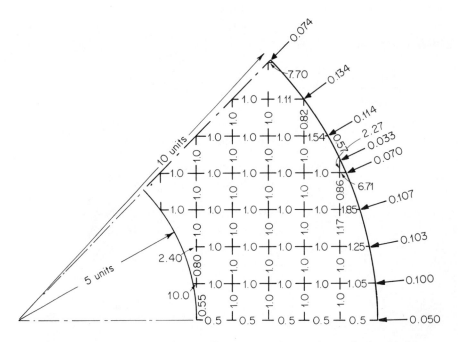

Figure 3.9. Coefficients of finite-difference equations and magnitude of inflow

Analytical values can be calculated from Eq. 3.25. For zero potential at the boundary, r_w, the analytical expression in non-dimensional form becomes

$$\frac{h2\pi T}{Q} = \log_e \left(\frac{r}{r_w}\right) \tag{3.34}$$

The analytical values are recorded below the line in Fig. 3.10. Small differences do occur; these are partly due to the finite-difference approximation and partly due to the approximation introduced in the representation of the curved boundaries. However, the maximum error of 2 per cent on the outer boundary does indicate that this finite-difference approach is adequate.

3.8 Time-variant Problems

When the seepage or groundwater flow is time-variant, the equation is usually of the form

$$\nabla^2 h = c\, \partial h / \partial t \tag{3.35}$$

where c may be a function of space and also of the potential h. This is an elliptic differential equation (Smith, 1965).

Many alternative approaches are possible when representing the time dimension. One approach when using an *electrical analog* is to model the physical time by an analogous electrical time. This method is introduced and discussed in detail in Sections 3.9 and Chapter 7.

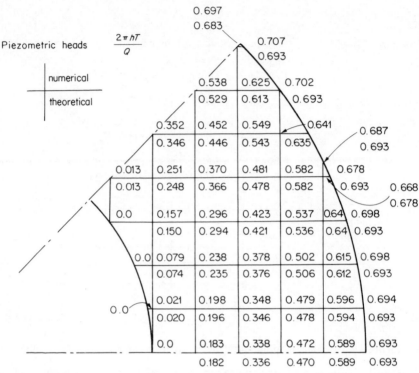

Figure 3.10. Comparison of numerical and theoretical groundwater potentials

In a second approach which is particularly suitable to *digital computer* methods, the continuous time is divided into discrete time steps at certain predetermined times in a similar manner to the discrete representation of continuous space.

Discrete time steps

Since in a discrete-time solution, the function is only defined at particular

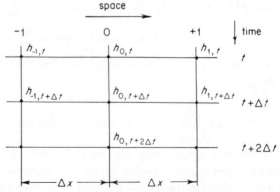

Figure 3.11. Representation of time dimension

times, discrete time steps can be represented pictorially as a series of horizontal lines (Fig. 3.11). Increasing time is represented in the downwards direction; this convention is adopted since it is convenient to print the results of a digital computer solution at succeeding times down a page.

It is usually adequate to take the simplest finite-difference approximation, namely

$$c\frac{\partial h}{\partial t} = c\frac{h_{0,t+\Delta t} - h_{0,t}}{\Delta t} \tag{3.36}$$

where the first suffix refers to the position in space and the second suffix refers to the time. The various finite-difference approximations depend on the time at which $\nabla^2 h$ is defined.

Approximation 1

In the case where $\nabla^2 h$ is defined at time t, this is called a *forward-difference approximation*. Taking the one-dimensional case where

$$\frac{\partial^2 h}{\partial x^2} = c\frac{\partial h}{\partial t}$$

the forward-difference finite-difference approximation in the space dimension with regular intervals Δx is

$$h_{-\Delta,t} - 2h_{0,t} + h_{1,t} = (c\Delta x^2/\Delta t)(h_{0,t+\Delta t} - h_{0,t})$$

or

$$h_{0,t+\Delta t} = \frac{\Delta t}{c\Delta x^2}\left[h_{-1,t} - \left(2 - \frac{c\Delta x^2}{\Delta t}\right)h_{0,t} + h_{1,t}\right] \tag{3.37}$$

Since the function at time $t + \Delta t$ can be expressed explicitly in terms of the known values of the function at time t, this is termed an *explicit* formulation. The calculation is described pictorially in Fig. 3.12(a), in which the symbol $\times$ refers to unknowns, the symbol $\bigcirc$ refers to values of the function calculated at the previous time step, and the heavy lines indicate terms involved in the calculation.

Approximation 2

When $\nabla^2 h$ is defined at time $t + \Delta t$, then the approximation is termed *backward difference*. The unknown functions at time $t + \Delta t$ are calculated from

$$h_{-1,t+\Delta t} - (2 + c\Delta x^2/\Delta t)h_{0,t+\Delta t} + h_{1,t+\Delta t} = -(c\Delta x^2/\Delta t)h_{0,t} \tag{3.38}$$

This leads to simultaneous equations as illustrated in Fig. 3.12(b); the scheme is therefore said to be *implicit*.

Approximation 3

When the term $\nabla^2 h$ is defined both at times t and $t + \Delta t$,

$$0.5(\nabla^2 h)_t + 0.5(\nabla^2 h)_{t+\Delta t} = (c\Delta x^2/\Delta t)(h_{0,t+\Delta t} - h_{0,t}) \tag{3.39}$$

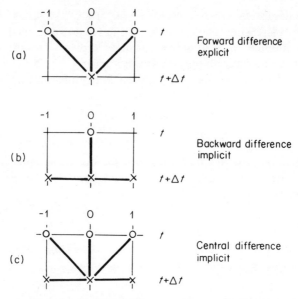

Figure 3.12. Finite-difference time approximations

This is a *central-difference approximation*, often called the Crank–Nicholson method. The dependence of the unknown functions $h_{0,t+\Delta t}$, $h_{1,t+\Delta t}$, $h_{-1,t+\Delta t}$ on the known functions is illustrated in Fig. 3.12(c). From this diagram it is clear that this is also an *implicit* approximation.

Stability and convergence

From the above description it is apparent that the most convenient technique is the forward-difference explicit procedure since it does not involve the solution of simultaneous equations. However, it suffers from two serious limitations. The first limitation is that if the time step exceeds a critical value unstable results are obtained, with the calculated functions showing wild fluctuations. The stability of a finite-difference scheme can be studied using a von Neumann analysis (Smith, 1965, p. 70). The stability restriction for a one-dimensional problem with constant coefficients using a constant mesh interval Δx is

$$\Delta t / c\, \Delta x^2 \leqslant 0.5 \tag{3.40}$$

There is no stability criterion for the backward-difference approximation.

The second limitation of finite-difference schemes relates to the convergence. Differences between the analytical and numerical results can be caused by the finite-difference approximation. The space discretization leads to certain errors, but the nature of the forward-difference time approximation, where the current value of the function depends only on the behaviour at the previous time step, can lead to more serious errors. For example, the effect of a change in a condition at a boundary has no effect within the field until later time steps. Therefore the convergence of the forward-difference approximation is usually inferior to the backward-difference approximation.

Difficulties arising from stability and convergence will be discussed in greater detail in subsequent chapters, but it should be noted that the occurrence of sudden changes in flow leads to severe constraints on finite-difference schemes.

3.9 Solution of Equations

Finite-difference approximations for both steady-state and time-variant problems lead to sets of simultaneous equations. There are three main methods of solving these equations: electrical analogs; direct matrix solution; and iterative methods.

Electrical analogs

Arrays of resistances and capacitances can be devised which, when Kirchhoff's current law is written for a nodal point, lead to an equation which is directly analogous to the finite-difference equation. If the boundary conditions are simulated as equivalent voltages or currents the network automatically gives a solution to the set of finite-difference equations.

Table 3.4 summarizes electrical circuits which are analogous to a number of the finite-difference equations. Three of the examples are discussed in detail below.

Example 1

Consider the differential equation which describes *one-dimensional flow* in an aquifer, Eq. 3.1,

$$T\frac{d^2h}{dx^2} = -q$$

A finite-difference form of the differential equation was derived as Eq. 3.9; this will be rearranged as

$$\frac{T}{\Delta x^2}(h_{-1} - 2h_0 + h_1) = -q \tag{3.41}$$

Now consider the electrical network of Fig. 3.13(a). Voltages at the nodal points are V_{-1}, V_0, and V_1 with both resistors of R ohms. Kirchhoff's current law states that the total current entering a nodal point should equal zero. Thus at node 0,

$$I_a + I_b + I_0 = 0$$

Since

$$I_a = \frac{V_{-1} - V_0}{R} \quad \text{and} \quad I_b = \frac{V_1 - V_0}{R}$$

Table 3.4 Resistance networks for selected finite-difference equations

Eq. no.	Dimensions	Cartesian or radial	Finite-difference equations
3.9	1	C	$\dfrac{T}{\Delta x^2}(h_{-1} - 2h_0 + h_1) = -q$
3.13	3	C	$\dfrac{1}{\Delta x^2}(h_1 - 2h_0 + h_3) + \dfrac{1}{\Delta y^2}(h_2 - 2h_0 + h_4)$ $+ \dfrac{1}{\Delta z^2}(h_5 - 2h_0 + h_6) = 0$
3.19	1	R	$T\left[\dfrac{h_{-1} - 2h_0 + h_1}{\Delta r^2} + \dfrac{1}{r_0}\dfrac{h_1 - h_{-1}}{2\Delta r}\right] = -q$
3.24	1	R	$\dfrac{m_b k_b(h_{-1} - h_0)}{\Delta a_b} + \dfrac{m_a k_a(h_1 - h_0)}{\Delta a_a}$ $= -0.5\, q r_0 (\Delta a_a + \Delta a_b)$
3.33 irregular boundary	2	C	$\dfrac{0.81\,\Delta y}{0.88\,\Delta x}(h_1 - h_0) + \dfrac{0.99\,\Delta x}{0.55\,\Delta y}(h_2 - h_0) + \dfrac{\Delta y}{\Delta x}(h_3 - h_0)$ $+ \dfrac{\Delta x}{\Delta y}(h_4 - h_0) = \dfrac{-1.88\,\Delta x\, 1.55\,\Delta y}{4T}q$
6.5	1	C	$\dfrac{T}{\Delta x^2}(h_{-1} - 2h_0 + h_1) = -q + S\dfrac{\partial h}{\partial t}$

Resistance network	*Electrical equation*
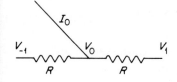	$$\frac{V_{-1} - 2V_0 + V_1}{R} = -I_0$$
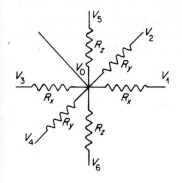	$$\frac{V_1 - 2V_0 + V_3}{R_x} + \frac{V_2 - 2V_0 + V_4}{R_y}$$ $$+ \frac{V_5 - 2V_0 + V_6}{R_z} = 0$$
	Resistance network cannot be devised
	$$\frac{V_{-1} - V_0}{R_b} + \frac{V_1 - V_0}{R_a} = -I_0$$
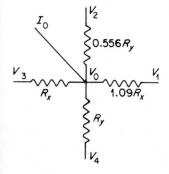	$$\frac{V_1 - V_0}{1.09\,R_x} + \frac{V_2 - V_0}{0.556R_y} + \frac{V_3 - V_0}{R_x}$$ $$+ \frac{V_4 - V_0}{R_y} = -I_0$$
	$$\frac{V_{-1} - V_0}{R} + \frac{V_1 - V_0}{R} = -I_0 + C\frac{\partial V_0}{\partial t_e}$$

then

$$\frac{V_{-1} - 2V_0 + V_1}{R} = -I_0 \qquad (3.42)$$

Equations 3.41 and 3.42 are of similar form. Therefore if:

(a) voltage is proportional groundwater potential;
(b) electrical resistance is inversely proportional to transmissivity; and
(c) electrical current is proportional to recharge q;

then Eqs. 3.41 and 3.42 are of identical form. Consequently the finite-difference equations can be solved by an analogous resistance network. Details of the scaling factors between physical and electrical quantities are discussed in Section 3.10. This example is summarized in the first row of Table 3.4.

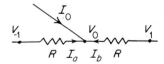

(a) Resistance network

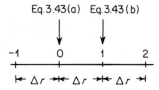

(b) Node - numbering system for radial example

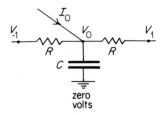

(c) Resistance – capacitance network

Figure 3.13. Derivation of resistance network analogs

Example 2

A different situation occurs when the *flow is radial*. For constant permeability and saturated depth, the differential equation is

$$T\left(\frac{d^2h}{dr^2} + \frac{1}{r}\frac{dh}{dr}\right) = -q \qquad (3.18)$$

If a constant mesh interval Δr is chosen the finite-difference equation becomes

$$T\left[\frac{h_{-1} - 2h_0 + h_1}{\Delta r^2} + \frac{1}{r_0}\left(\frac{h_1 - h_{-1}}{2\Delta r}\right)\right] = -q \qquad (3.19)$$

Note that this equation is centred at node 0.

To investigate whether an equivalent resistance network can be devised the equation must be rearranged as

$$T\left[\left(\frac{1}{\Delta r^2} - \frac{1}{2r_0\Delta r}\right)(h_{-1} - h_0) + \left(\frac{1}{\Delta r^2} + \frac{1}{2r_0\Delta r}\right)(h_1 - h_0)\right] = -q \quad (3.43a)$$

An electrical network can be devised for this equation and, in particular, the resistance between nodes 0 and 1 must be inversely proportional to

$$\left(\frac{1}{\Delta r^2} + \frac{1}{2r_0\Delta r}\right)$$

It is essential to check whether the same resistance is required between nodes 1 and 0 when the equation is written centred at node 1 with nodes 0 and 2 a distance Δr to the left and right of node 1 (Fig. 3.13(b)).

The equation becomes

$$T\left[\left(\frac{1}{\Delta r^2} - \frac{1}{2r_1\Delta r}\right)(h_0 - h_1) + \left(\frac{1}{\Delta r^2} + \frac{1}{2r_1\Delta r}\right)(h_2 - h_1)\right] = -q \quad (3.43b)$$

The required resistance between node 1 and 0 must now be inversely proportional to

$$\left(\frac{1}{\Delta r^2} - \frac{1}{2r_1\Delta r}\right)$$

Clearly this conflicts with the expression required by Eq. 3.43(a) and therefore this series of simultaneous equations cannot be solved by a resistance network.

However, as was shown in Section 3.5, if the mesh increases logarithmically, an equation similar to the one-dimensional cartesian equation is derived:

$$T\left(\frac{h_{-1} - 2h_0 + h_1}{\Delta a^2}\right) = -qr_0^2 \qquad (3.22)$$

where $a = \log_e r$. This can be solved by a resistance network similar to Example 1.

Example 3

The third example concerns a *one-dimensional time-dependent* equation based on Eqs. 3.1 and 3.35,

$$T\frac{\partial^2 h}{\partial x^2} = -q + S\frac{\partial h}{\partial t} \qquad (3.44)$$

Equation 3.44 is discretized in space in a similar form to example (1),

$$\frac{T}{\Delta x^2}(h_{-1} - 2h_0 + h_1) = -q + S\left(\frac{\partial h}{\partial t}\right)_0 \tag{3.45}$$

To devise an analogous electrical network to Eq. 3.45, a capacitor of capacitance C is added to the network. The capacitor has the property of absorbing or releasing an electrical current as the potential difference across it changes. Since one end of the capacitor is connected to zero voltage and the other end is connected to node 0, the current entering the network due to a change in voltage V_0 is

$$I_c = -C\left(\frac{\partial V}{\partial t_e}\right)_0$$

where t_e is the *electrical time*.

Kirchhoff's Law for node 0 leads to the equation

$$\frac{V_{-1} - 2V_0 + V_1}{R} = -I_0 + C\left(\frac{\partial V}{\partial t_e}\right)_0 \tag{3.46}$$

This equation is analogous to Eq. 3.45.

There are three scaling factors relating physical to electrical quantities; see Section 7.2. Note that in this analogy the physical time is represented by continuous electrical time.

The application of electrical networks in the solution of a large number of other field problems is discussed by Karplus (1958). The particular merit of the electrical-analog approach is that immediate solutions to the finite-difference equations are obtained. Resistance and resistance–capacitance networks are used for many of the examples in Parts II and III. Practical details of a typical resistance network solution are given in Section 3.10.

Direct matrix solution

For all but the simplest problems, a direct solution of the simultaneous equation is uneconomical even when high-speed digital computers are used. However, for one-dimensional problems, efficient solutions can be obtained by means of Gaussian elimination of the tri-diagonal matrix which is described in detail in Section 9.1.

Since efficient direct solutions to one-dimensional problems can be obtained, an attractive approach to two-dimensional problems is possible if the two-dimensional grid is considered as a series of interconnected one-dimensional strips. During the first sub-step, the grid is swept in the x-direction, row by row, leading to a series of one-dimensional solutions, each with a tri-diagonal matrix. In the second sub-step the equations are swept in the y-direction. This *alternating direction implicit* procedure, first introduced by Peaceman and Rachford (1955), is suitable for certain types of problem.

Iterative solutions

Most iterative solutions are developments of the relaxation method introduced by Southwell (1940). Initially relaxation was devised as a 'hand' method, and 'human skill' could be introduced to speed the convergence. With the advent of high-speed digital computers, the relaxation procedure is carried out automatically. The simplest technique to program is the *point successive over-relaxation* method (SOR); it has the advantage that the speed of convergence is not particularly sensitive to the choice of the over-relaxation parameter. Other variations of the SOR approach which can lead to a more rapid convergence are described by Smith (1965). One particular advantage of the iterative techniques is that non-linearities can be included without difficulty.

Choice of method

The choice of method of solving the finite-difference equations depends largely on the nature of the problem and the type of analog or digital computer available. Each of the above techniques is discussed in greater detail in the following chapters. However, since there is little published information on the practical approach to resistance network solutions, a particular example is described in the next section.

3.10 Practical Resistance Network Solutions

This section presents a detailed description of the solution of a typical problem using the resistance network method. Aspects to be covered include the choice of resistance values, layout of the network, and instrumentation.

The problem selected is concerned with the radial flow through an aquifer from an outer to an inner boundary. Though an efficient solution can be obtained using radial coordinates, for the purpose of demonstrating the techniques a cartesian (x, y) grid will be used. A non-dimensional numerical solution to this problem was presented in Section 3.7, but in this instance dimensional parameters will be used. This problem is representative of a large number of seepage and groundwater problems and is chosen because it illustrates many of the features of the resistance network technique.

Details of the problem are given in Fig. 3.14(a). A total quantity of $200\pi \, \text{m}^3/\text{d}$ enters the outer boundary of the aquifer which is at a radial distance of 100 m. This recharge is equivalent to $1.0 \, \text{m}^3/\text{d}$ for each unit length of the outer boundary. The water then flows through the homogeneous porous medium to an inner boundary at a radius of 50 m, along which the groundwater potential is maintained at a fixed value of 5.5 m. The aquifer is 2.5 m thick with a uniform permeability of 2.0 m/d.

A square mesh of side 10 m is selected. This is a relatively coarse mesh and is chosen for clarity of presentation of the results. However, for an accuracy of better than 1 per cent it would be advisable to reduce the mesh interval to 5 m, thereby

50

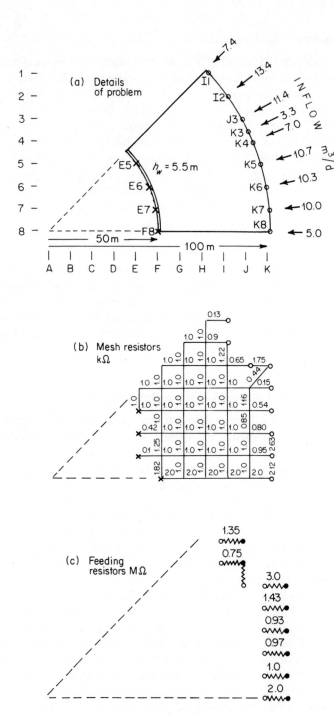

Figure 3.14. Details of problem and equivalent finite-difference network

providing 10 mesh subdivisions between outer and inner boundaries. Due to symmetry, only one-eighth of the field needs to be modelled.

Since a discrete model is used, the recharge must be distributed between the nodes according to the length of arc that each nodal point represents. Thus, for node K6 where the arc length associated with that node is 10.3 m, the recharge is 10.3 m³/d.

An analytical solution is available for this problem (Eq. 3.34):

$$h = h_w + \frac{Q}{2\pi km} \log_e \left(\frac{r}{r_w} \right) \tag{3.47}$$

where $h_w = 5.5$ m and $r_w = 50$ m.

Electrical analogy

The derivation of the analogy between electrical and physical terms is discussed in greater detail in Sections 7.2 and 9.1; the following simplified analysis is based on a regular mesh. The governing differential equation

$$mk \left(\frac{\partial^2 h}{\partial x^2} + \frac{\partial^2 h}{\partial y^2} \right) = -q \tag{3.48}$$

can be written in finite-difference form for the mesh of Fig. 3.15(a) as

$$mk \left[\frac{h_1 - 2h_0 + h_3}{\Delta x^2} + \frac{h_2 - 2h_0 + h_4}{\Delta y^2} \right] = -q$$

Multiplying through by $\Delta x \Delta y$ gives

$$\frac{mk\Delta y}{\Delta x}(h_1 - 2h_0 + h_3) + \frac{mk\Delta x}{\Delta y}(h_2 - 2h_0 + h_4) = -q\Delta x \Delta y = -Q \tag{3.49}$$

where Q is the total recharge at that node.

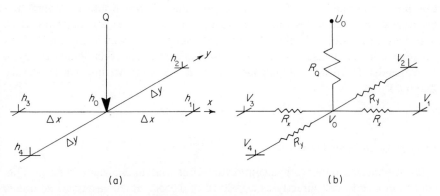

(a)   (b)

Figure 3.15. Finite-difference grid and equivalent resistance network

Identical equations are obtained for the resistance network of Fig. 3.15(b). Kirchhoff's current law for node 0 leads to the equation

$$\frac{V_1 - 2V_0 + V_3}{R_x} + \frac{V_2 - 2V_0 + V_4}{R_y} = \frac{-(U_0 - V_0)}{R_Q} \tag{3.50}$$

Equations 3.49 and 3.50 are analogous and the following relationships can be defined:

$$V = F(1)h$$

$$R_x = \frac{F(2)\Delta x}{km\Delta y} \qquad R_y = \frac{F(2)\Delta y}{km\Delta x} \tag{3.51}$$

where $F(1)$ and $F(2)$ are scaling factors.

From these a further relationship can be derived:

$$\frac{U_0 - V_0}{R_Q} = \frac{F(1)}{F(2)}Q \tag{3.52}$$

Choice of scaling factors

Factors $F(1)$ and $F(2)$ are selected so that the resistors and the voltages take convenient values. For this steady-state problem a wide range of parameters could be chosen; in this instance the mesh resistors are around $1\,k\Omega$, the mesh voltages are about 0.1 volt and the feeding resistors which provide the inflows should be in the region of $1\,M\Omega$. The procedure is to take trial values of $F(1)$ and $F(2)$; if these lead to unsuitable parameters, then modifications need to be made.

A value of $F(1) = 0.01$ is assumed, and if $F(2) = 5 \times 10^3$, the mesh resistors calculated from Eqs. 3.51 for internal nodes become

$$R_x = R_y = \frac{5 \times 10^3 \times 10}{10 \times 2.5 \times 2} = 10^3\,\Omega$$

On the horizontal axis of symmetry the resistors are doubled. Where the boundaries cut the mesh lines, the values of Δx and Δy are determined as described in Section 3.7. The resistance values are given in Fig. 3.14(b). It is possible to fit all the resistors in the standard square grid (further details are given in Appendix 1), though in one instance it is necessary to insert a resistance diagonally.

Fixed potential condition

On the inner boundary, the groundwater potential takes a value of 5.5 m. This condition is enforced directly as a voltage of 0.055 V at the appropriate nodes (E5, E6, E7, F8).

Inflow condition

On the outer boundary, the inflows per node, which are proportional to the arc length, are indicated in Fig. 3.14(a). These inflows are simulated by injecting electrical currents. Taking, for example, the mesh point K5, the quantity to be injected is $10.7\,m^3/d$. According to Eq. 3.52, this quantity can be represented as a voltage difference across a large resistance. Choosing $U_0 - V_0$ as 20.0 volts, then

$$R_Q = \frac{F(2)(U_0 - V_0)}{F(1)Q} = \frac{5 \times 10 \times 20}{0.01 \times 10.7} = 0.93\,M\Omega.$$

Other resistances can be calculated in the same manner; they are recorded in Fig. 3.14(c). In general it is advisable to arrange for the feeding resistances to be about 1000 times the value of the mesh resistances.

Electrical circuits and equipment

A schematic diagram of the electrical circuit is given in Fig. 3.16. In addition to the resistance network, the equipment required includes a digital voltmeter and a stabilized power supply. A specification of inexpensive equipment for this problem is presented in Table 3.5. The voltage on the inner boundary is set to 0.055 V using a 100-Ω helical potentiometer in series with a dropping resistance. Details of the construction of the network can be found in Appendix 1.

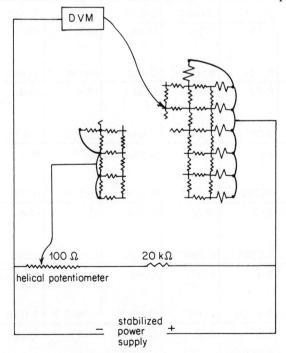

Figure 3.16. Schematic circuit diagram of resistance network

c

54

Table 3.5 Specification of equipment for simple experiment

Item	Minimum specification
Resistors	$\frac{1}{2}$ watt, 5% tolerance, E24 range
Digital voltmeter	sensitivity $100\,\mu$V, three digit, input impedance greater than $10\,M\Omega$
Power supply	30 V, 1 A, variable
Potentiometer	helical, 3 turn, $100\,\Omega$ 2 W

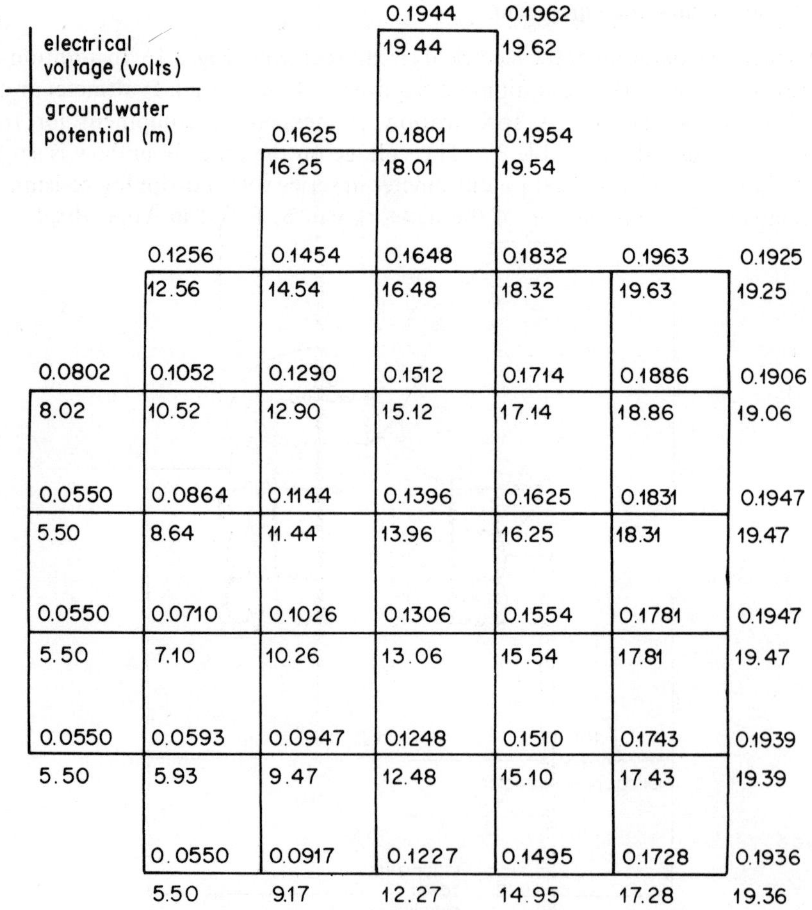

electrical voltage (volts) / groundwater potential (m)

| | | | | | 0.1944 | 0.1962 |
| | | | | | 19.44 | 19.62 |

| | | | 0.1625 | 0.1801 | 0.1954 | |
| | | | 16.25 | 18.01 | 19.54 | |

| | 0.1256 | 0.1454 | 0.1648 | 0.1832 | 0.1963 | 0.1925 |
| | 12.56 | 14.54 | 16.48 | 18.32 | 19.63 | 19.25 |

| 0.0802 | 0.1052 | 0.1290 | 0.1512 | 0.1714 | 0.1886 | 0.1906 |
| 8.02 | 10.52 | 12.90 | 15.12 | 17.14 | 18.86 | 19.06 |

| 0.0550 | 0.0864 | 0.1144 | 0.1396 | 0.1625 | 0.1831 | 0.1947 |
| 5.50 | 8.64 | 11.44 | 13.96 | 16.25 | 18.31 | 19.47 |

| 0.0550 | 0.0710 | 0.1026 | 0.1306 | 0.1554 | 0.1781 | 0.1947 |
| 5.50 | 7.10 | 10.26 | 13.06 | 15.54 | 17.81 | 19.47 |

| 0.0550 | 0.0593 | 0.0947 | 0.1248 | 0.1510 | 0.1743 | 0.1939 |
| 5.50 | 5.93 | 9.47 | 12.48 | 15.10 | 17.43 | 19.39 |

| | 0.0550 | 0.0917 | 0.1227 | 0.1495 | 0.1728 | 0.1936 |
| | 5.50 | 9.17 | 12.27 | 14.95 | 17.28 | 19.36 |

Figure 3.17. Measured voltages and equivalent groundwater potentials

Experimental procedure

The first step in the experiment is to set the voltages at the ends of the current-feeding resistances (indicated by solid circles in Fig. 3.14(c)) to 20.0 V. Next, the voltages on the nodes representing the outer boundary are measured and found to be approximately 0.19 V. The applied voltage is then increased to 20.19 V, thereby ensuring that the voltage across the current-feeding resistances $(U_0 - V_0)$ equals 20.0 V.

After making this adjustment, the potentials can be measured at as many nodes as required. Measurements at all the nodes are recorded in Fig. 3.17; they can be converted into groundwater potentials by multiplying by 100. As was explained in Section 3.7, the potentials compare well with analytical values.

This type of steady-state problem with known boundary conditions is particularly convenient for solution using resistance networks. For problems with moving boundaries, iterative techniques are required (Sections 4.8 and 5.4), whilst for other time-variant problems resistance–capacitance networks are selected (Sections 7.2 and 9.1).

PART II

Seepage

CHAPTER 4

Steady-state Seepage

4.1 Introduction

The fundamental equations of groundwater flow were discussed in Chapter 2 and in this chapter the particular problem of steady-state seepage will be considered. The analysis of seepage has been described by many writers, notably Harr (1962), Cedergren (1977), and Scott (1963). The main emphasis by these writers has been on solutions either by analytical techniques or by constructing flow nets.

Any seepage problem involves three principal factors: the soil media, the type of flow, and the boundary conditions.

If the coefficient of permeability is independent of the direction of the flow, the soil is said to be *homogeneous* and *anisotropic*. In such soils the coefficient of coefficient of permeability at all points within the flow region, the soil is described as being *homogeneous* and *isotropic*. If the coefficient of permeability is dependent on the direction of the flow and if it is the same at all points in the flow region, the soil is said to be homogeneous and *anisotropic*. In such soils the coefficient of permeability is dependent on the direction of the flow but it is independent of the space coordinates. Finally, if the soil is composed of random diverse elements it is described as being *heterogeneous*.

The flow may be either *steady* or *time-variant*; in the latter case its characteristics are affected by the lapse of time. In order to preserve a steady state of flow within a region, the conditions at the boundaries of the region must remain unchanged in time. In some cases an unsteady flow may ultimately settle down to a steady-state condition.

The correct definition of the conditions at the boundaries of the region is very important. In flow domains where all the boundaries are fixed and therefore known initially, the flow is said to be *confined* but where one boundary is a free surface the flow pattern is said to be *unconfined*.

Fixed boundaries may be caused, for example, by the retention of water in a reservoir, behind a dam or due to the presence of a cut-off wall. If the soil media change from one type of soil to another, the plane of contact between the formations becomes an internal boundary which requires certain conditions to be satisfied when the boundary conditions are formulated.

The complexity of a seepage analysis is increased in circumstances of unconfined flow by the occurrence of a free streamline surface. This streamline surface is a boundary whose actual position in space cannot, *a priori*, be specified but can be defined only as the position of the boundary at which the value of the working function and its normal derivative take prescribed values. For example, when studying the flow of water through a porous soil whose top surface is open to the atmosphere, the position of the water surface will not necessarily coincide with the surface of the soil, and thus the position of this boundary cannot be specified directly. Though the position of the boundary cannot be specified, two conditions do have to be satisfied simultaneously on this boundary. The first is that the pressure is atmospheric and the second is that no flow crosses the boundary. This boundary is called a free streamline surface. In this book neither the capillary effect above the water table nor flow in the unsaturated zone are included in the analysis.

Confined flow problems in which the soil is isotropic, or even anisotropic, are relatively easy to solve, often by graphical methods (Cedergren, 1977) or, in certain idealized cases, by a mathematical analysis using complex-variable methods (Harr, 1962). Even so, unacceptable assumptions have often to be made in order to simplify the boundary conditions.

In other cases, particularly those involving a soil of heterogeneous composition, where the soil properties may vary randomly from point to point, the only practicable method of solution is to resort to a numerical method using either a digital or an electrical analog computer.

The primary concern of this chapter is the numerical solution of steady-state seepage problems for various soil formations and boundary conditions.

Emphasis has been placed on using the groundwater potential as the working function and it will be shown that the use of this function, in conjunction with a digital or electrical analog computer, provides a numerical grid analysis which can be applied without undue effort and gives all the required information on pressure and quantity of flow. This approach is an extension of the classical methods of analysis, in which either the groundwater potential or the velocity potential is selected as the working function, coupled with the conjugate stream function; this leads to the flow-net concept. This latter technique, although producing solutions to certain cases of importance, lacks the generality of the numerical grid analysis and cannot be applied satisfactorily in many practical situations, particularly those involving heterogeneous soil.

The basic theory, appropriate to each seepage condition, will now be presented followed by a description of the practical numerical solution of a number of illustrative examples.

4.2 Equations of Seepage

This section summarizes the various equations describing steady-state flow. It shows how the general three-dimensional equations can be simplified in certain situations.

Isotropic soil

It will be recalled from Chapter 2 that the general equation for steady state flow is (Eq. 2.7)

$$\frac{\partial}{\partial x}\left(k_x\frac{\partial h}{\partial x}\right) + \frac{\partial}{\partial y}\left(k_y\frac{\partial h}{\partial y}\right) + \frac{\partial}{\partial z}\left(k_z\frac{\partial h}{\partial z}\right) = 0 \tag{4.1}$$

For homogeneous isotropic soils the permeability is independent of the direction of flow, thus

$$k_x = k_y = k_z = k$$

Then the basic equation becomes

$$\frac{\partial^2 h}{\partial x^2} + \frac{\partial^2 h}{\partial y^2} + \frac{\partial^2 h}{\partial z^2} = 0 \tag{4.2}$$

This equation is often written in operational form as $\nabla^2 h = 0$, where the operator ∇^2 is used to stand for

$$\frac{\partial^2}{\partial x^2} + \frac{\partial^2}{\partial y^2} + \frac{\partial^2}{\partial z^2}$$

This is the well-known equation of Laplace which governs a large number of physical phenomena and arises in almost every branch of mathematical physics.

It is often convenient, particularly when solving problems involving radial flow, to use polar instead of cartesian coordinates. In polar cylindrical coordinates (r, θ, z), Eq. 4.2 becomes

$$\nabla^2 h \equiv \frac{\partial^2 h}{\partial r^2} + \frac{1}{r}\frac{\partial h}{\partial r} + \frac{1}{r^2}\frac{\partial^2 h}{\partial \theta^2} + \frac{\partial^2 h}{\partial z^2} = 0 \tag{4.3}$$

where

$$r = (x^2 + y^2)^{\frac{1}{2}} \quad \text{and} \quad \theta = \tan^{-1}\left(\frac{y}{x}\right)$$

For radial axi-symmetrical flow in the horizontal plane there is no variation with θ, and Eq. 4.3 reduces to

$$\frac{1}{r}\frac{\partial}{\partial r}\left(r\frac{\partial h}{\partial r}\right) = 0 \tag{4.4}$$

This equation can be solved directly by integration to give

$$h = C_1 \log_e r + C_2 \tag{4.5}$$

where C_1 and C_2 are constants of integration to be determined from the boundary conditions.

For homogeneous isotropic soils the groundwater potential h is a harmonic potential function of x, y, z or r, θ, z satisfying the Laplace equation.

Mathematically, the function is entirely determined in a unique fashion in the interior of a domain if the values of the function h or the values of its normal derivative $\partial h / \partial n$ are specified at the boundary. These two boundary conditions are known as Dirichlet and Neumann conditions, respectively. It will be shown later that there are other conditions at the boundary which can be encountered.

It is most important to realize that the permeability of the soil, k, does not figure in the Laplace equation. Thus the distribution of the groundwater potential does not depend on the permeability in the case of isotropic soils. The solution depends uniquely on the geometric form of the flow domain and on the boundary conditions.

Similar conditions hold for two-dimensional flow, Laplace's equation reducing, in cartesian coordinates, to

$$\nabla^2 h \equiv \frac{\partial^2 h}{\partial x^2} + \frac{\partial^2 h}{\partial z^2} = 0 \qquad (4.6)$$

Anisotropic soil

In certain cases sedimentary deposits allow the flow to proceed more easily along the planes of deposition than across them. A soil which exhibits this characteristic is defined as being *anisotropic*, although if the condition of anisotropy is the same from point to point in the stratum it is still homogeneous.

Situations involving non-orthogonal anisotropy are very difficult to study by analytical methods but the simpler case of orthogonal anisotropy and, in particular, the case in which the permeability is the same for all directions of a given plane and is different in a perpendicular direction ($k_x = k_y \neq k_z$) can be solved without undue difficulty.

Consider the general case of three-dimensional anisotropic flow. Commencing, as before, with the general equation for steady-state flow and in accordance with the definition of anisotropic soil, the directional permeabilities may be assumed to be invariant during the flow process and so they can be placed outside the derivatives, giving

$$k_x \frac{\partial^2 h}{\partial x^2} + k_y \frac{\partial^2 h}{\partial y^2} + k_z \frac{\partial^2 h}{\partial z^2} = 0 \qquad (4.7)$$

This equation may be reduced to the Laplace equation by means of a simple affine transformation.

Assume that k_z is smaller than both k_x and k_y and introduce the following changes in the coordinates

$$\bar{x} = x \sqrt{\frac{k_z}{k_x}}$$

$$\bar{y} = y \sqrt{\frac{k_z}{k_y}} \qquad (4.8)$$

which implies that distances in the direction of greater hydraulic conductivity are reduced.

Then

$$\frac{\partial h}{\partial x} = \frac{\partial h}{\partial \bar{x}} \frac{\partial \bar{x}}{\partial x}$$

$$= \sqrt{\frac{k_z}{k_x}} \frac{\partial h}{\partial \bar{x}}$$

and

$$\frac{\partial^2 h}{\partial x^2} = \sqrt{\frac{k_z}{k_x}} \frac{\partial}{\partial x}\left(\frac{\partial h}{\partial \bar{x}}\right)$$

$$= \sqrt{\frac{k_z}{k_x}} \frac{\partial}{\partial \bar{x}}\left(\frac{\partial h}{\partial \bar{x}}\right) \frac{\partial \bar{x}}{\partial x}$$

$$= \frac{k_z}{k_x} \frac{\partial^2 h}{\partial \bar{x}^2}$$

Similarly

$$\frac{\partial^2 h}{\partial y^2} = \frac{k_z}{k_y} \frac{\partial^2 h}{\partial \bar{y}^2}$$

Using these results the general equation for homogeneous anisotropic steady-state flow transforms to

$$\frac{\partial^2 h}{\partial \bar{x}^2} + \frac{\partial^2 h}{\partial \bar{y}^2} + \frac{\partial^2 h}{\partial z^2} = 0 \tag{4.9}$$

which is the Laplace equation in the variables $\bar{x}$, $\bar{y}$, and z. It is then possible to apply a scale transformation.

In the case of two-dimensional anisotropic flow, Eq. 4.9 reduces to

$$\frac{\partial^2 h}{\partial \bar{x}^2} + \frac{\partial^2 h}{\partial z^2} = 0 \tag{4.10}$$

Therefore, once the transformation has been effected, a problem involving an anisotropic medium reduces to that obtaining for an isotropic medium; that is to say, the solution of the Laplace equation in either two or three dimensions subject to the required boundary conditions. Finally, the results can be re-plotted to the true dimensional scale.

Stratified soil

Although in many cases the flow medium may be defined as being either isotropic or anisotropic, circumstances arise where sediments are deposited in layers. In most stratified soils the restriction to flow is less along the planes of deposition than across them, although this is not always the case.

Generally, the various layers may be either isotropic or anisotropic or a mixture of both, and in addition they may be either homogeneous or heterogeneous in composition.

If the soil is homogeneous, although composed of a mixture of isotropic and anisotropic layers, the problem may be posed as a series of Laplace equations but with involved conditions on the interfaces. A number of problems involving stratified soil have been considered by Harr (1962).

Heterogeneous soil

In some soil deposits the degree of heterogeneity is so great that it is inaccurate to idealize the situation as being homogeneous. Not only may the soil have been naturally deposited in layers of different compositions, but the layers themselves may contain inclusions of a different composition.

To appreciate the problem, consider again the general equation for steady-state three-dimensional flow which was established in Chapter 2 (Eq. 2.7)

$$\frac{\partial}{\partial x}\left(k_x\frac{\partial h}{\partial x}\right) + \frac{\partial}{\partial y}\left(k_y\frac{\partial h}{\partial y}\right) + \frac{\partial}{\partial z}\left(k_z\frac{\partial h}{\partial z}\right) = 0 \tag{4.11}$$

Now, as the permeabilities k_x, k_y, and k_z are functions of the spatial variables x, y, and z, it is impossible to reduce this equation to the three-dimensional form of Laplace's equation and hence normal analytical and flow-net methods for obtaining a solution cannot be applied.

Boundary conditions

All the usual boundary conditions which occur in a typical problem are illustrated in Fig. 4.1, which represents a permeable dam resting on an impermeable base and retaining water which seeps through the dam, running down the downstream face into a reservoir. These conditions will apply for both homogeneous and heterogeneous soils. Further discussion of boundary conditions can be found in Section 2.3 and Table 2.2.

The boundary conditions are as follows:

(a) There is no flow across the impermeable base and so this boundary is a flow line.

(b) On the upstream and downstream faces the pressure head is solely due to the water pressure and varies with the height above the datum; the groundwater potential and likewise the velocity potential assume appropriate constant values on these faces, which are therefore equipotential lines.

(c) On the free water surface (or water table) there are two conditions to be satisfied: the first is the flow line condition that there is no flow across the surface; the second condition results from the pressures on the boundary being zero

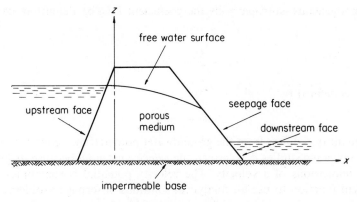

Figure 4.1. Typical seepage problem

(atmospheric). Thus the groundwater potential must equal the potential head at every point on the surface. The pressure head will be zero.

(d) The fourth type of boundary, the seepage face, is open to the atmosphere hence the groundwater potential equals the potential head. This condition allows a flow across the seepage face.

4.3 Velocity Potential and Stream Function

In the previous section it was shown that the steady flow in homogeneous isotropic or anisotropic soils reduces to the solution of the Laplace equation over the interior of the region. For these problems it is also possible to formulate the problem in terms of the velocity potential or stream function. The derivation of the relevant equations will be discussed in this section but, at the outset, it must be stressed that the alternative formulation can not be applied to heterogeneous soils.

Velocity potential

Two-dimensional flow with constant permeability has been shown to lead to Laplace's equation

$$\nabla^2 h \equiv \frac{\partial^2 h}{\partial x^2} + \frac{\partial^2 h}{\partial z^2} = 0 \qquad (4.12)$$

In two-dimensional flow, Darcy's law gives

$$v_x = -k\frac{\partial h}{\partial x} \qquad v_z = -k\frac{\partial h}{\partial z} \qquad (4.13)$$

where v_x and v_z are the Darcy velocities (see Section 2.2).

In homogeneous isotropic soils the coefficient k is, by definition, constant. Then

$$v_x = \frac{\partial \phi}{\partial x} \qquad v_z = \frac{\partial \phi}{\partial z} \qquad (4.14)$$

where ϕ is defined by

$$\phi = -kh \qquad (4.15)$$

The quantity ϕ differs from the groundwater potential by a constant factor k. In hydrodynamic studies ϕ is termed the *velocity potential* because its gradient has the dimensions of a velocity. The velocity potential is sometimes a very convenient function to use for analytical solutions of seepage problems.

Like the groundwater potential h, the velocity potential ϕ satisfies Laplace's equation

$$\nabla^2 \phi \equiv \frac{\partial^2 \phi}{\partial x^2} + \frac{\partial^2 \phi}{\partial z^2} = 0 \qquad (4.16)$$

and the equation of continuity, Eq. 2.6, becomes

$$\frac{\partial v_x}{\partial x} + \frac{\partial v_z}{\partial z} = 0 \qquad (4.17)$$

In Chapter 2, Eq. 2.4, the groundwater potential h was defined as

$$h = \frac{p}{\rho g} + z \qquad (4.18)$$

Then from Eq. 4.15

$$\phi = -k\left(\frac{p}{\rho g} + z\right) \qquad (4.19)$$

Hence

$$\nabla^2 \left(\frac{p}{\rho g}\right) = 0 \qquad (4.20)$$

by reason of Eqs. 4.16 and 4.19.

It will be appreciated that Eq. 4.20 expresses another Laplace equation which is written in terms of the pressure head, $p/\rho g$.

Stream function

It is well known (Sokolnikoff and Sokolnikoff, 1941) that Laplace's equation is satisfied by two conjugate harmonic functions which may be defined as ϕ and ψ and that the curves $\phi(x, z) = $ constant are the orthogonal trajectories of the curves $\psi(x, z) = $ constant. In groundwater flow and in fluid flow generally, the function $\psi(x, z)$ is known as the *stream function*; it is defined as

$$v_x = \frac{\partial \psi}{\partial z} \qquad v_z = -\frac{\partial \psi}{\partial x} \tag{4.21}$$

Substituting Eq. 4.21 into the equation of continuity, Eq. 4.17, yields

$$\frac{\partial^2 \psi}{\partial x\, \partial z} - \frac{\partial^2 \psi}{\partial z\, \partial x} = 0 \tag{4.22}$$

Equating the respective potential and stream functions of v_x and v_z, Eqs. 4.14 and 4.21, gives

$$\frac{\partial \phi}{\partial x} = \frac{\partial \psi}{\partial z} \qquad \frac{\partial \phi}{\partial z} = -\frac{\partial \psi}{\partial x} \tag{4.23}$$

These are the well-known Cauchy–Riemann differential equations. It is seen that $\psi(x, z)$ satisfies identically the equation of continuity and hence Laplace's equation, namely

$$\nabla^2 \psi \equiv \frac{\partial^2 \psi}{\partial x^2} + \frac{\partial^2 \psi}{\partial z^2} = 0 \tag{4.24}$$

The stream function ψ is of great importance in the understanding of groundwater flow and some of its properties will now be discussed.

Considering two-dimensional incompressible flow and referring to Fig. 4.2, let A be a fixed point in the plane of motion of the fluid and ABP, ACP two curves also in the plane joining A to an arbitrary point P. On the supposition that no fluid is created or destroyed within the region bounded by these curves, the condition of continuity may be expressed as follows.

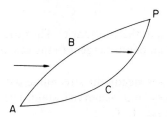

Figure 4.2. Arbitrary curves in the
flow field joining points A and P

The rate at which fluid flows into the region from left to right across the curve ABP is equal to the rate at which it flows out from left to right across ACP. The direction left to right is relative to an observer who proceeds along the curve from the fixed point A in the direction in which the arc of the curve measured from A is increasing.

Using the term *flux* to denote the rate of flow, it is seen that the flux from left to right across ACP is equal to the flux from left to right across any curves joining A to P. Once the datum point A has been fixed, this flux, denoted by ψ, therefore depends solely on the position of P and the time t. The existence of the stream function ψ is merely a consequence of the assertion of the continuity of an incompressible fluid.

Now consider two points P_1 and P_2 and let ψ_1 and ψ_2 be corresponding values of the stream function (Fig. 4.3). Then the flux across AP_2 is equal to the flux across AP_1 plus that across $P_1 P_2$. Hence the flux across $P_1 P_2$ from left to

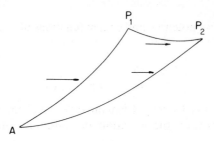

Figure 4.3. Flux across arbitrary boundaries

right $= \psi_2 - \psi_1$. It follows that if a different datum point is taken, A' say, the stream function merely changes by a constant, the flux from left to right across A'A.

Furthermore, if P_1 and P_2 are points on the same streamline, the flux from left to right across $P_1 P_2$ is equal to the flux from left to right across the streamline on which P_1 and P_2 lie. Thus $\psi_1 - \psi_2 = 0$ and therefore the stream function is constant along a streamline. (The locus of the path of an individual particle of fluid is described as a streamline or flowline.)

The equations of the streamline are therefore obtained from

$$\psi = C \tag{4.25}$$

by giving arbitrary values to the constant C. Thus the curves $\psi(x, y)$ equal to a series of constants are at all points tangent to the velocity vectors and define the path of flow.

When the fluid motion is steady the streamline pattern is fixed; in unsteady motion the pattern changes from instant to instant. The dimensions of the stream function are represented by $[L^2 T^{-1}]$.

An important property of the stream function can be obtained by considering the flow between the two adjacent streamlines ψ_1 and ψ_2 shown in Fig. 4.4. If the quantity of discharge through the line ab is q per unit length normal to the plane of flow then

$$q = \int_{\psi_2}^{\psi_1} v_x \, dz$$

$$= \int_{\psi_2}^{\psi_1} d\psi$$

$$= \psi_1 - \psi_2 \tag{4.26}$$

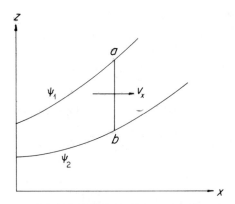

Figure 4.4. Flow in a stream tube

It is important to note that Eq. 4.26 states that the quantity of flow between two streamlines, termed a *flow channel*, is constant.

Properties of the velocity potential

Returning to a consideration of the velocity potential, which is analogous to a force potential whose directional derivative is the force in that direction, the velocity potential is a scalar function of space such that its derivative with respect to any direction is the velocity of the fluid in that direction. Like the stream function, the velocity potential has the dimensions $[L^2T^{-1}]$.

The velocity potential and stream functions differ in one important respect, inasmuch as the velocity potential function exists only for irrotational flow. (A particle of fluid is described as being irrotational if the circulation, which is the line integral of the tangential velocity taken around the particle, is zero.)

Consider the example shown in Fig. 4.5 in which ABCD represents a rectangular element in two-dimensional flow. The circulation for it is

$$v_x\, dx + \left(v_z + \frac{\partial v_z}{\partial x}\, dx\right) dz - \left(v_x + \frac{\partial v_x}{\partial z}\, dz\right) dx - v_z\, dz$$

from which, for the circulation to be zero,

$$\frac{\partial v_z}{\partial x} - \frac{\partial v_x}{\partial z} = 0 \qquad (4.27)$$

Using Eq. 4.14 in conjunction with Eq. 4.27 yields

$$\frac{\partial^2 \phi}{\partial x\, \partial z} - \frac{\partial^2 \phi}{\partial z\, \partial x} = 0 \qquad (4.28)$$

showing that the existence of the velocity potential implies that the flow is irrotational.

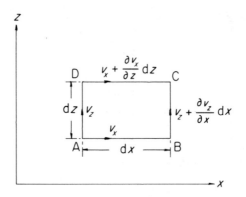

Figure 4.5. Rectangular element in two-
dimensional flow

Within a defined region of flow the streamlines and equipotential lines are
unique. Therefore, in solving homogeneous seepage problems it is only necessary
to be concerned with the determination of one of the functions, subject to the
imposed boundary conditions; the other function follows directly from the
Cauchy–Riemann equations, Eq. 4.23.

These properties of the potential function and the stream function lead on to
the well established method of analysis using flow nets.

4.4 Flow Nets

It has been shown that the solution of a problem of steady-state flow in
isotropic or anisotropic soils involves the solution of the Laplace equation
subject to certain boundary conditions. If the problem is solved in terms of one of
the functions, a family of equipotential lines will be derived which can be plotted
on a suitable diagram. In the case of two-dimensional flow the streamlines form
another family of curves which are orthogonal to the equipotential lines as they
intersect them at right angles. The grid of lines so plotted is called a *flow net*.

There is, however, one important difference, as far as the streamlines are
concerned, between two- and three-dimensional flow. For three-dimensional
fields of flow the velocity potential still satisfies Laplace's equation but the stream
function at a given point is no longer defined except in the special case of axi-
symmetrical flow. The reason for this is that a point specifies a certain streamline;
in the two-dimensional case this line is sufficient to divide the flow into regions,
whereas in the three-dimensional case a surface is required. For axially symmetric
flow the surface formed by the revolution of the streamline about the axis of
symmetry is used; as might be expected, even here ψ is no longer a solution of
Laplace's equation although streamlines are still normal to equipotential
surfaces (Bear, 1972).

The basic characteristics of flow nets are as follows:

(a) The flow net is based upon the assumption of irrotational flows, the voids

being filled with fluid; it is not necessary that the flow shall be steady since time-dependent seepage can be represented by the steady equation (see Section 2.5).
(b) The fluid is incompressible and of constant density and viscosity.
(c) Darcy's law must be valid.
(d) For a given set of boundary conditions there is only one possible pattern of flow.
(e) The equipotential lines intersect the streamlines and the fixed boundaries normally.
(f) No limitation is placed upon the number of equipotential lines and streamlines which may be drawn. The grid is composed of an assembly of curvilinear quadrilaterals which if sufficient lines are inserted will, in the limit, reduce to rectangles or squares according to the relative number of ϕ- and ψ-lines. Special consideration must be given to stagnation points or to points where the flow is theoretically infinite.

The flow-net concept was originally used by Forchheimer (1930) and extended by Casagrande (1940); its development stemmed from the original trial-and-error graphical method of determining the pattern of flow for a given set of boundary conditions. Numerous examples of flow-net techniques have been given by Cedergren (1977) and Scott (1963).

When using the graphical method, which has been well described by Cedergren (1977), it is advantageous to select the number of equipotential lines and streamlines as nearly equal as possible, avoiding as far as feasible fractional flow tubes and potential drops. In this way the net will consist of curvilinear squares, thus assisting the freehand sketching process. Although it has been claimed by some experienced operators that this trial-and-error process is simple and rapid, nevertheless many investigators have found the method to be both tedious and inaccurate, and it is for this reason that numerical methods have been developed.

Returning to the conventional flow net and referring to Fig. 4.6(a), consider the curvilinear quadrilateral element *abcd* extracted from the flow net. Let Δn denote the distance between the pair of adjacent streamlines and Δs the distance between the pair of adjacent equipotential lines; then the potential drop Δh across the two equipotential lines is the difference between the potential values along the two lines. The quantity of water Δq flowing through the cross-sectional area of the quadrilateral per unit width of the cross-section, normal to the diagram, is given by

$$\Delta q = k \frac{\Delta h}{\Delta s} \Delta n \tag{4.29}$$

As the quantity of flow between any two streamlines is constant, the total flow between the two fixed boundaries is obtained by summing the results for each streamline in accordance with Eq. 4.29. It should be noted that the answer must be the same whatever line joining the fixed boundaries is selected for the summation.

It will be opportune, at this stage, to discuss the use of the flow net in seepage problems, remembering that for practical purposes a seepage analysis is

72

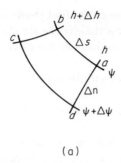

(a)

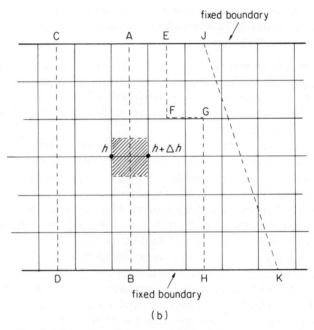

(b)

Figure 4.6. (a) Curvilinear quadrilateral element extracted
from flow net of Fig. 4.11. (b) Square grid in flow field

performed either to estimate the quantity of seepage or to determine pressures in order to undertake a stability analysis. For these reasons it is only necessary to determine one of the potential functions over the flow region; once one of these functions has been determined the others can be derived, if required, by a simple arithmetical calculation. In some, but not all cases it may be desirable to obtain the streamlines which have a very real value in providing a visualization of the flow pattern. It must be appreciated that the stream function is the conjugate function of the velocity potential and also of the groundwater potential which is related to the velocity potential by the definition $\phi = -kh$, Eq. 4.15. The pressure head is an analytic function that possesses a conjugate function which is not the stream function and is not of any practical significance. If a problem were solved in terms of the pressure head and if the streamlines were required, then it would be necessary for the easily deduced values of the velocity potential to be obtained before calculating the streamlines.

4.5 Numerical Grid Analysis

Graphical and analytical methods for flow-net construction can only be applied with any measure of success in cases where the homogeneous soil is either isotropic or anisotropic and the problem can be considered as being in two dimensions. For three-dimensional problems and those involving heterogeneity, these classical methods are impractical and some other method must be sought.

A numerical grid analysis will now be proposed as an alternative to the graphical flow net. Instead of using a flow net with curvilinear elements as illustrated in Fig. 4.6(a), suppose that, as shown in the Fig. 4.6(b), a square grid is drawn over the region of interest and assume that the values of the groundwater potential at the nodal points of the grid have been obtained by some numerical method. Then, as the same boundary conditions have been imposed, the quantity of seepage under the dam may be calculated by the identical method described for the flow-net method. In Eq. 4.29, $\Delta n/\Delta s$ will equal unity for a square grid and Δh will equal the difference between two adjacent recorded potentials. When carrying out the summation, any path between the fixed boundaries such as AB, CD or even the zig-zag path EFGH could be chosen and the result be the same. A slanting path such as JK could not be used without altering the values of Δn and Δs. This result, slightly surprising at first sight, will be understood when it is appreciated that the flow along a stream tube is constant and that the region between the fixed boundaries, which are themselves streamlines, is filled with stream tubes.

Therefore, in very many cases a flow net need not be drawn. Instead, if the values of the potential function can be determined at the nodal points of a square or rectangular mesh, the necessary information is provided by a numerical solution of groundwater potentials alone.

The popularity of the flow net appears to stem from the early approximate graphical techniques of constructing them and also to the use of conformal mapping in certain idealized cases. The numerical grid analysis method is far more versatile and can be applied to any case, however complicated, either dimensionally or due to the heterogeneity of the flow medium.

4.6 Finite-Difference Formulation

Finite-difference numerical solutions can include all the problems that can be studied by conventional methods of analysis as well as problems involving heterogeneous soils for which the conventional methods are unsuitable.

Reference will therefore be made to the case of heterogeneous flow, Eq. 4.1,

$$\frac{\partial}{\partial x}\left(k_x \frac{\partial h}{\partial x}\right) + \frac{\partial}{\partial y}\left(k_y \frac{\partial h}{\partial y}\right) + \frac{\partial}{\partial z}\left(k_z \frac{\partial h}{\partial z}\right) = 0$$

Suppose a three-dimensional orthogonal mesh is drawn to cover the region of interest and, to be quite general, the mesh separation in the directions x, y, and z may be variable. Fig. 4.7(a) shows one node of the grid, node 0, surrounded by the

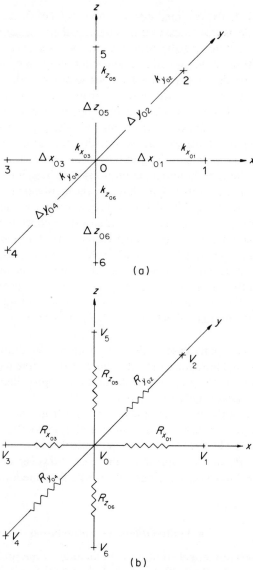

Figure 4.7. Electrical analogy for flow in heterogeneous medium: (a) three-dimensional orthogonal mesh; (b) equivalent resistance network

adjacent nodes, 1, 2, 3, 4, 5, and 6. Then, denoting $\Delta x_{01}, \Delta y_{02}, \Delta z_{05}$, etc., and $k_{x01}, k_{y02}, k_{z05}$, etc., as the distances between nodal points and the average permeabilities between points respectively, the finite-difference form of Eq. 4.1 for node 0 will be as follows:

$$\frac{2}{\Delta x_{01} + \Delta x_{03}}\left[k_{x01}\frac{h_1 - h_0}{\Delta x_{01}} + k_{x03}\frac{h_3 - h_0}{\Delta x_{03}}\right]$$
$$+ \frac{2}{\Delta y_{02} + \Delta y_{04}}\left[k_{y02}\frac{h_2 - h_0}{\Delta y_{02}} + k_{y04}\frac{h_4 - h_0}{\Delta y_{04}}\right]$$
$$+ \frac{2}{\Delta z_{05} + \Delta z_{06}}\left[k_{z05}\frac{h_5 - h_0}{\Delta z_{05}} + k_{z06}\frac{h_6 - h_0}{\Delta z_{06}}\right] = 0 \qquad (4.30)$$

This expression is based on the finite-difference approximations introduced in Section 3.4.

Multiplying throughout by the volume represented by node 0,

$$\tfrac{1}{8}(\Delta x_{01} + \Delta x_{03})(\Delta y_{02} + \Delta y_{04})(\Delta z_{05} + \Delta z_{06})$$

gives

$$\frac{1}{4}(\Delta y_{02} + \Delta y_{04})(\Delta z_{05} + \Delta z_{06})\left[k_{x01}\frac{h_1 - h_0}{\Delta x_{01}} + k_{x03}\frac{h_3 - h_0}{\Delta x_{03}}\right]$$
$$+ \frac{1}{4}(\Delta x_{01} + \Delta x_{03})(\Delta z_{05} + \Delta z_{06})\left[k_{y02}\frac{h_2 - h_0}{\Delta y_{02}} + k_{y04}\frac{h_4 - h_0}{\Delta y_{04}}\right]$$
$$+ \frac{1}{4}(\Delta x_{01} + \Delta x_{03})(\Delta y_{02} + \Delta y_{04})\left[k_{z05}\frac{h_5 - h_0}{\Delta z_{05}} + k_{z06}\frac{h_6 - h_0}{\Delta z_{06}}\right] = 0 \quad (4.31)$$

At each node of the mesh, an equation of the form of Eq. 4.31 will hold. Solutions of these equations can be obtained either using analog or digital computers. For the present the analog computer approach will be considered.

Analogous resistance network

Consider a single node, 0, of a three-dimensional orthogonal grid of electrical resistors, as shown in Fig. 4.7(b). Then, by Kirchhoff's current law, the net current flowing into node 0 must be zero; thus

$$\frac{V_1 - V_0}{R_{x01}} + \frac{V_3 - V_0}{R_{x03}} + \frac{V_2 - V_0}{R_{y02}} + \frac{V_4 - V_0}{R_{y04}}$$
$$+ \frac{V_5 - V_0}{R_{z05}} + \frac{V_6 - V_0}{R_{z06}} = 0 \qquad (4.32)$$

where V and R represent electrical potential and resistance respectively. Setting

$$R_{x01} = \frac{F(2)\Delta x_{01}}{\frac{1}{4}(\Delta y_{02} + \Delta y_{04})(\Delta z_{05} + \Delta z_{06})k_{x01}} \qquad (4.33)$$

where $F(2)$ is a convenient scaling factor, with similar expressions for the other resistances, Eq. 4.32 may be rewritten as

$$\frac{1}{4}(\Delta y_{02} + \Delta y_{04})(\Delta z_{05} + \Delta z_{06}) \left[k_{x01} \frac{V_1 - V_0}{\Delta x_{01}} + k_{x03} \frac{V_3 - V_0}{\Delta x_{03}} \right]$$

$$+ \frac{1}{4}(\Delta x_{01} + \Delta x_{03})(\Delta z_{05} + \Delta z_{06}) \left[k_{y02} \frac{V_2 - V_0}{\Delta y_{02}} + k_{y04} \frac{V_4 - V_0}{\Delta y_{04}} \right]$$

$$+ \frac{1}{4}(\Delta x_{01} + \Delta x_{03})(\Delta y_{02} + \Delta y_{04}) \left[k_{z05} \frac{V_5 - V_0}{\Delta z_{05}} + k_{z06} \frac{V_6 - V_0}{\Delta z_{06}} \right] = 0$$

$$(4.34)$$

Clearly Eqs. 4.31 and 4.34 are analogous if, in addition to the identities defined by Eq. 4.33, it is accepted that a groundwater potential h is represented by an electrical potential V such that

$$V = F(1)h$$

An examination of the type of equation defined by Eq. 4.33 reveals that an electrical resistance has been equated to the distance between mesh points divided by the product of a cross-sectional area and a permeability. In other words, throughout the region under consideration the effective hydraulic resistance has been represented by an electrical resistance. This derivation is similar to that introduced in Chapter 3.

Clearly, the electrical analogy provides a tool for the numerical solution of the most complicated case of seepage, the determination of the flow in a heterogeneous medium.

It is of interest to note that if a uniform cubical mesh is used and the permeability is constant, Eqs. 4.31 and 4.34 reduce to

$$\sum_{n=1}^{6} h_n - 6h_0 = 0 \qquad (4.35)$$

and

$$\sum_{n=1}^{6} V_n - 6V_0 = 0 \qquad (4.36)$$

respectively; both equations represent the three-dimensional form of Laplace's equation and are appropriate for the solution of a three-dimensional seepage problem in an isotropic medium.

Clearly, by similar reasoning, it can be shown that Eqs. 4.35 and 4.36 reduce to

$$\sum_{n=1,5,3,6} h_n - 4h_0 = 0 \qquad (4.37)$$

and

$$\sum_{n=1,5,3,6} V_n - 4V_0 = 0 \qquad (4.38)$$

respectively, for a similar two-dimensional problem in the (x, z)-plane.

It should be appreciated that although, in the foregoing reasoning, a three-dimensional mesh with variable mesh separation was assumed, it would be quite possible to use a uniform cubical mesh for the solution of a three-dimensional seepage problem involving heterogeneity. However, from a practical point of view, it is often preferable to use a variable mesh so that a fine mesh separation can be used for regions of particular interest, such as regions of flow change; and a coarse mesh elsewhere. The use of variable and graded meshes will be described in detail later when practical seepage examples are presented.

Digital computer solution

The simultaneous finite-difference equations can be solved equally well using a digital computer routine. Since there are usually a large number of equations, direct solution using a library subroutine is not usually the best approach. Instead, iterative techniques are more economical.

Chapter 8 contains detailed discussions of various methods of solving sets of finite-difference equations. Though it is the time-variant form of the flow equation that is under examination in Chapter 8, most of the methods can be adapted to solve the steady-state equation (Eq. 4.31). Of all the methods available, successive over-relaxation (SOR) is probably the simplest to program and the most reliable.

For further details of the SOR method reference should be made to Section 8, and to Smith (1965). Appendices 3 and 4 contain two programs designed for seepage problems; the programs will be introduced in Section 4.7 and 4.8.

4.7 Solution of Confined-Flow Problems

The following very simple example, illustrated in Fig. 4.8, has been selected in order to describe the procedure for obtaining a numerical solution to a confined-flow problem.

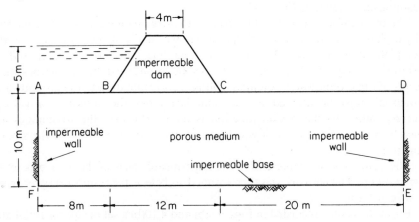

Figure 4.8. Example of confined flow

A dam, whose walls and base are impervious, rests on a homogeneous isotropic soil which has confined impermeable boundaries as shown. The dam retains water at a constant height of 5 m, the downstream level being zero. The flow may be considered to be predominantly two-dimensional with negligible flow in the lateral direction.

This problem may be solved by working in terms of the stream function, the groundwater potential, the velocity potential, or the pressure head.

The boundary conditions are that the groundwater potential is specified as 5 units along the line AB and zero along the downstream surface CD. On the other boundaries the condition is that no flow shall cross the boundary, thus defining them as streamlines. Solutions using both analog and digital computers will be described below.

Resistance network solution

A numerical solution to this problem may be obtained easily by using the electrical resistance network analogy and it will be advantageous to select the groundwater potential as the working function.

An electrical resistance network of the type described in Appendix 1 and shown diagrammatically in Fig. 4.9(a) was used. For convenience, a mesh interval of 2 m was chosen in the x- and z-directions.

The nodes of the network along the permeable boundaries AB and CD were short-circuited with electrical potentials of 5.0 V and 0.0 V respectively, corresponding to the groundwater potential. The impermeable boundaries BC, DE, EF, and FA were set up as free edges, or *selvedges*, in which case the resistors along them take twice the value of the resistors within the interior of the network. This arrangement satisfies all the necessary boundary conditions.

The network was energized from a stabilized power supply and the electrical potentials were measured with a digital voltmeter at all the nodes. The recorded readings are shown in Fig. 4.9(b).

The problem was also solved in terms of the stream function, which required a slight change in the electrical circuit, as shown in Fig. 4.10(a). In this case the complete boundary AFED, which is a streamline, was short-circuited and an arbitrary voltage of 7 V applied to it. It must be emphasized that when working in terms of the groundwater potential, voltages corresponding to the groundwater potential must be applied to the boundaries but the choice of a voltage corresponding to the stream function is quite arbitrary; the streamlines are merely contour lines, each of which represents a constant value of the stream function.

The boundary BC, representing the impermeable base of the dam, was set to zero voltage. The network was then scanned, as before, the results being recorded in Fig. 4.10(b).

From the results recorded in Figs. 4.9(b) and 4.10(b) a set of equipotential and streamlines were plotted, and these are shown in the form of a flow net in Fig. 4.11.

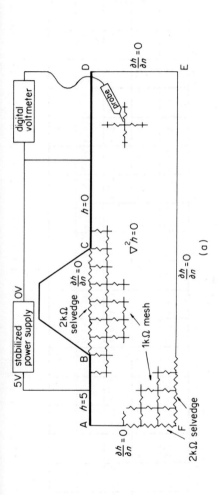

digital
volt meter

stabilized power supply

5V   0V

A $h=5$

$\frac{\partial h}{\partial n}=0$

B

$2\,k\Omega$
selvedge $\frac{\partial h}{\partial n}=0$

C $h=0$

D

$\nabla^2 h=0$

$1\,k\Omega$ mesh

$2\,k\Omega$ selvedge

F

probe

$\frac{\partial h}{\partial n}=0$

$\frac{\partial h}{\partial n}=0$ E

(a)

Head h

5.00	5.00	5.00	5.00	5.00	5.00	3.86	3.11	2.46	1.82	1.09	0.0	0.0	0.0	0.0	0.0	0.0	0.0	0.0	0.0	0.0
4.73	4.72	4.66	4.54	4.28	3.66	3.05	2.46	1.87	1.27	0.67	0.40	0.27	0.19	0.13	0.10	0.07	0.06	0.05	0.04	0.04
4.50	4.47	4.38	4.21	3.92	3.47	2.97	2.46	1.94	1.43	0.99	0.68	0.48	0.34	0.25	0.19	0.14	0.11	0.09	0.08	0.07
4.32	4.29	4.18	4.00	3.71	3.34	2.91	2.45	1.99	1.55	1.16	0.85	0.63	0.46	0.34	0.25	0.19	0.15	0.12	0.11	0.10
4.22	4.18	4.07	3.88	3.61	3.26	2.86	2.45	2.02	1.61	1.25	0.95	0.71	0.53	0.40	0.30	0.23	0.18	0.14	0.13	0.12
4.18	4.14	4.03	3.84	3.57	3.24	2.85	2.44	2.03	1.63	1.28	0.98	0.74	0.55	0.41	0.31	0.24	0.19	0.15	0.13	0.13

(b)

Figure 4.9. (a) Electrical analog circuit for determination of the groundwater potential h. (b) Values of groundwater potential

80

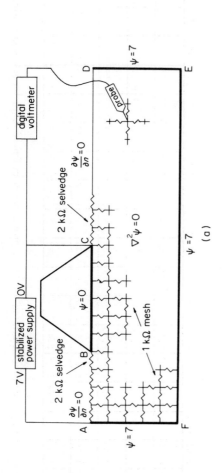

(a)

Stream function ψ

7.00	6.86	6.71	6.53	6.29	5.99	5.58	5.00	4.14	2.74	0.0	0.0	0.0	0.0	0.0	0.0	3.10	4.79	5.98	7.00
7.00	6.87	6.72	6.55	6.33	6.05	5.66	5.14	4.41	3.43	2.24	1.82	1.65	1.63	1.68	1.88	2.39	3.80	5.04	6.07
7.00	6.89	6.76	6.62	6.43	6.20	5.89	5.48	4.96	4.34	3.70	3.33	3.16	3.12	3.20	3.42	3.87	4.66	5.49	6.27
7.00	6.92	6.83	6.72	6.60	6.43	6.21	5.94	5.62	5.24	4.88	4.65	4.52	4.49	4.56	4.72	5.03	5.49	6.00	6.50
7.00	6.96	6.91	6.86	6.79	6.70	6.60	6.46	6.30	6.13	5.97	5.85	5.79	5.76	5.80	5.89	6.04	6.26	6.50	6.75
7.00	7.00	7.00	7.00	7.00	7.00	7.00	7.00	7.00	7.00	7.00	7.00	7.00	7.00	7.00	7.00	7.00	7.00	7.00	7.00

(b)

Figure 4.10. (a) Electrical analog circuit for determination of the stream function ψ. (b) Values of stream function

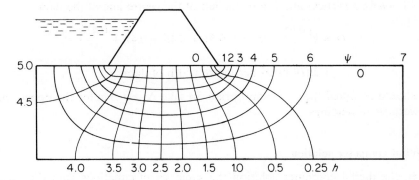

Figure 4.11. Equipotentials and streamlines

Strictly speaking, only one experiment need have been undertaken. Once either the equipotentials or the stream function have been determined, the other function can be plotted, bearing in mind the properties of conjugate functions in which the equipotential and streamlines form a family of orthogonal curves. However, the experimental procedure is so simple that, in this instance, it was quicker to carry out the two experiments instead of deriving the second set of curves from the first.

Although the flow net is useful in providing a visualization of the flow pattern, all the information required for the calculation of flow quantities and pressure heads is provided by Fig. 4.9(b).

The quantity of water seeping underneath the dam can be calculated from the results given in Fig. 4.9(b) and is

$$Q = \sum k \frac{\Delta h}{\Delta x} \Delta z \, \Delta y \qquad (4.39)$$

where Q = quantity of water

       k = coefficient of permeability

       Δh = groundwater potential increment

$\Delta x, \Delta y, \Delta z$ = mesh spacing

Now as

$$\Delta x = \Delta z = 2 \, \text{m}$$

the quantity of water flowing per unit width ($\Delta y = 1$ m) of the dam is given, in this case, by

$$Q = k \sum \Delta h$$

It will be remembered that the quantity flowing between two fixed boundaries, which are streamlines, must be constant, and therefore the line between boundaries along which the summation is made is immaterial.

Choosing a vertical line 1 m to the left of the centre line of the dam

$$Q = k\left(\frac{0.65}{2} + 0.59 + 0.51 + 0.46 + 0.41 + \frac{0.41}{2}\right)$$

$$= 2.50k\,\mathrm{m}^3/\mathrm{d} \text{ per metre width of dam.}$$

It should be noted that it is appropriate to take $\Delta z = 0.5\,\mathrm{m}$ for the upper and lower sets of readings.

Digital computer solution

In the digital computer solution the same set of equations with the same boundary conditions must be solved. The successive over-relaxation technique (SOR) has proved to be satisfactory and a program for confined seepage is included as Appendix 3. Comment cards are included in the program to show how it is used.

This steady-state program is derived from the time-variant groundwater flow program of Appendix 2. Reference should be made to Appendix 2 and Section 8.3 for further details.

The data corresponding to this particular problem is listed at the end of Appendix 3. One item that requires some explanation is the variable 'ERROR'. Since SOR is an iterative technique, some form of convergence criterion must be introduced. ERROR is a flow rate per unit volume; if the flow balance at every node is to a better accuracy than the specified error, a satisfactory convergence has been achieved. This error is set to 0.01 per cent of the maximum flow rate in the aquifer.

Values of the groundwater potential obtained from the digital solution are not quoted since they differ from the analog values of Fig. 4.9(b) by less than 0.02 m. The total flow under the dam according to the digital computer program is $2.52k\,\mathrm{m}^3/\mathrm{d}$.

Variable permeability

The example described above has been concerned with a homogeneous isotropic medium. However, as far as the numerical technique is concerned any variation in permeability can be used, provided that the solution is obtained in terms of the groundwater potential. Variable permeability is reflected by a range of resistance values in the analog solution or variable coefficients $A(I,J)$, $B(I,J)$, $C(I,J)$, $D(I,J)$ in the digital computer program.

Care must be taken in the calculation of the flows. In Eq. 4.39 each head difference must be multiplied by the appropriate permeability. It is important to note that in heterogeneous soils it is not possible to obtain a solution in terms of the velocity potential or stream function.

For the particular case of anisotropic flow, certain interesting properties can be noted. These are discussed below.

Anisotropy

A case in which the soil is anisotropic will now be considered as a second example of confined flow.

An impermeable dam rests on anisotropic soil which has confined impermeable boundaries, as shown in Fig. 4.12. The coefficient of permeability in the longitudinal direction is 16 times that in the vertical direction; thus $k_x/k_z = 16$. Again, the flow will be considered to be predominantly two-dimensional and so only a unit width of the dam in the lateral direction has to be considered.

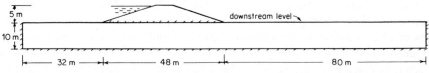

Figure 4.12. Example of anisotropic confined flow

The first step is to carry out the simple affine transformation described in Section 4.3. As the present problem is two-dimensional, Eq. 4.9 reduces to

$$\frac{\partial^2 h}{\partial \bar{x}^2} + \frac{\partial^2 h}{\partial z^2} = 0 \tag{4.40}$$

where

$$\bar{x} = x\sqrt{k_z/k_x}$$

$$\bar{x} = x/4$$

Hence, according to the transformation, all dimensions in the longitudinal direction are reduced to one-quarter of their actual values and the problem to be solved has been transformed to one in which the soil is isotropic. The next step is, in fact, the obtaining of the previous solution, Figs. 4.9(b) and 4.10(b).

The quantity of water seeping under the dam will in this case be

$$Q = k_x \sum \left(\frac{0.65}{2} + 0.59 + 0.51 + 0.46 + 0.41 + \frac{0.41}{2} \right) \frac{\Delta z}{\Delta x}$$

As $\Delta x = 4\Delta z$,

$$Q = \frac{2.50k_x}{4}$$

Finally, the results recorded in Figs. 4.9(b) and 4.10(b) are used to plot, to the *true* scale, the flow net as shown in Fig. 4.13. It should be noted that, in this case, the equipotential lines and the flowlines are not orthogonal except when the equipotentials or the flow lines coincide with the axes of principal permeability. Consequently, the equipotentials and streamlines are orthogonal at the boundaries. Flow nets for anisotropic conditions are discussed by Bear (1972) and Harr (1962).

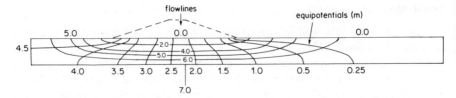

Figure 4.13. Equipotentials and flowlines for anisotropic problem

4.8 Solution of Unconfined-flow Problems

The complexity of the analysis of seepage problems is increased in cases of unconfined flow, where one boundary is a free surface. This surface is a boundary whose actual position in space cannot be predetermined but can be defined only as the position of the boundary at which the value of the working function and its slope take prescribed values. When studying the flow of water through a porous soil whose top surface is open to the atmosphere, the position of the water surface will not necessarily coincide with the surface of the soil, and thus the position of the boundary cannot be specified directly. Though the position of the boundary cannot be specified, *two conditions* do have to be *satisfied simultaneously* on this boundary. The first is that the pressure is atmospheric and the second is that no flow crosses the boundary. This boundary is called a *free flow surface* (or *water table*) and is a streamline.

As before, the governing equation which has to be solved is the second-order equation

$$\frac{\partial}{\partial x}\left(k_x \frac{\partial h}{\partial x}\right) + \frac{\partial}{\partial y}\left(k_y \frac{\partial h}{\partial y}\right) + \frac{\partial}{\partial z}\left(k_z \frac{\partial h}{\partial z}\right) = 0 \qquad (4.1)$$

and to appreciate the complexity of the present problem the boundary conditions must be fully understood.

It is commonly stated that to obtain the solution to Eq. 4.1 over a certain region it is necessary to know either the value of the function, or its normal derivative, around the boundary. The position of the boundary is implied, and thus the specification of its location does, in effect, imply a second boundary condition. Now, in confined-flow problems the location of the boundary is known, thus satisfying one boundary condition. However, in cases where a free surface occurs, the boundary position of that surface is not known and hence *two other conditions* must be imposed, namely the value of the function and its normal derivative.

Any technique chosen for the solution of free-surface problems must be sufficiently flexible to allow for changes in the position of the unknown boundary and to provide a rapid and sensitive method for determining whether convergence to the correct boundary position has been achieved.

In general, the problem is in three dimensions although there are many cases in which a two-dimensional solution obtains, or can be accepted as a close approximation to the three-dimensional case. Nevertheless, it is highly desirable that the method of solution shall be sufficiently versatile to be applied to both two- and three-dimensional problems.

A two-dimensional example, chosen to demonstrate the method of solution, is illustrated in Fig. 4.14. A dam constructed of isotropic permeable material rests on an impermeable stratum and retains water to a height of 10 m; water seeps through the dam and emerges at an initially unknown height on the downstream face draining freely down the wall into a reservoir, the height of the water there being maintained at 2 m. Though the example is concerned with an isotropic medium, the techniques described below are equally applicable to heterogeneous permeabilities.

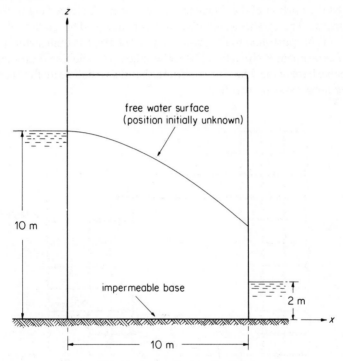

Figure 4.14. Example of unconfined flow

It will be convenient, for reasons which will be explained later, to work in terms of the groundwater potential. If the datum point is located at the heel of the dam the groundwater potential, h will be constant at a value of 10 m on the upstream face and 2 m on the downstream face where the water is in contact with the dam. The boundary condition on the impermeable base is that there shall be no flow across it; at points along the seepage face and the undetermined free-flow line the pressure is atmospheric, thus requiring the groundwater potential to equal the height above datum, $h = z$. Additionally, as previously stated, there must be no flow across the free surface.

Solution on resistance network

An electrical resistance network provides a very simple method of solving free-surface problems and the use of the analog will now be described for the solution of this two-dimensional example.

The electrical resistance network required for the solution of this problem is essentially the same as that required for the previous problem but with one important difference; it must be capable of being 'cut' as the experiment proceeds. The construction of a suitable network is described in Appendix 1.

The network is set up as shown in Fig. 4.15. The line corresponding to the upstream face of the dam is short-circuited and set to an electrical potential of 10 V whilst the portion of the downstream face, over which the height of water is maintained at 2 m, is short-circuited and set at 2 V. The choice of voltage, provided it is proportional to the groundwater potential, is quite arbitrary and a matter of convenience. No action has to be taken with the line representing the impermeable base of the dam; the condition that there shall be no flow across this line being automatically satisfied.

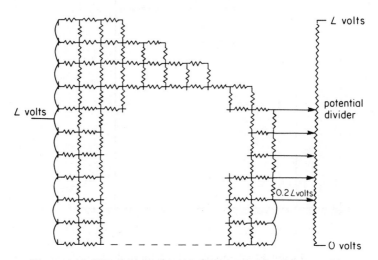

Figure 4.15. Electrical analog circuit for unconfined flow problem

On the downstream face, above the reservoir level, the groundwater potential must equal the height above datum and therefore electrical potentials, proportional to the potential head, have to be set at nodes along this boundary. Herein lies a difficulty, because the point of emergence of the free-water surface on the downstream face is, as yet, unknown. Therefore, as a starting point for the analysis the point at which the water emerges must be assumed. It would be quite possible to assume that the free-water surface coincides with the upstream water level and to proceed with the iterative process from this position. However, it will generally be found that time can be saved by making an intelligent guess at the location of the free-water surface and then isolating the nodes of the network nearest to this line from the network above it.

Having commenced the experiment in this manner, it is then necessary to measure the electrical potentials at the assumed free-surface boundary nodes to discover whether the free-surface boundary condition, that the potential there equals the height above datum, has been satisfied. The condition that there shall be no flow across the free-surface boundary is satisfied by the 'cutting' of the network. If the voltage corresponding to the groundwater potential on the assumed free-surface boundary equals the height above datum, then the correct boundary position has been found. If this condition has not been satisfied (and this will probably be the case), successive trial positions of the boundary surface must be made until the condition of atmospheric pressure is approximately achieved.

A simple, *systematic manual procedure* for determining the free surface may be followed by studying the present example. The electrical potentials at the nodes corresponding to the first assumed position of the free surface are recorded in Fig. 4.16. It should be noted that until the correct position for the free surface has been

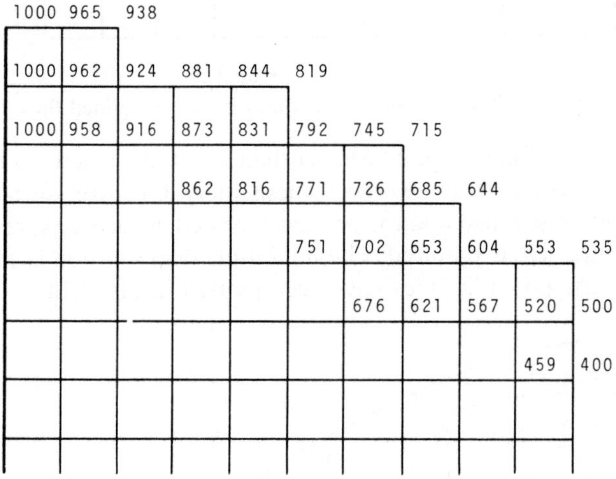

```
1000 965  938
    ┌────┐
1000│962 │924  881  844  819
    │    │
1000│958 │916  873  831  792  745  715
    │    │    ┌────┐
    │    │    │862 │816  771  726  685  644
    │    │    │    │    ┌────┐
    │    │    │    │    │751 │702  653  604  553  535
    │    │    │    │    │    │    ┌────┐
    │    │    │    │    │    │    │676 │621  567  520  500
    │    │    │    │    │    │    │    │    ┌────┐
    │    │    │    │    │    │    │    │    │459 │400
    │    │    │    │    │    │    │    │    │    │
    │    │    │    │    │    │    │    │    │    │
    └────┴────┴────┴────┴────┴────┴────┴────┴────┘
```

Figure 4.16. Initial approximation to free-surface position

found it is unnecessary to scan the complete network. An examination of Fig. 4.16 shows that the assumed line for the free surface is too high, the potentials recorded all being lower than the corresponding potential head; equally well they might have been higher had the assumed free-surface line been located lower. The next step is to disconnect resistors in the top line of nodes, from the downstream face, until the potential at the remaining nodes is in reasonable agreement with the potential head, in this case 9.5 m. The value of 9.5 m is used because the line of nodes at 10 m represents a layer down to a level of 9.5 m. This procedure is repeated for each successive layer downwards until a new emergence point for the free-flow surface, on the downstream face, is determined. For each layer the

1000	978									
1000	967	931	905							
1000	961	921	881	831	791	764				
1000	957	913	868	820	776	735	693			
1000	953	905	856	806	756	705	651	586	547	
1000	949	898	845	791	736	679	619	559	509	473
1000	946	891	835	777	717	654	588	521	456	400
1000	943	885	826	764	700	632	559	480	394	300
1000	941	881	820	755	687	614	535	446	340	200
1000	940	879	816	749	679	603	520	427	321	200
1000	940	878	844	747	676	599	515	421	316	200

Section

Section

Figure 4.17. Final numerical solution for unconfined flow

groundwater potential must not be lower than $z - 0.5\Delta z$, where z is the elevation of the line of nodes and Δz is the vertical mesh interval. The free-surface boundary nodes are then re-scanned and the process repeated, if necessary, resistors being added or subtracted as required. The final numerical solution is shown in Fig. 4.17 and the equipotential lines have been plotted in Fig. 4.18.

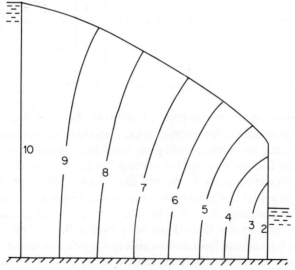

Figure 4.18. Equipotential lines for unconfined flow problem

The representation of a free surface on a resistance analog computer has been discussed in detail by Herbert and Rushton (1966) and Redshaw (1967), who described both an automatic transistor switching device as well as purely manual procedures. Usually it will be found that the manual procedures are simple, rapid, and quite satisfactory for most seepage problems.

Digital computer solution

A digital computer program which can solve this and similar problems is presented in Appendix 4. The program is a development of that for a confined aquifer, Appendix 3. The technique used is similar to that of the resistance network solution, though the convergence to the correct finite-difference position tends to be slower. Another limitation is that the computer program may fail to find the correct free-surface position. This can occur if the free surface falls by more than one vertical mesh length within a horizontal mesh interval. Then a message will be printed out and the computation will stop. The program should be run again with more mesh divisions in the vertical direction. It must be emphasized that this is a simple program and care should be taken in using it for more complicated problems.

The digital computer program does not converge to exactly the same free-surface position as that for the analog technique. This arises because slightly different criteria are used to check for the condition of zero atmospheric pressure. However, the free-surface profiles for the analog and digital solutions are generally within 0.1 m of each other. The total flows agree to better than 0.5 per cent.

In both the analog and digital simulations the free surface has been approximated as a series of steps. It would be more correct to use a curved boundary with parameters calculated as described in Section 3.7. However, this leads to only small improvements in the predicted free-surface position and total flow.

Quantity of flow

The quantity of water seeping through the dam can be calculated by summing the flow through any section between the free surface and the impermeable base, such as the section indicated in Fig. 4.17. Then, from Eq. 4.37, if k is the coefficient of permeability for unit lateral length of the dam, the quantity of water is given by

$$Q = k \times 1.0(0.40 + 0.44 + 0.50 + 0.55 + 0.60 + 0.64 + 0.68 + 0.70 + 0.71/2)$$

$$= 4.86k$$

$\Delta z = 0.5$ for the bottom set of readings but not for the top pair. Strictly speaking, the free-flow line should be represented on the network as a selvedge, which would entail doubling the resistances on this upper boundary; but if a fine network mesh is used this refinement is not really necessary.

It is of interest to compare the result of this calculation of the quantity of seepage with the result which would be obtained by applying the Dupuit theory of unconfined flow. As early as 1863 Dupuit propounded his theory which has been concisely summarized by Harr (1962). Dupuit assumed:

(a) For small inclinations of the line of seepage, the streamlines can be taken as horizontal.
(b) The hydraulic gradient is equal to the slope of the free surface and is invariant with depth.

The first assumption implies that the equipotential lines approach the vertical, and it will be seen that the validity of the second assumption depends on the steepness of the line of seepage.

Dupuit's formula for the quantity of seepage per unit width of aquifer is

$$Q = k\frac{z_1^2 - z_2^2}{2L} \qquad (4.41)$$

where z_1 and z_2 represent the elevations of two known points on the free surface, and L is the horizontal distance separating them.

In the present example, considering the elevations at the two faces of the dam, $z_1 = 10\,\mathrm{m}$, $z_2 = 2\,\mathrm{m}$, and $L = 10\,\mathrm{m}$; inserting these values into Eq. 4.41 yields

$$Q = 4.8k$$

which is in close agreement with the result from the numerical solutions.

Charni (1951) has shown that an exact analytical solution to the two-dimensional problem leads to an equation for the total flow which is identical to the Dupuit formula, Eq. 4.41, provided that the potentials at the upstream and downstream faces are used in the formula. Nevertheless, it must be emphasized that, although a good estimate for the quantity of seepage may be obtained, it is inherent in the derivation based on the Dupuit assumptions that the free-surface line is parabolic; an examination of Fig. 4.18 indicates that this is not the case. Also, the implied assumption that the equipotential lines approach the vertical, that is to say that there is no vertical component of flow, is incorrect as an examination of the equipotential values recorded in Fig. 4.17 and the contours of equipotential lines in Fig. 4.18, clearly show.

Choice of working potential

The contour lines shown in Fig. 4.18 are lines of constant groundwater potential, not lines of flow. Had the free-surface line emerged at the reservoir level, streamlines could have been obtained by adopting the procedure used in the confined-flow example. That is to say, an arbitrary electrical potential would be set up on the network between the short-circuited free-flow line and the impermeable base. In the present example a difficulty arises because the streamline condition on the downstream face, between the point of emergence of the free-flow line and the reservoir level, is indeterminate. Water will seep from the dam and run down the downstream face, not emerging at right angles to it.

The choice of the working function, in this case the groundwater potential, now calls for comment. Clearly, as the position of the free surface was initially unknown, it would have been impossible to have worked in terms of the stream function, but the problem could have been solved by either using the velocity potential (for isotropic soils) or the pressure head.

It would be quite feasible to use the pressure head as the working function, although less convenient since the boundary condition on the upstream face would require a different potential to be set at each node of this boundary. The potentials would, of course, vary linearly with height. The boundary condition on the free surface and on the downstream face above the level of the reservoir requires zero pressure, and therefore the condition of zero potential would have to be enforced on these boundaries. A pressure condition is required for the boundary on the downstream face below the reservoir level and the boundary condition for the impermeable base becomes $\partial p/\partial z = 1.0$. It is interesting to note that when a solution to this type of problem was first sought by the exercise of the relaxation method it was found to be preferable to work in terms of the pressure head (Southwell, 1946).

4.9 Radial Flow to a Well

As a second example of unconfined flow, the seepage of water into a well will be considered. This problem was first analysed by Boulton (1951) but he did not use the logarithmic radial mesh described below. A typical small well, such as occurs in practice, is illustrated in Fig. 4.19. Radial flow takes place through an aquifer of constant radial permeability k_r from a fixed head H at a radius r_{max} towards a well of radius r_{well} discharging at a constant rate Q.

This problem, like the previous one, can be solved very easily by a numerical method. A choice of coordinate system exists but as the problem involves radial flow it will be more convenient to work in terms of polar cylindrical coordinates with h, the groundwater potential, as the working function.

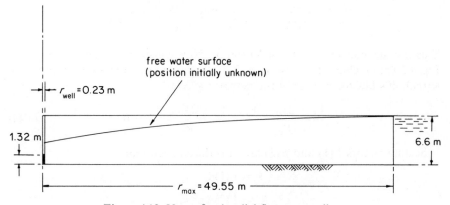

Figure 4.19. Unconfined radial flow to a well

The basic equation to be solved is the one derived in Chapter 2 as Eq. 2.8, that is,

$$\frac{\partial}{\partial r}\left(k_r\frac{\partial h}{\partial r}\right) + \frac{k_r}{r}\frac{\partial h}{\partial r} + k_z\frac{\partial^2 h}{\partial z^2} = 0 \tag{4.42}$$

where k_r and k_z are the radial and vertical coefficients of permeability respectively. It will be observed that Eq. 4.42 implies radial symmetry, that is to say the coefficient of permeability does not vary with the angle of rotation θ about the vertical axis, which obviously is conveniently taken as the axis of the well.

It will be seen that Eq. 4.42 expresses the two-dimensional form of Laplace's equation in polar coordinates, for the solution of the groundwater potential h. It is, in fact, a pseudo–three-dimensional equation as it expresses the reduction from a three-dimensional problem to one in two dimensions by taking advantage of axial symmetry.

The particular example chosen to illustrate the numerical technique concerns radial flow between a vertical boundary at a radial distance of 49.55 m where the water level is at a constant height of 6.6 m to a well of radius 0.23 m. The upper horizontal boundary coincides with the undisturbed water level 6.6 m above the horizontal impermeable base. The height of water in the well is 1.32 m.

It is now necessary to devise a suitable finite-difference approximation. In Section 3.5 it was shown that the differential equation can be simplified by using a logarithmic scale in the radial direction. This not only has the merit of allowing the coefficients of the finite-difference equations in the radial direction to be constant, but also has the very desirable feature of providing a finer mesh in the vicinity of the well.

Thus, rewriting Eq. 4.42 in terms of $a = \log_e r$, leads to

$$k_r\frac{\partial^2 h}{\partial a^2} + k_z r^2\frac{\partial^2 h}{\partial z^2} = 0 \tag{4.43}$$

In finite-difference form, with the notation of Fig. 4.20(a), this equation becomes

$$k_r\frac{h_1 - 2h_0 + h_3}{\Delta a^2} + k_z r^2\frac{(h_2 - 2h_0 + h_4)}{\Delta z^2} = 0$$

This assumes constant values of Δa and Δz. Now consider the network shown in Fig. 4.20(b) in which the resistances R_a in the horizontal direction are constant. Kirchhoff's law for a node of the network gives

$$\frac{V_1 - 2V_0 + V_3}{R_a} + \frac{V_2 - 2V_0 + V_4}{R_z} = 0 \tag{4.44}$$

Scaling factors $F(1)$ and $F(2)$ are introduced such that

$$V = F(1)h$$

$$R_a = F(2)\Delta a^2/k_r$$

$$R_z = F(2)\Delta z^2/k_z r^2$$

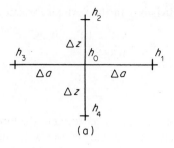

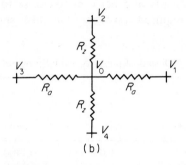

Figure 4.20. (a) Finite-difference network.
(b) Equivalent resistance network

If the aquifer is isotropic $k_r = k_z$. Thus for Eqs. 4.43 and 4.44, stated in physical and electrical terms respectively, to be analogous,

$$R_z = \frac{\Delta z^2 R_a}{\Delta a^2 r^2} \tag{4.45}$$

If the network has ten mesh divisions in the vertical and fourteen in the horizontal direction,

$$\Delta z = 0.66 \, \text{m}$$

and

$$\Delta a = \frac{\log_e 49.55 - \log_e 0.23}{14}$$

$$= 0.383\,76$$

which, owing to the logarithmic scale, implies that there are six mesh intervals for a tenfold increase in radius.

Bearing in mind the availability of resistors, it was found to be convenient to choose

$$R_a = 9.1 \, k\Omega$$

and so

$$F(2) = \frac{9.1 \times 10^3}{0.383\,76^2}$$

With ten equal mesh divisions in the vertical direction

$$R_z = \frac{9.1 \times 10^3}{0.383\,76^2} \times \frac{0.66^2}{r^2}$$

$$= \frac{2.72 \times 10^4}{r^2}$$

Table 4.1 lists the resistance values according to the foregoing expressions; the values of the radius r between the wall of the well and the outer boundary, correspond to equal intervals of $\Delta a = 0.383\,76$. Finally, resistors were selected from commercial components of 5 per cent tolerance to match, as closely as possible, the calculated required resistance. The table shows that a very close

Table 4.1 Design of mesh for radial flow network

Logarithmic radius a	r metres	r^2 metres^2	$R_z = \dfrac{2.72 \times 10^4}{r^2}$ ohms	R_z chosen ohms
− 1.469 6	0.23	0.052 9	513 400	510 000 × 2
− 1.086 0	0.337 6	0.113 97	239 360	240 000
− 0.702 2	0.495 5	0.245 5	110 610	110 000
− 0.318 5	0.727 3	0.529	51 340	51 000
0.065 3	1.067 6	1.1397	23 936	24 000
0.449 1	1.567	2.455	11 061	11 000
0.832 8	2.30	5.29	5 134	5 100
1.216 6	3.376	11.397	2 393.6	2 400
1.600 3	4.955	24.55	1 106.1	1 100
1.984 1	7.273	52.9	513.4	510
2.367 9	10.676	113.97	239.36	240
2.751 6	15.67	245.5	110.61	110
3.135 4	23.0	529.0	51.34	51
3.519 1	33.76	1 139.7	23.94	24
3.902 9	49.55	2 455.0	11.06	11 × 2

match was obtained: the reason for this is that the preferred values of resistors have a nominal logarithmic increase with twenty-four steps for a tenfold increase in value, and so the six mesh interval for a tenfold increase in radius enables an excellent fit to be obtained.

It should be noted that the resistance values on the four boundaries are doubled in value.

Having set up the network, the procedure is identical to that of the previous example (Section 4.8). The electrical potentials, corresponding directly to the groundwater potentials, are recorded in Fig. 4.21. The results, plotted to the true scale, are shown in Fig. 4.22, on which the equipotential lines have been drawn. Note the steepness of the free surface as the well wall is approached.

													6.37	6.60
											5.87	5.12	6.37	6.60
							4.95	5.12	5.36	5.63	5.87	6.12	6.37	6.60
4.32	4.32	4.33	4.36	4.41	4.51	4.67	4.91	5.11	5.36	5.63	5.87	6.12	6.37	6.60
3.96	4.04	4.10	4.20	4.30	4.44	4.63	4.87	5.10	5.35	5.63	5.87	6.12	6.37	6.60
3.30	3.51	3.70	3.90	4.11	4.32	4.56	4.82	5.08	5.35	5.63	5.87	6.12	6.36	6.60
2.64	2.96	3.26	3.57	3.87	4.18	4.48	4.78	5.07	5.35	5.62	2.87	6.12	6.36	6.60
1.98	2.41	2.83	3.24	3.64	4.04	4.40	4.74	5.06	5.35	5.62	5.87	6.12	6.36	6.60
1.32	1.90	2.45	2.97	3.46	3.92	4.34	4.71	5.05	5.34	5.62	5.87	6.11	6.36	6.60
1.32	1.84	2.37	2.88	3.37	3.85	4.30	4.69	5.04	5.34	5.62	5.87	6.11	6.36	6.60
1.32	1.83	2.34	2.86	3.35	3.83	4.28	4.68	5.04	5.34	5.62	5.87	6.11	6.36	6.60

Figure 4.21. Numerical values of groundwater potential with logarithmic spacing in radial direction

Calculation of flow

According to the Dupuit theory, which gives the correct total flow, the quantity of flow (Kirkham, 1964) is given by

$$Q = \frac{\pi k(H^2 - h_w^2)}{\log_e(r_0/r_w)} \tag{4.46}$$

where

$$H = \text{height of water at outer boundary}$$
$$h_w = \text{height of water in the well}$$
$$r_w = \text{radius of well}$$
$$r_0 = \text{radius to outer boundary}$$

Then

$$Q = \frac{(6.6^2 - 1.32^2)\pi k}{\log_e(49.55/0.23)}$$

$$= 7.7\pi k$$

Using the electrical analog results, the groundwater potential at a radius of 33.76 m is 6.367 m; and at 49.55 m it is 6.6 m. Thus the average fall in head over the last mesh interval is 0.233 m. The mesh length is $49.55 - 33.76 = 15.79$ m. The area through which flow takes place is $2\pi r \times$ depth $= 2\pi \times 40.9 \times 6.6$, where r is calculated from the mean logarithmic radius of $a = 3.711\,0$. Thus from Darcy's law

$$Q = \frac{0.233k}{15.79} \times 2\pi \times 40.9 \times 6.6$$

$$= 7.96\pi k$$

As with the previous example, a good comparison between the flow as obtained by the use of the Dupuit assumption and the finite-difference results has been obtained. Again, however, the Dupuit assumption that the equipotential lines approach the vertical, is not correct as an examination of the measured equipotential values (Fig. 4.22) shows.

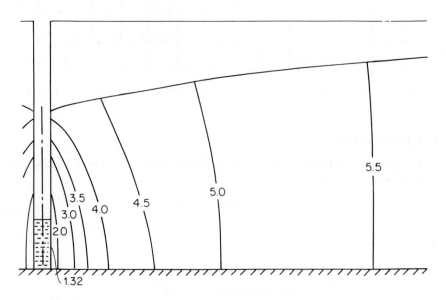

Figure 4.22. Equipotentials for radial flow to well plotted to true scale

Instead of using a radial network, this problem could have been solved with a three-dimensional cartesian coordinate network as demonstrated by Redshaw (1967). Although the cartesian network possesses the possible advantage of being a general-purpose network, it loses in accuracy compared to the logarithmic mesh spacing used in the polar coordinate network.

In the foregoing problem, the level of water in the well was maintained constant, but for the condition of a specified total discharge it is necessary to use a trial-and-error method to find the equivalent well-water level.

4.10 Solution of a Series of Seepage Problems

Introduction

The previous examples describe the manner in which a numerical solution to problems in confined and unconfined flow could be obtained. There are, however, certain complications which can arise and on which some comment is necessary. To illustrate the various points, an example of a permeable dam resting on an aquifer of different permeability has been chosen and the treatment of a drain and a cut-off wall will be described in stages. Table 4.2 summarizes the cases considered.

Table 4.2 Examples considered and resultant flows

		Flow through $1 \text{ m } width \times 10^{-3} \text{ m}^2/\text{d}$		
Example	Results	Across upstream face	Through aquifer	Total flow
1. Aquifer and dam with seepage face	Fig. 4.26	—	—	33.4
2. Aquifer only	Fig. 4.28(a)	0	26.8	26.8
3. Aquifer only with cut-off wall	Fig. 4.28(b)	0	20.3	20.3
4. As (3) but with singularity correction	Fig. 4.28(c)	0	19.3	19.3
5. Dam only with seepage face	Fig. 4.29(a)	4.3	0	4.3
6. Dam only with drain	Fig. 4.29(b)	5.8	0	5.8
7. Dam and aquifer with drain and cut-off wall	Fig. 4.30	—	—	33.0

Referring to Fig. 4.23, a dam having sloping faces rests on an aquifer and retains water to a height of 12 m. Water seeps through and under the dam, the

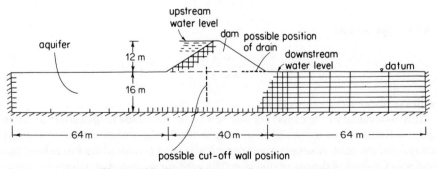

Figure 4.23. General layout of dam and aquifer

water level being maintained constant at the toe of the dam. The aquifer has a constant depth of 16 m. It will be assumed that the aquifer extends to impermeable boundaries 64 m forwards and backwards from the toe and heel of the dam respectively, giving an aquifer of total length 168 m. Although the aquifer has thus been treated as being of limited length, the results show that it is effectively of infinite length since the vertical impermeable boundaries are at such a remote distance from the dam. As water can percolate through and under the dam, the problem is one of combined confined and unconfined flow.

The dam was constructed of sandy clay and the aquifer consists of a compacted sandstone, the coefficients of permeability being 1.3×10^{-3} m/d and 7.4×10^{-3} m/d respectively.

The effect of a drain in the dam and a cut-off wall below it located as shown in Fig. 4.23 was also studied. The lateral dimensions of the dam are such that the problem may be considered as being two-dimensional.

Aquifer and dam with seepage face

For the first case (example 1 of Table 4.2), neither a drain nor a cut-off wall is included. This introduces only one feature which has not been encountered in the previous examples, namely the sloping walls of the dam. Solutions to this problem can be obtained using either an analog or a digital computer.

If the electrical analogy method is used, the first step is to assemble a suitable electrical resistance network. Choosing a square mesh of 2 m separation, $\Delta x = \Delta z = 2$ m, and in the lateral direction $\Delta y = 1$ m, the required mesh resistance R_x is given by

$$R_x = F(2)\frac{\Delta x}{k_x \Delta y \Delta z} \tag{4.47}$$

where $F(2)$ is a scaling factor.

Selecting $F(2) = 20$, the mesh resistances are for the dam:

$$(R_x)_{dam} = \frac{20 \times 2}{1.3 \times 1 \times 2} = 15.38 \text{ k}\Omega$$

and for the aquifer:

$$(R_x)_{aquifer} = \frac{20 \times 2}{7.4 \times 1 \times 2} = 2.703 \text{ k}\Omega$$

with corresponding values of R_z.

It is unnecessary to use this very fine mesh over the whole field of study and it is quite satisfactory to use a mesh, graded in the horizontal direction, in regions away from the dam. A simple graded mesh was used commencing 8 m behind the heel and forward of the toe of the dam. The grading consisted of two steps each 4, 8, and 16 m in length.

The method of calculating the various graded mesh resistances is shown in Table 4.3. Resistors having preferred values closest to the required values were chosen and assembled. The sloping faces of the dam only coincided with some of

Table 4.3 Calculation of resistance values for graded mesh of aquifer using equation $R = F(2) \times$ length/(cross-sectional area $\times$ permeability) with $k = 7.4 \times 10^{-3}$ m/d, $\Delta y = 1$ m, $F(2) = 20$

	2 m	2 m	4 m	8 m	16 m	
2 m		a	c	e	g	i
2 m		b b	d	f	h	

Resistance	Length (m)	Cross-sectional area (m²)	Resistance (kΩ)	Preferred value (kΩ)
a	2	2	2.703	2.7
b	2	2	2.703	2.7
c	2	3	1.802	1.8
d	4	2	5.406	5.6
e	2	6	0.901	0.91
f	8	2	10.814	11.0
g	2	12	0.45	0.47
h	16	2	21.63	22.0
i	2	8	0.676	0.68

the mesh nodal points, and where the boundary cuts a mesh line a resistor having a proportional resistance value was inserted. The resistance values on part of the upstream and downstream faces of the dam are indicated in Fig. 4.24. It should be noted that where a selvedge exists the resistance values are doubled and that the base of the dam and the aquifer below it are both selvedges; the internal boundary condition requires corresponding mesh nodes to be connected.

Having set up the network as shown in Fig. 4.25 and choosing the groundwater potential as the working function, the experiment proceeded on the lines already described for the problems in confined flow (Sections 4.7 and 4.8) and no additional comment is necessary. The recorded values of the groundwater potential are shown in Fig. 4.26.

Aquifer only, with and without cut-off wall

In the next three cases (examples 2, 3, and 4 of Table 4.2) it is assumed that the permeability of the dam is zero and therefore the faces and the base of the dam are effectively impermeable, thus presenting a case of confined flow. The effect of introducing a cut-off wall and the corrections necessary to take account of the singularity at the bottom of the wall are then considered.

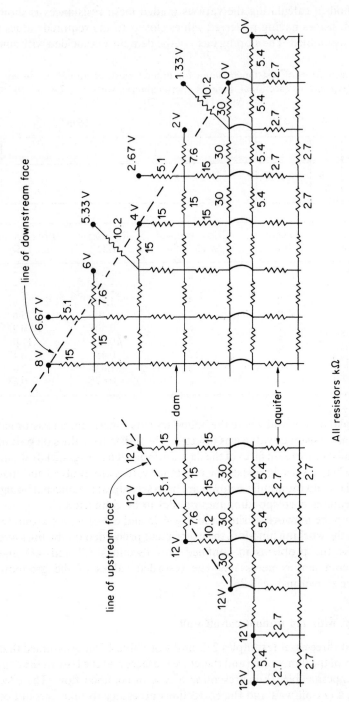

All resistors kΩ

Figure 4.24. Representation of inclined faces of dam and connections to aquifer

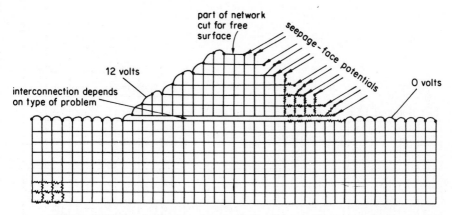

Figure 4.25. Circuit diagram for aquifer and dam with seepage face

The basic network remains unchanged and the electrical set-up is similar to that illustrated in Fig. 4.9(a). The only new feature concerns the introduction of the cut-off wall. To take account of a cut-off wall, extending downwards to a distance three-quarters the depth of the aquifer, the network had to be 'cut' and boundary resistors inserted as shown in Fig. 4.27. The method of correcting for the singularity has been fully described in Section 3.6, and simply requires that all the resistances connected to the lowest node of the cut-off wall are multiplied by 1.64.

Groundwater equipotentials estimated from the numerical results at discrete points for the three cases studied in this section are shown for the central region of interest in Fig. 4.28.

Dam only, with and without drain

For these cases (examples 5 and 6 of Table 4.2) it was assumed that the dam rested on an extensive horizontal impermeable stratum. Without the drain the problem is a simple one in unconfined flow, and the basic technique previously described in Section 4.8 was used.

The introduction of a drain in the base of the dam simply requires that the network nodes, along the line of the drain, are connected to zero potential.

Groundwater equipotentials constructed from the numerical results are plotted in Fig. 4.29 for these two cases. The singularity correction has been applied to the resistors at the end of the drain. Note that when a drain is provided, the free surface does not extend to the downstream face of the dam, and consequently there is no seepage face.

Dam and aquifer with drain and cut-off wall

In the final case (example 7 of table 4.2) the dam and aquifer are modelled with both the drain and the cut-off wall present. The circuit diagram is similar to that

																	120			
																1200	1171	110		
															1200	1147	1091	102		
														1200	1175	1125	1073	1017	956	
													1200	1150	1101	1052	1001	948	891	
												1200	1172	1122	1074	1027	980	931	881	829
1200	1200	1200	1200	1200	1200	1200	1200	1200	1200	1200	1135	1086	1041	997	954	909	863	816	768	
1200	1200	1199	1197	1194	1190	1184	1180	1174	1165	1147	1108	1067	1025	983	941	897	852	806	758	
1200	1199	1198	1194	1188	1181	1170	1162	1152	1137	1116	1085	1049	1010	970	929	886	842	796	749	
1200	1199	1196	1192	1182	1172	1158	1147	1134	1117	1094	1066	1033	997	958	918	876	833	788	742	
1200	1199	1196	1190	1177	1165	1147	1135	1120	1101	1078	1050	1019	985	948	909	868	826	782	736	
1199	1199	1194	1188	1173	1159	1139	1125	1109	1089	1066	1039	1009	976	940	902	862	820	777	732	
1199	1199	1194	1186	1170	1155	1133	1119	1102	1082	1059	1032	1002	970	935	897	857	816	773	728	
1199	1199	1194	1186	1168	1153	1130	1115	1098	1077	1054	1028	998	966	931	894	855	814	771	726	
1199	1198	1194	1185	1168	1152	1129	1114	1097	1076	1052	1026	997	965	930	893	854	813	770	726	

Figure 4.26. Groundwater potential values for aquifer and dam with seepage face

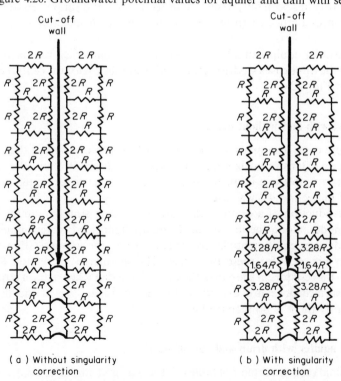

(a) Without singularity
correction

(b) With singularity
correction

Figure 4.27. Resistance network for cut-off wall in aquifer

951																			
880	787	734																	
821	749	680	603	557															
768	704	639	573	512	441	400													
716	658	598	537	476	411	345	256												
666	612	558	501	443	381	314	237	141	0	0	0	0	0	0	0	0	0	0	0
657	605	551	496	439	378	314	244	168	92	59	42	31	25	15	10	4	2	0	0
650	599	547	493	437	380	320	258	197	141	103	78	60	47	30	19	9	4	1	0
645	595	543	491	437	383	327	272	219	172	134	105	83	66	43	28	12	5	2	0
641	592	541	491	438	387	335	284	236	193	156	126	102	82	54	35	16	7	2	1
638	589	540	491	440	390	341	293	248	208	172	142	116	95	63	41	18	8	2	1
636	588	539	491	441	393	346	300	257	218	183	152	126	103	69	47	21	9	2	1
634	587	539	491	442	395	348	304	262	223	188	158	131	108	73	49	22	9	2	1
634	587	539	491	443	395	349	305	263	225	190	160	133	110	74	49	22	9	2	1

of Fig. 4.25 apart from the condition of zero potential along the line of the drain and a 'cut' network as shown in Fig. 4.27 on the cut-off wall. The singularity correction has been applied to both the drain and the cut-off wall. Figure 4.30 contains a plot of the groundwater potentials constructed from the numerical results. It is of value to compare the equipotentials of Fig. 4.30 with those of Figs. 4.28(c) and 4.29(b); there are significant differences between the aquifer and dam treated separately and the combined behaviour.

Flow through the dam and the aquifer

In the above description attention has been confined to methods of obtaining the groundwater potentials, but the practical significance of such an example concerns the effect on the magnitudes of the flows. The third, fourth, and fifth columns of Table 4.2 list the individual flows through the upstream face of the dam, through the aquifer, or through the combined system. These results repay careful study. Only two of the many significant points will be commented on here. The cut-off wall extending for 0.75 of the aquifer depth reduces the flow to 72 per cent of the value when no cut-off wall is present. Secondly, summing the results of cases 4 and 6 gives a total flow of $25.1 \times 10^{-3} \, \text{m}^2/\text{d}$ through the dam and the aquifer when they are considered separately. This compares with a total flow of $33.0 \times 10^{-3} \, \text{m}^2/\text{d}$ for case 7, where the dam and aquifer are analysed as a single unit.

These results indicate that a reliable method is essential to obtain accurate information for the flows as well as the groundwater potentials. The finite-difference method provides a reliable method of modelling the complex flow patterns in the vicinity of cut-off walls and drains.

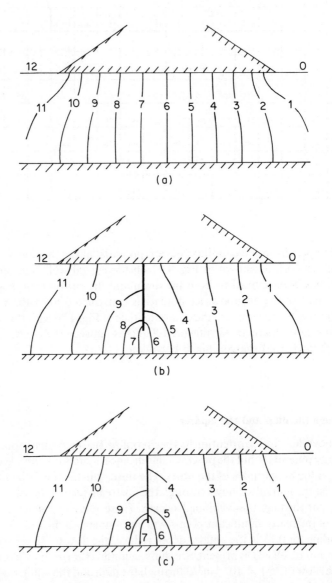

Figure 4.28. Groundwater potentials for the aquifer: (a) aquifer only; (b) aquifer only with cut-off wall; (c) aquifer only with cut-off wall and correction for singularity

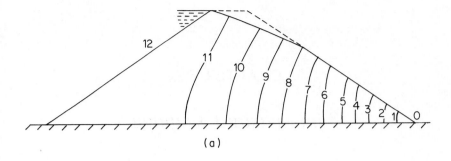

(a)

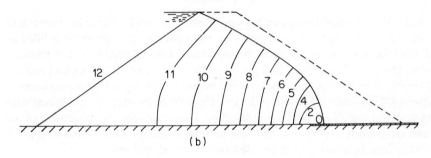

(b)

Figure 4.29. Groundwater potentials for dam: (a) dam only; (b) dam only with drain and singularity correction

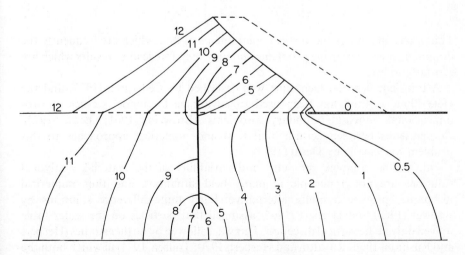

Figure 4.30. Groundwater potentials for dam and aquifer with drain and cut-off wall

CHAPTER 5

Time-variant seepage

5.1 Principles

The time-variant nature of seepage problems is generally due to the movement of the water table (or free water surface). For example, as a reservoir is filled or emptied the upstream water level changes, and this is reflected by a movement of the water table. The change in position of the water table is not instantaneous; indeed a considerable time may elapse before a new equilibrium condition is achieved. This delay is a result of the specific yield of the material combined with the slow seepage of water through the dam. Consequently, the movement of the water table tends to dominate the flow pattern within the dam. A similar pattern of behaviour is noted during the dewatering of excavations.

The prediction of the movement of water tables in seepage problems has received little attention in the literature. This is primarily due to the difficulty in obtaining reliable solutions. For simple problems involving constant permeability, approximate results can be obtained using the transient form of the Dupuit–Forchheimer equation (Jacob, 1950):

$$\frac{k}{2}\left[\frac{\partial^2(h^2)}{\partial x^2} + \frac{\partial^2(h^2)}{\partial y^2}\right] = S_y\frac{\partial h}{\partial t} \tag{5.1}$$

This equation ignores the vertical components of flow which are frequently the dominant effect in seepage problems and may therefore lead to results which are seriously in error.

A transient flow-net technique was described by Cedergren (1977) and the Hele–Shaw method has been used to advantage for model studies of two-dimensional moving boundary problems (Browzin, 1961; Bear, 1972). Comparisons between certain analytical and numerical approaches to this problem were made by Desai (1973).

Due to the complex geometry and variability of the material, analytical solutions are not applicable in many field situations, and thus numerical techniques provide a useful alternative. Early finite-difference solutions by Bouwer (1962) and Herbert (1965) assumed that particles on the water table moved only in the vertical direction. This was followed by further studies (Herbert and Rushton, 1966; Rushton and Herbert, 1970; Todsen, 1971) in which both the horizontal and vertical components of flow of a particle on a water table were considered. Taylor and Luthin (1969) used an alternative finite-difference approach for radial flow to a well in which the region above the water table was represented as an unsaturated zone. Finite element (Neuman and Witherspoon, 1971; Cheng and Li, 1973) and boundary integral equation (Liggett, 1977) techniques are also available. For further information on the developments of methods of solution, the paper by Todsen (1971) should be consulted.

Mathematical formulation

In Section 2.5, where the idealizations involved in various types of problems were introduced, it was noted that in seepage problems the volume of water released from storage due to the compressibility of the aquifer can be neglected when compared with the volume of water which is released from storage as the water table is displaced. As a consequence, the specific storage is set to zero and the differential equation takes the same form as for steady flow, namely

$$\frac{\partial}{\partial x}\left(k_x\frac{\partial h}{\partial x}\right) + \frac{\partial}{\partial y}\left(k_y\frac{\partial h}{\partial y}\right) + \frac{\partial}{\partial z}\left(k_z\frac{\partial h}{\partial z}\right) = 0 \qquad (5.2)$$

Numerical solutions to this equation can be obtained by a variety of techniques using either digital or analog computers (see Sections 4.7, 4.8, and Appendices 3 and 4).

The distinctive feature of the time-variant seepage problem is the moving boundary. In Section 2.4, an equation describing the movement of the water table was derived (Eq. 2.12), from which the change in elevation of the moving water table on a vertical ordinate during a time increment δt is shown to be

$$\delta H = \frac{\delta t}{S_y}(v_z + v_x \tan\alpha + v_y \tan\beta)$$

The angles α and β are defined in Fig. 5.1. This equation can be rewritten in terms of the groundwater potentials as

$$\delta H = -\frac{\delta t}{S_y}\left(k_z\frac{\partial h}{\partial z} + k_x\frac{\partial h}{\partial x}\tan\alpha + k_y\frac{\partial h}{\partial y}\tan\beta\right)$$

First it must be recognized that at any point within the saturated aquifer the groundwater potential is a function of the space coordinates x, y, z and the time coordinate t, thus

$$h = F(x, y, z, t)$$

On the water table the pressure is atmospheric, thus

$$z = H(x, y, t)$$

The symbol H is used to signify the groundwater potential on the water table. Consequently, on the water table surface at any time t,

$$h = H(x, y, t) = F(x, y, H(x, y, t), t)$$

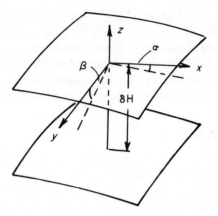

Figure 5.1. Movement of water table
during time step δt

By the rules of partial differentiation

$$\frac{\partial H}{\partial x} = \frac{\partial F}{\partial x} + \frac{\partial F}{\partial z}\frac{\partial H}{\partial x}$$

$$\frac{\partial H}{\partial y} = \frac{\partial F}{\partial y} + \frac{\partial F}{\partial z}\frac{\partial H}{\partial y}$$

Also $\partial H/\partial x = -\tan\alpha$, $\partial H/\partial y = -\tan\beta$; consequently

$$\frac{\partial F}{\partial x} = \left(\frac{\partial F}{\partial z} - 1\right)\tan\alpha; \qquad \frac{\partial F}{\partial y} = \left(\frac{\partial F}{\partial z} - 1\right)\tan\beta$$

Since $h = F(x, y, z, t)$,

$$\frac{\partial h}{\partial x} = \left(\frac{\partial h}{\partial z} - 1\right)\tan\alpha; \qquad \frac{\partial h}{\partial y} = \left(\frac{\partial h}{\partial z} - 1\right)\tan\beta$$

Finally,

$$\delta H = \frac{-\delta t}{S_y}\left[k_z\frac{\partial h}{\partial z} - k_x\left(1 - \frac{\partial h}{\partial z}\right)\tan^2\alpha - k_y\left(1 - \frac{\partial h}{\partial z}\right)\tan^2\beta\right] \qquad (5.3)$$

Note that the movement of the water table depends on the specific yield S_y and on the permeabilities; these may all be functions of the elevation. The three-dimensional movement of the water table is determined, partly, by the angles α and β which are the inclinations below the horizontal of the water table in the xz and yz planes respectively. The dominant effect is, however, the potential gradient in the vertical direction, $\partial h/\partial z$.

Solutions at discrete time steps

This moving boundary is represented as values of the water table groundwater potential H at discrete time steps. The first result is obtained at a time $t = 0$, by solving numerically the differential equation, Eq. 5.2, with the water table at its initial position and all the other boundaries taking their appropriate values. From the groundwater potentials of this first solution, the free-surface position at a time Δt after the start of the calculation can be predicted using Eq. 5.3.

For the second solution, this new free-surface position together with the conditions on all the other boundaries, certain of which may have changed from the initial conditions, are used with the finite-difference form of Eq. 5.2 to obtain the distribution of groundwater potentials throughout the field at time Δt. This process is repeated many times and may be continued until a steady state is reached.

Clearly the choice of the time step is of great importance. Too large a time step leads to inaccurate solutions and perhaps instability, whereas too small a time step is wasteful in effort. No simple expression can be given for the size of the time step since individual problems differ significantly. However, as a general guideline the time step should be selected so that the water table movement is between 5 and 10 per cent of the maximum possible movement. For many problems this will mean that the time steps increase as the calculation proceeds.

The procedure outlined above is an explicit scheme since the water table positions at the new time are calculated only from the groundwater potentials at the previous time step. However, the water table position at a particular time really depends both on the potentials at previous times and on the current groundwater potentials throughout the field. With an explicit procedure there is a risk of instability (Todsen, 1971) but, provided that the time step is small and some form of smoothing is introduced, the explicit technique is usually acceptable. In an alternative approach Herbert (1966) showed that larger time steps can be used when the water table position is modified in the light of the current groundwater potentials.

Since a step-by-step technique is adopted, variable permeability and specific yield can be included. The modifications required to include such variable coefficients were discussed by Herbert (1966).

The technique used to represent a moving boundary in a finite-difference solution will be introduced by considering a two-dimensional problem involving a falling water table. Reference will then be made to a situation in which the water table rises. In a later section, the dewatering of excavations will be discussed.

All the examples of this chapter relate to cartesian coordinates. Solutions can be obtained for radial flow (Herbert, 1968b), but it is difficult to represent the behaviour close to any abstraction well. This difficulty occurs because the angle of the water table becomes very steep close to the well and the radial form of Eq. 5.3 tends to be unstable.

5.2 Example of Falling Water Table

Consider the rectangular dam of uniform permeability k, sketched in Fig. 5.2(a). The flow in the y-direction is sufficiently small to be neglected. Initially, the upstream and downstream water levels are both at $H_0 = H_1 = 1.0L$. At time $t = 0$, the downstream level is suddenly lowered to $H_1 = 0.2L$ and maintained at that value. The water table falls (Fig. 5.2(c)), eventually reaching a steady state (Fig. 5.2(d)). A steady-state solution to this problem was described in Section 4.8 and values of groundwater potentials are recorded in Fig. 4.17.

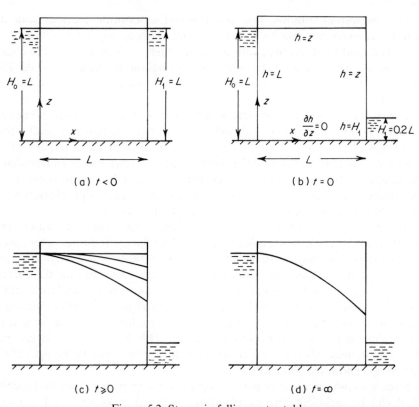

Figure 5.2. Stages in falling water table

The water table fall is traced at a series of discrete times. For the first step in the solution, the Laplace equation is solved with the boundary conditions as shown in Fig. 5.2(b). These conditions correspond to the instant when the downstream water level to the right has just fallen. The upper water surface of the dam is horizontal; thus it is held at $h = L$. On the downstream face ($x = L$), for $z > 0.2L$ the condition becomes that of a seepage face. Since the seepage face is at atmospheric pressure the condition which must be enforced is $h = z$.

Finite-difference solutions to all of the problems described in this chapter can be obtained using either analog or digital computers. On a digital computer an iterative method of solving the finite-difference equations, such as successive over-relaxation, is convenient, using a program similar to that of Appendix 4. Alternatively the resistance network analog can be used. A schematic circuit diagram is presented in Fig. 5.3. The electrical circuit is similar to that of Fig. 4.15 apart from the water table, where the condition is that of enforced fixed potentials.

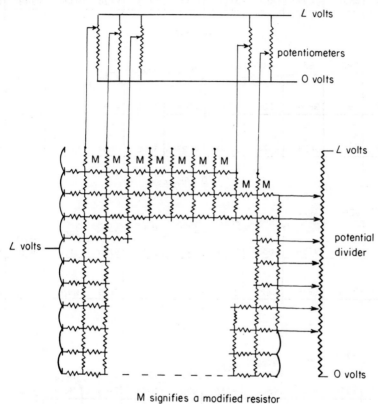

M signifies a modified resistor

Figure 5.3. Circuit for analog representation of moving water table

For the purpose of demonstrating the technique, mesh intervals in the x- and z-directions are all chosen to equal $0.1L$. If variations occurred in the permeability, then more mesh intervals would be advisable. The distribution of groundwater potentials on the horizontal lines $z = 1.0$, 0.9, and $0.8L$ for the initial step are recorded in Fig. 5.4(a).

From these groundwater potentials the position to which the water table will move after a time increment Δt can be estimated using Eq. 5.3 which, for this problem of uniform permeability and no flow in the y-direction, becomes

$$\frac{\Delta H}{L} = \frac{-k\,\Delta t}{S_y L}\left[\frac{\partial h}{\partial z} - \left(1 - \frac{\partial h}{\partial z}\right)\tan^2\alpha\right] \qquad (5.4)$$

112

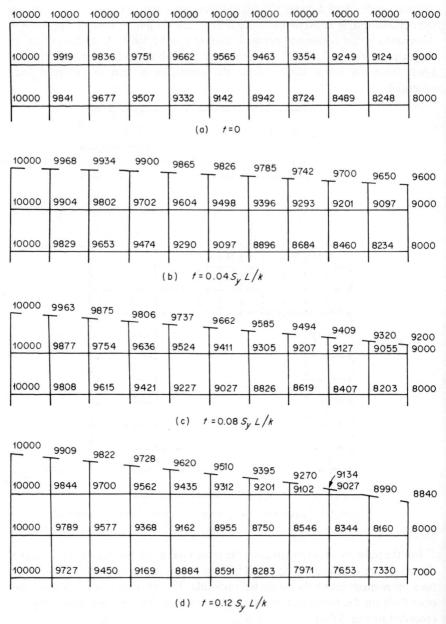

Figure 5.4. Groundwater potentials (times 10 000) for the first four stages of the analysis

The calculation on the vertical section $x = 0.6L$ is summarized in the upper part of Table 5.1. The simplest finite-difference approximations are used. Higher-order approximations do lead to some improvement but the most straightforward approach will be used for this demonstration problem. On other

Table 5.1 Evaluation of Eq. 5.4 on vertical section $x = 0.6L$; $t' = k\Delta t/S_y L$

Function	Equation	Expression	Value
1st time step $t' = 0$ to $t' = 0.04$			
$\partial h/\partial z$	$(H_{z=L} - h_{z=0.9L})/0.1L$	$(1.0 - 0.9463)/0.1$	0.537
$\tan\alpha$	initially horizontal		0.0
$\dfrac{\Delta H}{L}$	$\dfrac{-k\Delta t}{S_y L}\left[\dfrac{\partial h}{\partial z} - \left(1 - \dfrac{\partial h}{\partial z}\right)\tan^2\alpha\right]$	-0.04×0.537	$-0.021\,48$
new H	$H_{t'=0.04} = H_{t'=0.0} + \Delta H$	$1.0L - 0.021\,48L$	$0.9785L$
2nd time step $t' = 0.04$ to $t' = 0.08$			
$\partial h/\partial z$	$(H_{z=0.9785L} - h_{z=0.9L})/0.0785L$	$(0.9785 - 0.9396)/0.0785$	0.496
$\tan\alpha$	$(H_{x=0.5L} - H_{x=0.7L})/0.2L$	$(0.9826 - 0.9742)/0.2$	0.042
$\dfrac{\Delta H}{L}$	$\dfrac{-k\Delta t}{S_y L}\left[\dfrac{\partial h}{\partial z} - \left(1 - \dfrac{\partial h}{\partial z}\right)\tan^2\alpha\right]$	$-0.04(0.496 - (1 - 0.496)(0.042)^2)$	-0.0199
new H	$H_{t'=0.08} = H_{t'=0.04} + \Delta H$	$0.9785L - 0.0199L$	$0.9586L$

vertical sections, the calculation takes an identical form. The manner in which the calculation is modified for variable permeability has been described by Herbert (1966).

A suitable value for the time step can be estimated as follows. The total fall, towards the right-hand side of the aquifer, is roughly $0.5L$, so that if ten time steps are used, the fall for each time step is roughly $\Delta H/L = 0.05$. Now the initial value of $\partial h/\partial z$ is approximately 0.9, thus from Eq. 5.4 with $\tan \alpha = 0$

$$\frac{k \, \Delta t}{S_y L} = \left(\frac{\Delta H}{L}\right) \bigg/ \left(\frac{\partial h}{\partial z}\right) = \frac{0.05}{0.9} = 0.055$$

For the first three steps the time step is chosen to be $\Delta t = 0.04\, S_y L/k$. With a decreasing vertical velocity at the later stages of the calculation, the size of the time steps can be increased.

The new water table positions from the above calculations are used as the conditions on the upper boundary for the next time step. However, the upper boundary has been displaced downwards corresponding to the fall in the water table. For example, on the line $x = 0.6L$, the vertical mesh extends to $z = 0.9785L$ and at the uppermost point on the mesh, the condition is enforced that $H = 0.9785L$. Therefore the finite-difference equation at node $x = 0.6L$, $y = 0.9L$ has to be modified. Referring to Fig. 5.5, and applying the formula for irregular mesh spacings, Section 3.7,

$$(h_1 - h_0) + \frac{1}{0.785}(h_2 - h_0) + (h_3 - h_0) + (h_4 - h_0) = 0 \qquad (5.5)$$

where $h_2 = 0.9785L$. Similar modifications are made on all the other vertical sections.

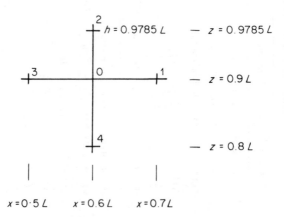

Figure 5.5. Modification to finite-difference network
due to moving water table

If a resistance network is used to solve the finite-difference equations, an alternative approach using a potential divider (Herbert and Rushton, 1966) can be adopted. This satisfies the condition of a mesh line of reduced length without physically changing the resistors.

When the new boundary potentials are applied to the modified mesh, with all the other boundary conditions remaining unchanged, results are obtained for the groundwater potentials at a time $t = 0.04S_yL/k$. Values for the upper part of the media are recorded in Fig. 5.4(b). The calculation of the next estimated water table position is illustrated by the lower half of Table 5.1. A similar calculation is repeated for all the vertical sections. At the boundary $x = 1.0L$, where both the seepage face and the water table conditions apply, a singularity occurs (Polubarinova-Kochina, 1962). As a consequence, Eq. 5.4 does not apply, yet the approximate position of the top of the seepage face can be obtained by extrapolation based on the water table profile for $x = 0.7L, 0.8L, 0.9L$.

Using these new water table positions, the upper boundary of the field is modified again and a new set of finite-difference equations of the form of Eq. 5.5 are obtained. Values of the groundwater potential corresponding to a time $t = 0.08S_yL/k$ are calculated from the full set of finite-difference equations; values in the upper part of the dam are recorded in Fig. 5.4(c).

For the next increment, the time step remains at $\Delta t = 0.04S_yL/k$. The calculation of the new water table position again follows the procedure of Table 5.1 but certain of the new water table positions now lie below $z = 0.9L$. Consequently, part of the upper boundary is below $z = 0.9L$ and the remainder is above. Groundwater potentials in the upper part of the aquifer are recorded in Fig. 5.4(d).

For the next step, the time increment is increased to $\Delta t = 0.06S_yL/k$ and the procedure is as before. As the slope of the water table increases, the rate of fall of the water table on a vertical line becomes less and therefore larger time steps are taken. After eight time steps when the elapsed time is $t = 0.72S_yL/k$, the water table position approaches the steady-state position. The fall in the water table through the various stages is plotted in Fig. 5.6.

Since this is an explicit technique, the rate of fall of the water table will tend to be overestimated. As a result the water-table profiles of Fig. 5.6 are slightly below their correct position. More detailed results with an implicit approximation suggests that the water-table fall plotted in Fig. 5.6 is an overestimate by about 10 per cent.

In the later stages of the calculation, there is a tendency to instability. As the steady state is approached, the terms in Eq. 5.4 almost cancel each other out, leaving a small net vertical movement, ΔH. Therefore, even small errors in either of the two terms will lead to erratic predictions of the water-table fall.

The results of this example have been given in non-dimensional terms. To give an idea of the magnitude of the quantities, particular numerical values may be substituted. If the dam consists of a silty sand with a permeability of 0.12 m/d, a length and height of 6 m, and a specific yield of 0.15, the time to fall to the final position plotted in Fig. 5.6 is

$$t = \frac{0.72S_yL}{k} = \frac{0.72 \times 0.15 \times 6}{0.12} = 5.4 \text{ days}$$

116

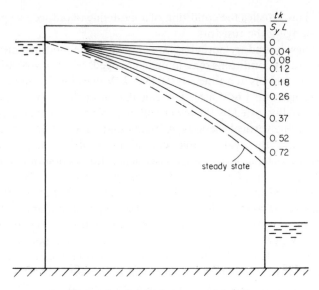

Figure 5.6. Stages in water-table fall

Alternative method of modelling

An alternative method of modelling unsteady, unconfined problems is possible if the moving water table is represented as a series of inflows rather than known groundwater potentials.

On a moving water table two conditions apply:

(a)   The pressure is atmospheric, hence $h = z$;
(b)   Water is released into the aquifer as the water table falls.

In the detailed experiment described above, the moving boundary was represented as a series of known groundwater potentials which satisfy directly the condition of zero atmospheric pressure. The second condition concerning the water released from storage is represented by Eq. 5.4 from which the water-table positions at successive time steps are calculated.

In the *alternative technique*, the water table is represented as a known flow rather than a known potential. Equation 5.4 can be rewritten as the rate of vertical movement of the water table using a similar approach to the derivation of Eq. 2.19 in Section 2.5.

$$\left(\frac{\delta H}{\delta t}\right)_z = \frac{-k}{S_y}\left[\frac{\partial h}{\partial z} - \left(1 - \frac{\partial h}{\partial z}\right)\tan^2\alpha\right] \tag{5.6}$$

where H signifies the groundwater potential on the water table. Thus the inflow at a nodal point on a vertical mesh line equals $\delta H/\delta t$ multiplied by the perpendicular cross-sectional area, that is

$$Q = \frac{-\delta H}{\delta t} \Delta x \, \Delta y$$

$$= \frac{k \, \Delta x \, \Delta y}{S_y} \left[\frac{\partial h}{\partial z} - \left(1 - \frac{\partial h}{\partial z} \right) \tan^2 \alpha \right] \qquad (5.7)$$

In two-dimensional problems in the xz plane, Δy equals unity.

This new procedure requires the following steps:

(a)   For the initial calculation when the downstream water level has suddenly fallen, the water table remains at its initial position, $H = 1.0L$. Figure 5.4(a) therefore represents the potentials in the upper region of the aquifer at a time $t = 0$. The evaluation of the terms within the square brackets of Eqs. 5.4 and 5.7 follows the same pattern as before and the predicted position of the water table at a time $t = 0.04 S_y L/k$ is also calculated as before. In addition, the rate of release of water due to the free-surface fall can be calculated from Eq. 5.7.

(b)   When determining the potentials throughout the aquifer at a time $t = 0.04 S_y L/k$, the water table is represented as inflows at the nodal positions on the water table. However, owing to the nature of the finite-difference approximation it is not necessary to form an irregular mesh since the flows can be injected at the nodal points directly above the water-table positions as shown in Fig. 5.7(a). Other boundaries are represented in the same manner as before. Groundwater potentials are then obtained by solving the finite-difference equations.

(c)   The procedure is repeated evaluating the new water-table position and the inflow rate using Eqs. 5.4 and 5.7 with the time steps increasing in the same form as with the fixed-potential technique. The boundary conditions at a later stage are as indicated in Fig. 5.7(b).

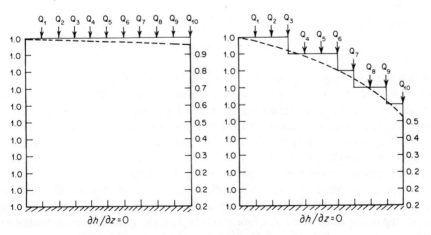

Figure 5.7. Moving water table represented as inflows

E

This approach has two major advantages. Firstly, the predicted potentials and flows from Eqs. 5.4 and 5.7 have a smoother distribution. Therefore smoothing of the results plays a less important part. Secondly, as the water table tends towards its steady-state position, the inflows at the water table tend to zero. Consequently it is easier to obtain a steady-state solution since, when the inflows are zero, the problem is posed in an identical form to the steady-state solution described in Section 4.8.

5.3 Rising Water Table

The situation of a rising water table is equally amenable to solution using the discrete time step method. In the following discussion the water table is simulated in the finite-difference solution as fixed potentials, yet the alternative method of specified flows can equally well be used. Certain new features are introduced: they will be described with reference to a particular problem.

Consider a rectangular dam of unit thickness (Fig. 5.8) having a uniform permeability k and a specific yield S_y. The dam rests on an impermeable base; the right-hand boundary, $x = L$, is also impermeable. On the left-hand boundary the water level is initially $H_1 = 0.4L$ and the water table within the dam is also at this level (Fig. 5.8(a)). Then the upstream water level is raised so that

$$H_1 = 0.4L + tk/S_y$$

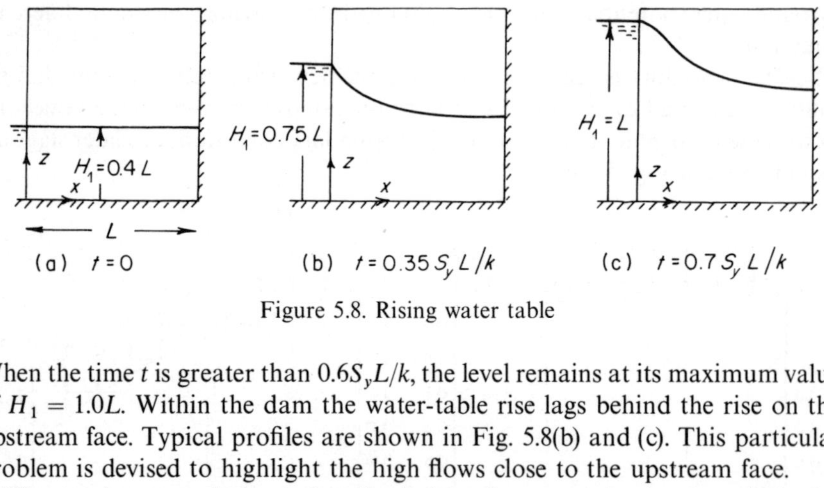

Figure 5.8. Rising water table

When the time t is greater than $0.6S_yL/k$, the level remains at its maximum value of $H_1 = 1.0L$. Within the dam the water-table rise lags behind the rise on the upstream face. Typical profiles are shown in Fig. 5.8(b) and (c). This particular problem is devised to highlight the high flows close to the upstream face.

The step-by-step solution is described with reference to the sketches of Fig. 5.9. A uniform time step of $\Delta t = 0.1S_yL/k$ is used; this is certainly too large for very accurate results in the region close to the upstream face.

There is no need to solve the differential equation for $t = 0$ since the groundwater potential throughout the aquifer is everywhere $h = 0.4L$. For the next step of the calculation at a time $t = 0.1S_yL/k$, the groundwater potential on the boundary $x = 0$, is $H_1 = 0.5L$. However, the water-table position within the

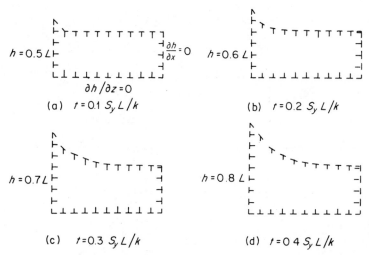

Figure 5.9. Finite-difference grid to represent rising water table

dam calculated using Eq. 5.4 from the potentials at $t = 0$, is that $H = 0.4L$ at all the internal water-table nodes. This occurs because at $t = 0$, $\partial h/\partial z = 0$ everywhere. Therefore the boundary conditions take the form indicated in Fig. 5.9(a). Within the dam the flow is described by the Laplace equation; numerical results are obtained using standard finite-difference techniques. Equipotential lines plotted from the numerical results are presented in Fig. 5.10(a).

The groundwater potentials for the time $t = 0.1S_yL/k$ are used to estimate the vertical movement of the water according to Eq. 5.4, from which the new water-table position at $t = 0.2S_yL/k$ can be predicted. The technique is exactly the same as for the falling water table. Since $\partial h/\partial z$ is negative, Eq. 5.4 predicts a rise in water-table position. Boundary conditions for the dam at a time $t = 0.2S_yL/k$ are as indicated in Fig. 5.9(b) and the boundary conditions and positions for the next two time steps are as sketched in Figs. 5.9(c) and (d).

For each successive step the calculation follows the same pattern. Certain difficulties arise in representing the curvature close to the upstream face, and the calculated values of both $\partial h/\partial z$ and $\tan \alpha$ often need to be smoothed. Smaller time steps and smaller mesh intervals close to the upstream boundary would lead to improved results.

An interesting feature of this problem is that all the boundary conditions change at each time step. Indeed, if it was not for the relatively slow movement of the water table, the groundwater potentials would be dominated by the rising upstream water level. The development of the flow pattern within the aquifer is apparent from the equipotential plots in Fig. 5.10.

5.4 Dewatering of Excavations

As a final example of the potentialities of finite-difference methods in solving seepage problems, an example concerning the dewatering of excavations is

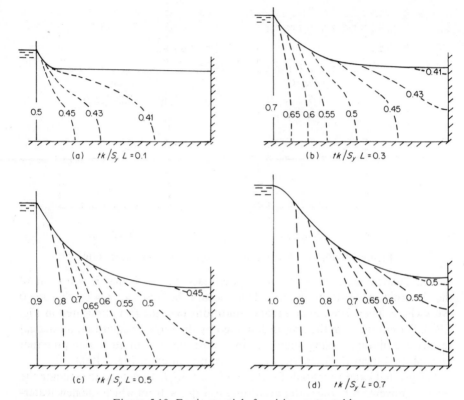

Figure 5.10. Equipotentials for rising water table

considered. If an excavation is to be made below the water-table level, some method of controlling the seepage of water into the excavation must be introduced.

One method commonly chosen involves the provision of wells around the excavation. These wells are pumped until the water surface is lowered below the lowest point of the excavation. The water surface is maintained at this position whilst the excavation and construction takes place. When examining this problem both a time-variant moving surface and a steady-state solution are required.

Following the detailed solution, an approximate method of analysis is suggested.

Three-dimensional solution

The dewatering of an excavation must be analysed as a three-dimensional problem. The governing equation is therefore

$$\frac{\partial}{\partial x}\left(k_x\frac{\partial h}{\partial x}\right) + \frac{\partial}{\partial y}\left(k_y\frac{\partial h}{\partial y}\right) + \frac{\partial}{\partial z}\left(k_z\frac{\partial h}{\partial z}\right) = 0 \qquad (5.8)$$

A finite-difference approximation to this equation was introduced in Section 4.6. Eq. 4.30 allows for random variations in permeability and a variable graded mesh.

On the moving water table, the full three-dimensional form of Eq. 5.3 is used. The change in water-table level ΔH on a vertical line during a time increment Δt is

$$\Delta H = \frac{-\Delta t}{S_y} \left[k_z \frac{\partial h}{\partial z} - k_x \left(1 - \frac{\partial h}{\partial z} \right) \tan^2 \alpha - k_y \left(1 - \frac{\partial h}{\partial z} \right) \tan^2 \beta \right] \qquad (5.9)$$

The technique of tracing the moving surface follows the same routine as that outlined in Section 5.2. The choice of the size and number of time steps is also based on similar considerations to those of the two-dimensional problems. Solutions can be obtained on either analog or digital computers. However, for a problem such as the dewatering of excavations where the number of nodal points exceeds 10 000 and a complex three-dimensional moving boundary has to be represented, there is considerable advantage in the choice of the resistance network analog technique.

Particular problem

The particular problem chosen to illustrate the technique is sketched in Fig. 5.11. A plan view of one-quarter of the excavation is shown by the hatched line;

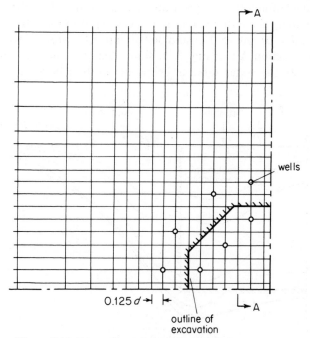

Figure 5.11. Example of dewatering of excavations. Because of symmetry only one-quarter of the field is included

due to symmetry it is only necessary to consider a quarter of the field. The total distance across the excavation is $1.625d$, where d is the original saturated depth of the aquifer. The maximum depth of the excavation is $0.35d$.

A total of twenty-eight wells are positioned as indicated in Fig. 5.11. These wells have a radius of $0.0125d$ and they are pumped so that the drawdown in each well is maintained at $0.875d$. Due to the presence of a seepage face in each well, the drawdown in the aquifer will never reach this magnitude.

An adequate design of the mesh is very important. Throughout the model the vertical mesh spacing is $z = 0.125d$. For the region in the vicinity of the excavation, the mesh spacing in the x- and y-directions equals $0.125d$ but at distances of $0.5d$ beyond the ring of wells the spacing is increased to $0.25d$ or more, so that the total plan area is approximately $18d$ by $18d$. Therefore, the mesh takes a complex rectangular shape.

On the network, the following *initial* and *boundary conditions* must be enforced. Around all the outer vertical boundaries of the grid, the assumption is made that the groundwater potentials are unaffected by the pumping. Thus the condition on this outer boundary is $h = d$. For each of the wells the well-water level is $0.125d$ above the base of the pervious stratum. Hence for $z = 0$ and $0.125d$ the groundwater potentials are $h = 0.125d$. Then on the seepage face of each well, $d \leqslant z \leqslant 0.25d$, $h = z$. The initial condition over the whole of the original water table surface is that for $z = d$, $H = d$.

The calculation follows a procedure similar to that described in Section 5.2. After obtaining the initial solution, the calculation proceeds with solutions at times $kt/S_y d = 20, 50, 95, 175$, and 300. Figure 5.12, which plots the water-table profiles on section AA of Fig. 5.11, shows that by a non-dimensional time of $kt/S_y d = 300$, the water table is below the bottom of the excavation. If, for

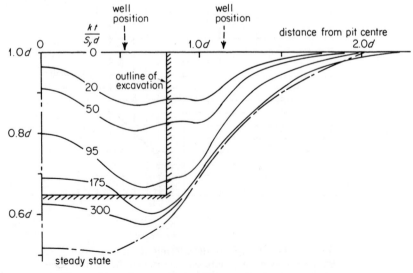

Figure 5.12. Water-table profiles on Section AA of Fig. 5.11

example, the excavation takes place in a relatively permeable sand with $k = 30$ m/d, $S_y = 0.15$, and $d = 15$ m, this corresponds to a time of 22.5 days. Contours of equal groundwater potential at a time of $kt/S_yd = 175$ are shown in Fig. 5.13.

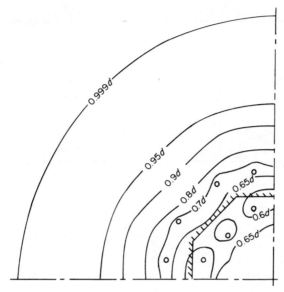

Figure 5.13. Contours of water table after time
$kt/S_yd = 175$

Approximate method

As an alternative to performing a full analysis, it is possible to gain sufficient information from just two solutions. The basis of the method is illustrated by Fig. 5.14. The diagram indicates that there are two main phases in the dewatering process. In the first phase the vertical potential gradients close to the well are high and the water table falls rapidly as shown in Fig. 5.14(a). In the second phase, Fig. 5.14(b), most of the flow to the wells comes from a greater distance, thus the rate of water-table fall is much lower. Only the first phase is of significance in the dewatering of excavations, although the second phase indicates the importance of maintaining the discharge from the wells.

This first phase can be studied by considering the initial state when the water table is horizontal, together with a steady-state solution. The drawdown of the steady-state free surface on any vertical line will be denoted by s_s. From the time–drawdown curve of Fig. 5.15 it is apparent that the average rate of fall of the water table during the first phase is roughly half of the initial rate of fall. Therefore, if a line is drawn corresponding to half of the initial rate of fall, it intersects the actual time–drawdown curve at roughly 0.8 of the total drawdown at the end of the first phase. The value of $0.8s_s$ has been confirmed in a number of different situations (Rushton and Herbert, 1970).

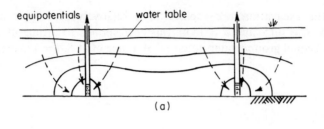

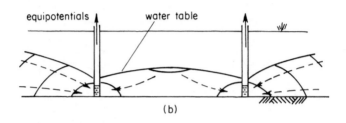

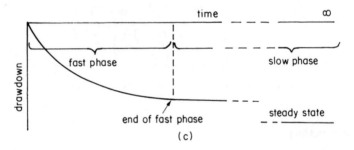

Figure 5.14. Stages in water-table fall: (a) initial stages;
(b) end of fast phase of fall; (c) time–drawdown curve

In order to identify this position of $0.8s_s$ a solution corresponding to the end of the first phase must be obtained; this will give values of the free surface drawdown s_s. An approximation to this position is given by a steady-state free-surface analysis with a zero drawdown boundary at a horizontal distance equal to the depth of the aquifer beyond the outermost wells. It is important that this boundary is used and not the outer boundary of the grid, which is about eight times the depth of the aquifer beyond the outermost wells. The profile of the steady-state solution is shown by the chain-dotted lines in Fig. 5.12. It is apparent that this gives a fair approximation to the groundwater potentials at the end of the first phase.

Though only a single example is described above, a range of problems are described by Rushton and Herbert (1970). From the results of these studies, the design method outlined above is found to be adequate in a variety of dewatering situations.

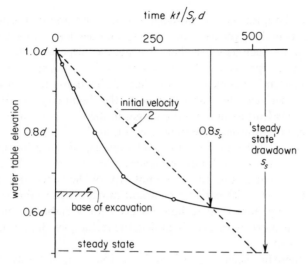

Figure 5.15. Approximate method of investigating dewatering

5.5 Usefulness of Numerical Methods for Seepage Problems

Chapters 4 and 5 have demonstrated that numerical solutions to a wide range of seepage problems can be obtained using analog or digital computers. Variable permeability can be included, as can variable storage coefficients. Many types of boundary can be simulated and techniques are available for cut-off trenches or walls. Free surfaces and seepage faces can be simulated using a simple iterative technique and the important condition of a moving water table can be represented by dividing the continuous time into a series of discrete steps.

All the calculations lead to values of groundwater potential on a regular grid. From these potentials both pressures and velocities can be calculated. Therefore the groundwater potential provides a complete picture of the flow mechanism.

There is often some hesitancy about using numerical techniques for the analysis of seepage problems. This occurs for a variety of reasons. Frequently, the lack of accurate information about the permeability is given as a reason for not attempting to obtain accurate solutions. Also it is suggested that the accuracy of the solution is no better than the accuracy of the data (Cedergren, 1977). This is not necessarily true. In many problems the groundwater potentials and the total discharge are changed only slightly by modifications to the permeability (see for example Desai, 1973). The uncertainties due to the lack of precise data may be quantified by investigating the *sensitivity* due to changes in permeability or other factors such as the nature of the boundary conditions. For example, a range of possible values of permeability can be introduced and solutions obtained for each case. The dependence of either the predicted groundwater potentials or flow on each of these parameters can then be investigated.

Another reason why numerical methods have not been used widely is that well-established techniques such as flow-net construction and conformal transformations are available; yet each of these techniques has severe limitations in certain situations, such as when there are significant permeability changes, or with time-variant problems. Furthermore, with conformal transformation techniques, a potential source of error is that idealizations often must be made in terms of the permeability variations and the boundary conditions. Such idealizations may lead to significant changes in the flow pattern with the result that incorrect groundwater potentials are obtained.

Time-dependent behaviour is of considerable significance in many seepage problems but comparatively little experience has been gained in the techniques of solution. This is partly due to the complexity of the time-variant numerical techniques. However, with care, time-variant moving water-table solutions can be obtained using numerical techniques even for variable permeability and specific yield. As more experience is gained, the techniques should become more firmly established.

In Part II there has been only limited discussion of the methods of solving the finite-difference equations. Where comments have been made they have usually referred to solutions on resistance network analogs, though basic digital computer programs are presented in Appendices 3 and 4. The authors believe that there is much to commend the resistance network method for solving seepage problems, since the networks give immediate solutions to the finite-difference equations. This is particularly valuable when devising iterative techniques for the free surface or moving boundaries. Once a technique has been thoroughly tested, a digital computer version should be prepared. In solving the simultaneous equations on a digital computer the successive over-relaxation method is reliable and relatively easy to program.

PART III

Analysis of Regional Groundwater Flow

CHAPTER 6

Regional Groundwater Flow

6.1 The Nature of Groundwater Flow

Introduction

Many of the rocks which lie beneath the ground surface contain water; they are called *aquifers*. In certain situations the water enters the aquifer by infiltration in areas where the strata crop out on the surface. Alternatively, the water may be transferred from one stratum to another. Not all aquifers contain recent water; sometimes the water has been trapped for many thousands of years. If this water is potable or if its quality is satisfactory for domestic, industrial or agricultural purposes, then the possibility of abstracting the water requires investigation.

The quantity and quality of the water abstracted does not depend only on the design of the well and the nature of the aquifer in the vicinity of the well. The total flow pattern within the aquifer may also be significant. Initially water is drawn from the vicinity of the well but as the abstraction continues, the resources further from the well are exploited. Therefore the groundwater flow in the whole region must be examined.

Two main physical responses may occur when abstraction takes place. If an aquifer is *confined*, there is an overlying impermeable layer which confines the aquifer and isolates it from the atmosphere. The water in the aquifer is then under pressure; if an observation borehole is taken through the impermeable layer to monitor conditions in the aquifer, water will rise up this borehole to a level, termed the *piezometric head*. When water is abstracted from a confined aquifer changes in the piezometric head spread rapidly to distant parts of the aquifer.

On the other hand, an *unconfined* aquifer is open to the atmosphere and at the water surface the pressure is atmospheric. Abstraction from an unconfined aquifer causes significant changes in groundwater potential only in the vicinity of the abstraction well.

Groundwater potential is the term that will be used in this chapter to describe both the piezometric head in a confined aquifer and the elevation of the free-water surface in an unconfined aquifer.

When analysing regional groundwater flow, certain assumptions and idealizations are made so that the solution of specific problems becomes possible (see Section 2.5). Vertical components of flow are assumed to be sufficiently small to be neglected, thus the problem reduces to one of two dimensions. However, if the vertical components of flow are significant, for example with partially penetrating rivers, then special techniques can be introduced; see Section 9.2.

Aquifers are not usually homogeneous; indeed most of the flow takes place through cracks and fissures so that flow through the pores may only be of minor significance. However, when examining an aquifer on a regional scale, it is usually acceptable to describe the flow mechanism in terms of an effective transmissivity and storage coefficient. In the vicinity of abstraction wells, particularly if solution channels have developed, then corrections may have to be made.

A further assumption is that velocities are sufficiently small for the flow to remain laminar and therefore Darcy's law is applicable (UNESCO, 1975; Childs, 1969). Most groundwater velocities are, indeed, very low. It is only in localized situations, such as close to wells, that the flow may be turbulent. On a regional scale, therefore, Darcy's law is assumed to hold and experience shows that this is an acceptable assumption.

Purpose of groundwater models

Mathematical models of groundwater systems provide a means of examining the flow processes within an aquifer. Each model is based on data obtained from a wide variety of sources. When all this information is incorporated in a single model it acts as a check on the consistency of the data. Deficiencies in the data become apparent and further investigation to obtain the required data can be initiated.

Groundwater models are demanding in the data that they require. Certain information may not be available, yet an estimate based on the probable ranges within which the parameters should lie must be made. Such an approach is acceptable since a number of solutions can be obtained with a range of possible values of the parameters; then the sensitivity of the results to the varying parameter values will indicate the overall accuracy of the model.

Extensive historical data are invaluable when devising and testing a groundwater model. Much useful information can also be gained from pilot scheme investigations (Thames Conservancy, 1971). Even when an acceptable model has been devised, it is important to continue monitoring the groundwater-potential variations and use this information to update the aquifer model.

Once a groundwater model has been proved against historical data it can be used to predict changes which are likely to occur when the aquifer is exploited in a different manner.

6.2 Data Requirements

Extensive information must be provided before a model investigation of groundwater flow in an aquifer can be carried out. The information can be deduced from geological, geophysical, hydro-geological, hydrological, and other evidence. Useful texts which describe some of the techniques for obtaining this information include Todd (1959), De Wiest (1965), Walton (1970), Bear (1972), Campbell and Lehr (1973), Cedergren (1977), and Bouwer (1978). Each of the following parameters must be considered.

Transmissivity and storage coefficient

The distribution of transmissivity and storage coefficient throughout the aquifer must be specified. In Section 2.5 the transmissivity was introduced as an integral of the permeability over the depth: thus the transmissivity in the x-direction at a particular location (x, y) is given by

$$T_x = \int_{z_b}^{z_t} k_x \, dz$$

where z_b is the elevation of the base of the aquifer and z_t is the top of a confined aquifer or the water-table elevation of an unconfined aquifer. For a permeability which remains constant with depth,

$$T_x = k_x m$$

where m, the saturated thickness of the aquifer, equals $z_t - z_b$. A discussion of the difference between the confined storage coefficient and the unconfined storage coefficient (or specific yield) can be found in Sections 2.4 and 2.5.

If pumping tests have been carried out, values of the transmissivity and storage coefficient may be available at various points throughout the aquifer. These pumping tests are particularly valuable since they are a measure of the actual hydraulic properties of the aquifer. Other sources of information include geological evidence and groundwater potential maps.

Sometimes, two pumping tests at the same site but at different times of the year provide conflicting evidence. This usually occurs because the tests were carried out when conditions within the aquifer differed. For example, an analysis in Chapter 12 shows that the transmissivity and storage coefficient can vary greatly with groundwater potential, and thus tests starting from different rest-water levels may indicate different aquifer parameters. Another reason why unreliable estimates of the aquifer parameters are often obtained from pumping tests is that the technique of analysis is inadequate. An alternative, more flexible approach to the conventional method of pumping-test analysis is described in Part IV.

There has been considerable interest in recent years in automatic methods of estimating aquifer parameters from the groundwater potentials. This is an inverse problem but, as was shown by Neuman (1975), the technique is very sensitive and may lead to unreliable estimates unless the field data are of a high quality. Furthermore, the techniques require accurate information about the recharge; this is usually difficult to obtain.

Boundaries

Along each of the boundaries of the aquifer under consideration, some form of boundary condition must be specified. Detailed information about the position of an aquifer boundary is often hard to obtain and further difficulties arise in determining the actual hydraulic conditions on these boundaries. The various types of boundary are discussed below; the range of possible conditions on boundaries are listed in Table 2.2.

Impermeable boundary

No flow crosses an impermeable boundary. A typical example of the occurrence of an impermeable boundary is a fault which isolates the aquifer from other pervious strata. In mathematical terms, the condition on an impermeable boundary is $\partial h/\partial n = 0$, where h is the groundwater potential and n is the direction normal to the boundary.

Groundwater divide

On a groundwater divide the groundwater potential reaches a maximum value. The flow perpendicular to a groundwater divide is zero and therefore the boundary can be described mathematically as $\partial h/\partial n = 0$.

Recharge boundary

For a recharge boundary, a certain volume of water enters the aquifer from adjacent strata. Usually it is difficult to determine the quantity of water that is transferred in this way; often an estimate can be made by considering the water balance of the adjacent strata. The relative transmissivities of the two rock types also give an indication of the likely order of the flows.

Fixed groundwater potential

When specifying a groundwater problem it is common practice to take a line along which the groundwater potential remains constant, and to enforce this as a boundary condition. This is a valid condition if the groundwater potential remains at this constant value because the aquifer is in hydraulic continuity with the sea or a large lake. However, a fixed potential infers that there is an infinite source of water on which the aquifer can draw. Therefore a small river should not be represented as a fixed potential, since increased abstraction in the vicinity of the river may be such that the river is unable to sustain the flows required to hold the groundwater potential at a fixed value. Thus the condition of a fixed groundwater potential should be introduced only if there is an adequate source of water.

Down-dip boundary

In many situations the flow is predominantly in the down-dip direction. The condition on the down-dip boundary is usually difficult to determine because the aquifer frequently extends below other impermeable strata as its depth below the ground surface increases. Little may be known of the flow mechanism in that region, other than that the quality of the water deteriorates down-dip.

In such a situation the only acceptable approach is to investigate the effect of different boundary conditions. For example, if the two alternative boundary conditions enforced in model solutions are impermeable and fixed potential, then the actual boundary condition must lie somewhere between these two.

Flowline boundary

It may not be convenient to analyse the whole of an aquifer, instead some arbitrary boundary is required to limit the extent of the study area. If a map of equipotential lines is constructed, it is then possible to identify flowlines approximately. Since no flow crosses a flowline, it can be treated as an impermeable boundary. This condition assumes that no water can be drawn from beyond the study area and therefore it is a safer condition than the assumption of known fixed groundwater potentials.

Inflows and outflows

Accurate information concerning the inflows and outflows of the aquifer must be obtained. Whenever possible, a check should be made to ascertain whether an overall water balance is achieved over a number of years. Sources of inflow and outflow are as follows.

Recharge

The process by which rainfall infiltrates through the ground and recharges the aquifer is not well understood, and therefore there is considerable difficulty in obtaining a good estimate of the recharge. There are a variety of methods of estimating the recharge; see for example Headworth (1970), Smith *et al.* (1970), Brustkern and Morel-Seytoux (1975), Nassif and Wilson (1975), Sukhija and Shah (1976), Kitching *et al.* (1977), and Collis-George (1977); the method described briefly below is the approach often used in the United Kingdom.

The following calculation should be carried out on a daily basis (Howard and Lloyd, 1979). Precipitation minus direct runoff is assumed to infiltrate into the soil. Depending on the state of the soil, a number of different mechanisms may apply.

(a) If the soil is saturated at field capacity then evapo-transpiration will take place at a rate equal to the potential rate. Values of the potential evaporation are usually obtained from an expression devised by Penman (1949). The quantity of water that remains when the potential evaporation and direct runoff are subtracted from the precipitation, is assumed to infiltrate through the soil to recharge the aquifer.

(b) If the soil is not saturated then there will be a soil-moisture deficit (SMD). Any of the precipitation which does not run off enters the soil and reduces the SMD. Meanwhile evaporation will take place. This will occur at the potential rate unless the SMD is greater than the root constant. The root constant is defined as the depth from which the roots can readily draw water multiplied by the porosity.

Plants and crops with different root systems have different root constants which may also differ with the time of year (Grindley, 1967). When the SMD exceeds the root constant, the roots have difficulty in gathering moisture from the soil and evaporation takes place more slowly than the potential rate. One relationship, which is simple to use, is that when the SMD exceeds the root constant, evaporation from depth takes place at 10 per cent of the potential rate when precipitation and runoff for that day have been subtracted. This is illustrated in Fig. 6.1.

(c) Even though there is no precipitation in any one day, the process described in (b) is still followed.

(d) No recharge to the aquifer takes place when there is a soil moisture deficit.

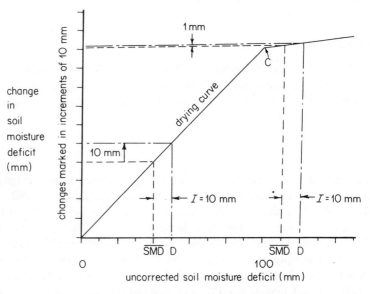

Drying curve for C =100 mm. Reference should be made to Table 6.1 for meaning of symbols.

Figure 6.1. Effect of root constant on change in soil moisture deficit. Root constant is 100 mm

Table 6.1 contains a typical calculation of recharge for a selected number of days for one period in the winter and another in the summer.

According to the Penman–Grindley theory, recharge to the aquifer takes place only when the SMD is zero. This is contrary to many field observations since recharge to certain aquifers has been observed after heavy summer rainfall, even when there is a large SMD. Further studies are required before the mechanism of aquifer recharge is fully understood. In the meantime, an acceptable approximation which appears to represent the observed behaviour of certain aquifers is to allow a fraction, say 15 per cent, of the precipitation minus the

Table 6.1 Calculation of recharge

Notation: P = precipitation; E_p = potential evaporation; RO = runoff; $\overline{SMD}$ = soil-moisture deficit at start of day; SMD = soil-moisture deficit at end of day; RC = root constant; $I = -$(quantity available for infiltration) $= -(P - E_p - RO)$; D = soil-moisture deficit without allowance for root constant = $\overline{SMD} + I$; RECH = recharge for that day; E_a = actual evaporation.

Calculation
(a) If $D < 0.0$; RECH $= -D$, SMD = 0.0
(b) If $I > 0.0$ and $0.0 \leqslant \overline{SMD} < RC$; RECH = 0.0, SMD = D
(c) If $I > 0.0$ and $\overline{SMD} \geqslant RC$; RECH = 0.0, SMD = $\overline{SMD} + 0.1I$
(d) If $I \leqslant 0.0$ and $D > 0.0$; RECH = 0, SMD = D

Examples Root constant = 100 mm; all values in mm.

Day	P	RO	E_p	I	D	SMD	RECH	E_a
39						2.0		
40	1.1	0.3	0.3	−0.5	1.5	1.5	0.0	0.3
41	8.1	3.2	0.0	−5.5	−4.0	0.0	4.0	0.0
42	5.3	2.3	0.1	−2.9	−2.9	0.0	2.9	0.1
43	0.2	0.1	0.4	0.3	0.3	0.3	0.0	0.4
44	0.1	0.0	0.5	0.4	0.7	0.7	0.0	0.5
45	3.3	1.3	0.2	−1.8	−1.1	0.0	1.1	0.2
46	10.6	4.7	0.1	−5.8	−5.8	0.0	5.8	0.1
179						96.0		
180	0.0	0.0	3.3	3.3	99.3	99.3	0.0	3.3
181	0.0	0.0	3.7	3.7	103.0	103.0	0.0	3.7
182	8.3	1.4	2.7	−4.2	98.8	98.8	0.0	2.7
183	2.3	0.2	3.2	1.1	99.9	99.9	0.0	3.2
184	0.5	0.1	3.8	3.4	103.3	103.3	0.0	3.8
185	0.0	0.0	3.7	3.7	107.0	103.67	0.0	0.37
186	2.1	0.4	3.4	1.7	105.37	103.84	0.0	1.87
187	15.6	3.6	2.5	−9.5	94.34	94.34	0.0	2.5

evaporation to enter directly as recharge; the remaining precipitation is included in the standard calculation (Rushton and Ward, 1979).

Abstraction

Abstraction is one parameter that is usually known accurately. It is important not to ignore small abstractions, for when they are totalled they may well amount to a significant part of the water balance.

Spring flow

Spring flow is usually a function of the difference between the groundwater potential and the elevation of the spring. The actual magnitude of the flow clearly depends on the geology of the region in the vicinity of the spring. Thus the spring should be represented as a flow condition, the actual flow depending on the groundwater potential. A typical flow relationship for a spring is sketched in Fig. 6.2(a). The spring flow is included in the differential equation as a negative recharge.

Springs have been represented by certain workers as a fixed potential which is maintained provided that the groundwater potential exceeds the elevation of the spring. Such a procedure can be very dangerous. If this condition is modelled on a finite-difference network with a mesh spacing of 1 km, then a single node with the condition of a fixed potential at ground level is equivalent to a lake of diameter 416 m in direct contact with the aquifer (see Eq. 9.14). The flow drawn into such a lake will clearly be far larger than that for a number of springs.

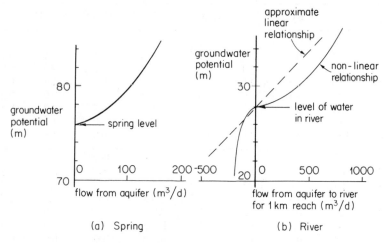

Figure 6.2. Typical relationships between flow from aquifer to rivers or springs and groundwater potential

River flow

River flow must be modelled with the same care as spring flow. A typical relationship between groundwater potential in the aquifer and flow from the aquifer into a river is shown in Fig. 6.2(b) by the full line. Once the groundwater potential falls below the river level, the river will feed the aquifer provided that there is sufficient water in the river. The actual flow from a river into an aquifer is usually considerably less than the flow in the reverse direction for a similar head difference; this is largely due to the sediment in a river bed.

Because of the difficulty in actually determining a relationship such as that sketched in Fig. 6.2(b), it is common practice to use a linear relationship as shown by the broken line. Methods of calculating this 'hydraulic river resistance' have been proposed by Herbert (1970) and Streltsova (1974) and will be discussed further in Section 9.2. Intermittent rivers clearly introduce additional complications: their behaviour is more like that of a spring.

Leakage

Leakage into an aquifer often occurs from overlying or underlying semi-permeable strata. The classical approach is that leakage, which acts as a recharge, is equal to $(k'/m')(H - h)$ where k' is the coefficient of permeability of the leaky aquifer, m' is its thickness, H is the groundwater potential in the leaky aquifer, and h is the unknown groundwater potential in the main aquifer. Relationships of this form can also be used for leakage between two aquifers; in this case both H and h may be unknown potentials which must be determined by a mathematical model.

Field data for comparisons

When a mathematical model has been devised, it is most important to examine its validity by modelling a historic period and checking the predicted behaviour against field records. A wide variety of field information is available for most aquifers. This may include:

(a) distribution of groundwater potential throughout the aquifer at certain times of each year;
(b) variation of groundwater potential with time at selected observation boreholes;
(c) base flows in rivers which originate from the aquifer;
(d) spring flow, particularly during very wet or very dry periods; loss of artesian pressure;
(e) presence of ponds;
(f) change in quality of the water;
(g) influence of new abstraction sites on existing abstraction sites.

As more field information becomes available, the model parameters should be kept continuously under review.

6.3 Differential Equations

Mathematical models of groundwater flow are based on Darcy's law and the condition of continuity of flow. The assumptions and idealizations introduced for regional groundwater flow are summarized in Section 2.5. When vertical components of flow are neglected Eq. 2.20 applies; thus the differential equation for regional groundwater flow is as follows:

$$\frac{\partial}{\partial x}\left(T_x\frac{\partial h}{\partial x}\right) + \frac{\partial}{\partial y}\left(T_y\frac{\partial h}{\partial y}\right) = S\frac{\partial h}{\partial t} - q(x, y, t) \qquad (6.1)$$

where x and y are the coordinates and h is the groundwater potential. The transmissivities, T_x and T_y, may themselves be functions of the groundwater potential. For example, with unconfined sandstone and gravel aquifers expressions of the following form may apply:

$$T_x = k_x(h - z_b), \qquad T_y = k_y(h - z_b) \qquad (6.2)$$

where z_b is the elevation of the base of the aquifer. Other aquifers such as limestone and chalk are not homogeneous but have a permeability that varies with depth, and relationships of the form of Eq. 6.2 are not valid. Frequently bands of higher permeability occur in an aquifer; if the positions of these bands are known then T_x and T_y can be modified as h varies. Similar variations may occur in the storage coefficient S.

The term $q(x, y, t)$ which has dimensions $[L/T]$, represents a whole range of physical phenomena such as recharge, abstraction, leakage, spring flow, and river flow. When the finite-difference approximation of Eq. 6.1 is used, the value of q which represent an abstraction of $-Q$ (m³/d) is

$$q = \frac{-Q}{A} \quad \text{(m/d)} \qquad (6.3)$$

where A is the plan area of aquifer represented by that node in square metres.

Associated with this differential equation there are boundary and initial conditions. Along every boundary, some form of boundary condition must be specified which will be one of two types, either a known potential or a known flow (see Table 2.2). On at least part of one of the aquifer boundaries it is advisable to arrange that the condition is a known groundwater potential.

When the aquifer response is time-dependent, initial conditions are required for the instant when the calculation commences. They must represent the previous history of the aquifer and take account of the dynamic changes within the aquifer. Different approaches for achieving satisfactory initial conditions are discussed later, in Sections 7.6 and 8.9.

The complexity of groundwater problems is apparent from the above description of the data, the differential equations, and the initial and boundary conditions. Analytical solutions are clearly impossible. If idealizations are made to simplify the problem, these may well mask significant features of the groundwater flow pattern. Numerical solutions are the only realistic approach; both analog and digital techniques based on finite-difference approximations are described below.

6.4 Steady-State Example

In this section a simple example of steady-state regional groundwater flow will be considered. Though the problem is concerned with a rectangular aquifer having uniform values of transmissivity, the basic principles involved can be applied to many of the complex time-variant situations that occur in practice.

The main purpose of this example is to demonstrate how a problem is specified and then indicate the physical significance of the results. The formulation of the finite-difference equations is considered later and the different techniques of solving the equations using analog and digital computers are discussed in detail in Chapters 7 and 8.

Specification of the problem

Consider the confined aquifer shown diagrammatically in Fig. 6.3. The problem can be specified as follows.

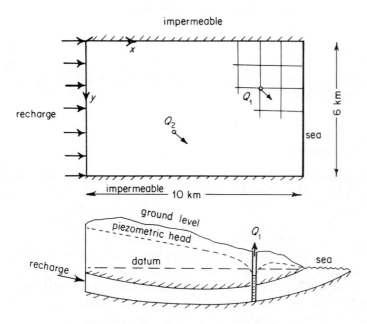

Figure 6.3. Simple steady-state example

Aquifer properties

Everywhere the transmissivities $T_x = T_y = 900 \, \text{m}^2/\text{d}$.

Boundaries

$0 < x < 10 \, \text{km}$, $y = 0$, impermeable

$0 < x < 10 \, \text{km}$, $y = 6 \, \text{km}$, impermeable

$x = 0$, $0 < y < 6 \, \text{km}$, recharge boundary, total recharge of $6500 \, \text{m}^3/\text{d}$ distributed uniformly.

$x = 10 \, \text{km}$, $0 < y < 6 \, \text{km}$, the aquifer is in direct contact with the sea, thus the groundwater potential is maintained at zero.

140

Inflows

Recharge boundary as described above. At the boundary with the sea, either inflow or outflow may occur (or both simultaneously).

Outflows

Well Q_1:$x = 8\,\text{km}$, $y = 2\,\text{km}$, flow $-4500\,\text{m}^3/\text{d}$

Well Q_2:$x = 4\,\text{km}$, $y = 4\,\text{km}$ flow $-1500\,\text{m}^3/\text{d}$

Mesh spacing

Select 1 km; this does not really give sufficient detail but it is adequate for this demonstration problem.

Numerical solution

With groundwater flow probems as specified above, the finite-difference solution for the potentials is given in Fig. 6.4(a). These results were obtained using a resistance network similar to that of Section 3.10, but a solution could equally well be obtained using the digital computer program of Appendix 2. The results from the analog and digital techniques should be identical since with steady-state problems the same finite-difference approximations are used. However, small differences of up to 0.04 m do occur due to the tolerances of the resistors.

In Fig. 6.4(b) a diagram is presented of the flows between nodal points. Since this model uses a square grid, these flows are calculated directly as the change in groundwater potential multiplied by the transmissivity.

Water balance

The flow into the aquifer, obtained by summing all the flows immediately to the right of the recharge boundary, equals $6450\,\text{m}^3/\text{d}$. The small error of 1 per cent between this calculated flow and the specified recharge of $6500\,\text{m}^3/\text{d}$ is a result of the tolerances of the resistances. The net flow out to the sea is $530\,\text{m}^3/\text{d}$, though over part of the boundary there is an inflow from the sea. At well Q_1, the four component flows total $-4530\,\text{m}^3/\text{d}$ and at well Q_2 they total $-1510\,\text{m}^3/\text{d}$. Thus the overall water balance is

$$6450 - 530 - 4530 - 1510 = -120\,\text{m}^3/\text{d}.$$

This overall error of $120\,\text{m}^3/\text{d}$ is roughly 2 per cent of the recharge; this error is due to the tolerance of the resistors and the inexpensive voltmeter used.

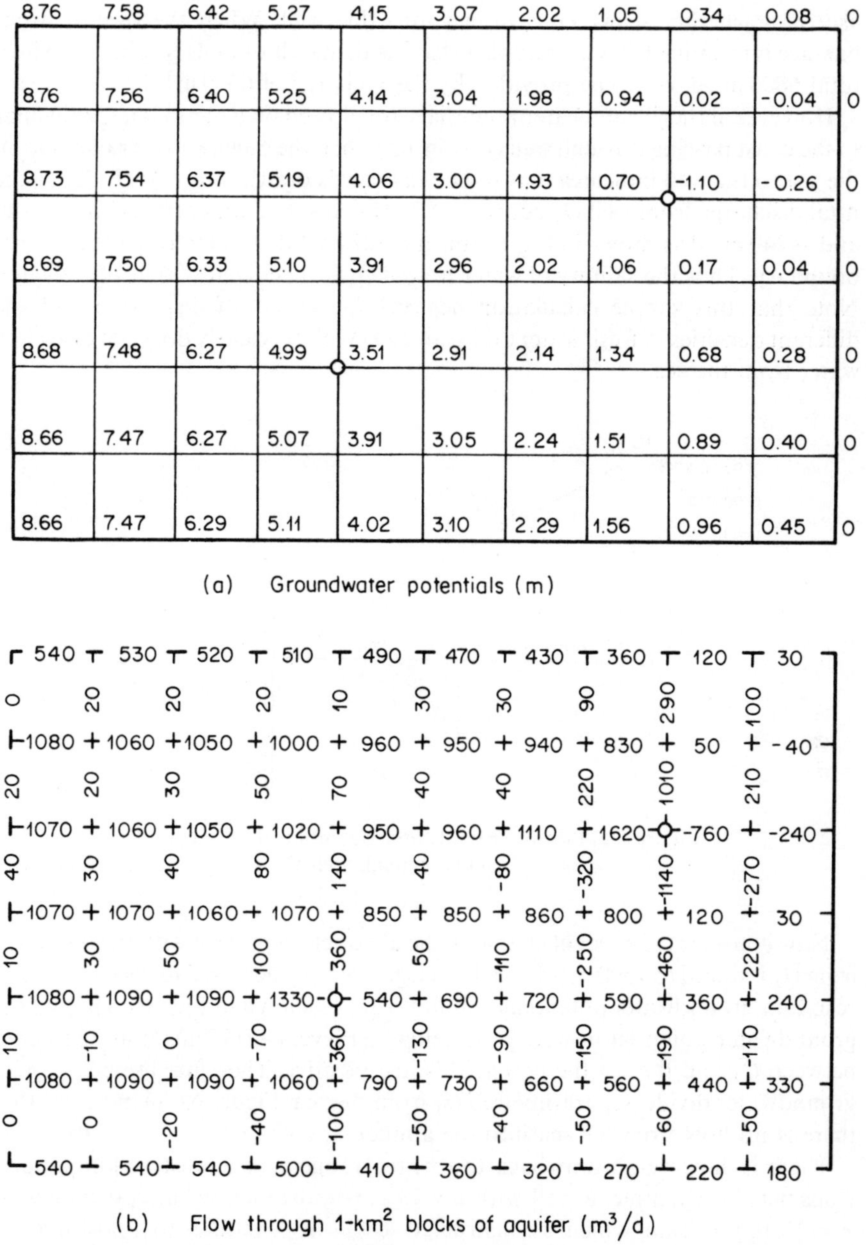

8.76	7.58	6.42	5.27	4.15	3.07	2.02	1.05	0.34	0.08	0
8.76	7.56	6.40	5.25	4.14	3.04	1.98	0.94	0.02	-0.04	0
8.73	7.54	6.37	5.19	4.06	3.00	1.93	0.70	-1.10	-0.26	0
8.69	7.50	6.33	5.10	3.91	2.96	2.02	1.06	0.17	0.04	0
8.68	7.48	6.27	4.99	3.51	2.91	2.14	1.34	0.68	0.28	0
8.66	7.47	6.27	5.07	3.91	3.05	2.24	1.51	0.89	0.40	0
8.66	7.47	6.29	5.11	4.02	3.10	2.29	1.56	0.96	0.45	0

(a)   Groundwater potentials (m)

```
  540 ┬ 530 ┬ 520 ┬ 510 ┬ 490 ┬ 470 ┬ 430 ┬ 360 ┬ 120 ┬ 30
 0     20     20     20     10     30     30     90    290    100
 ┝1080 +1060 +1050 +1000 + 960 + 950 + 940 + 830 +  50 + -40 ┤
 20     20     30     50     70     40     40    220   1010    210
 ┝1070 +1060 +1050 +1020 + 950 + 960 +1110 +1620 Ø-760 +-240 ┤
 40     30     40     80    140     40    -80   -320   1140    270
 ┝1070 +1070 +1060 +1070 + 850 + 850 + 860 + 800 + 120 +  30 ┤
 10     30     50    100    360     50   -110   -250   -460   -220
 ┝1080 +1090 +1090 +1330 Ø 540 + 690 + 720 + 590 + 360 + 240 ┤
 10    -10     0     -70   -360   -130    -90   -150   -190   -110
 ┝1080 +1090 +1090 +1060 + 790 + 730 + 660 + 560 + 440 + 330 ┤
 0      0     -20    -40   -100    -50    -40    -50    -60    -50
  540 ┴ 540 ┴ 540 ┴ 500 ┴ 410 ┴ 360 ┴ 320 ┴ 270 ┴ 220 ┴ 180
```

(b)   Flow through 1-km² blocks of aquifer (m³/d)

Figure 6.4. Groundwater potentials and flows; $Q_1 = -4500 \text{ m}^3/\text{d}$, $Q_2 = -1500 \text{ m}^3/\text{d}$

Interpretation

Saline intrusion

Since well Q_1 is positioned at a distance of 2 km from the sea, it is important to ascertain whether seawater is drawn into this well. When the overall water balance is examined, it would appear that the demands of wells Q_1 and Q_2, which total 6000 m³/d, could be provided by the recharge of 6500 m³/d.

However, if the groundwater potentials are plotted on a section perpendicular to the coast passing through well Q_1 (Fig. 6.5), then the downwards gradient from the sea to the well indicates that some seawater is drawn into well Q_1. Thus the total discharge from well Q_1 consists of 280 m³/d saline water (made up of -40 and -240 m³/d as shown in Fig. 6.4(b)) and 4220 m³/d of water from the recharge boundary. Thus the salinity of water in well Q_1 is 6 per cent of that of seawater. Note that this simple calculation neglects the effects of dispersion and the different densities of fresh water and seawater. Well Q_2 clearly does not draw any water from the sea.

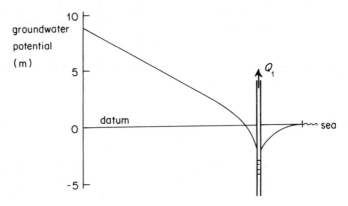

Figure 6.5. Plot of groundwater potentials on a section
$y = $ constant through well Q_1

Now consider the effect of changing the abstraction pattern with the discharge from Q_1 reduced to 1500 m³/d and discharge from Q_2 increased to 4500 m³/d. The resultant groundwater potentials and flows are as shown in Fig. 6.6. Though the groundwater potential in well Q_1 is below sea level (-0.17 m), at all the nodes between Q_1 and the sea the potential is just positive. Therefore there is a small groundwater divide, separating well Q_1 from the sea. Figure 6.6(b) indicates that there is no flow from the sea into the aquifer.

There is, however, a drawback to this rearrangement in abstraction pattern. Consider, for example, a well with a very small abstraction rate positioned at $x = 2$ km, $y = 2$ km, which is constructed so that it penetrates to a position 5 m above datum. With the initial arrangement where heavy abstraction occurs close to the coast, the groundwater potential is 6.37 m, hence the depth of water in the

6.68	5.51	4.45	3.42	2.61	1.82	1.23	0.98	0.34	0.13	0.0
6.66	5.48	4.39	3.35	2.48	1.75	1.22	0.66	0.23	0.10	0.0
6.58	5.39	4.26	3.15	2.22	1.55	1.04	0.52	-0.17	0.01	0.0
6.45	5.25	4.06	2.80	1.66	1.23	0.90	0.55	0.20	0.08	0.0
6.36	5.14	3.86	2.38	0.31	0.84	0.78	0.56	0.32	0.15	0.0
6.30	5.08	3.84	2.57	1.40	1.05	0.83	0.59	0.38	0.18	0.0
6.27	5.07	3.87	2.66	1.68	1.16	0.87	0.61	0.39	0.19	0.0

(a)   Groundwater potentials (m)

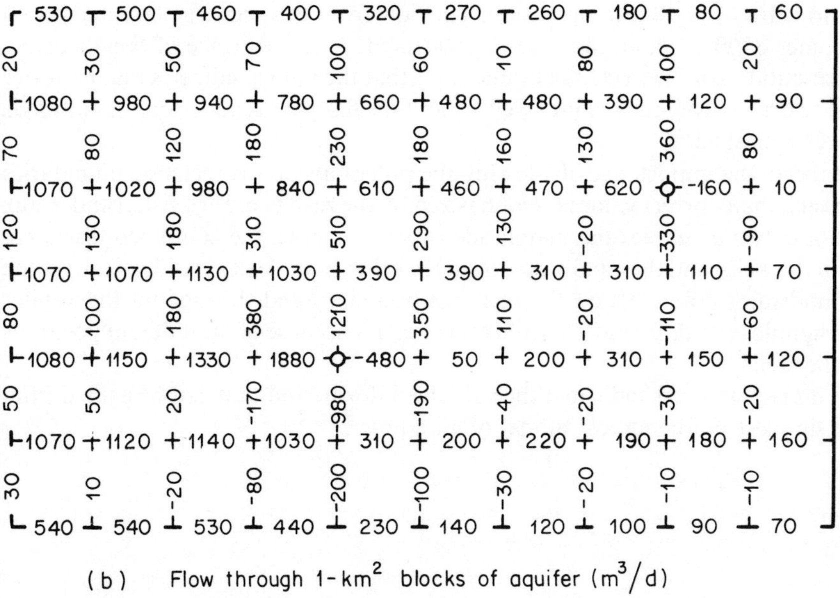

```
┌ 530 ┬ 500 ┬ 460 ┬ 400 ┬ 320 ┬ 270 ┬ 260 ┬ 180 ┬ 80 ┬ 60 ┐
  20    30    50    70   100    60    10    80   100   20
├1080 + 980 + 940 + 780 + 660 + 480 + 480 + 390 + 120 + 90 ┤
  70    80   120   180   230   180   160   130   360   80
├1070 +1020 + 980 + 840 + 610 + 460 + 470 + 620 ◇-160 + 10 ┤
 120   130   180   310   510   290   130   -20  -330  -90
├1070 +1070 +1130 +1030 + 390 + 390 + 310 + 310 + 110 + 70 ┤
  80   100   180   380  1210   350   110   -20  -110  -60
├1080 +1150 +1330 +1880◇-480 +  50 + 200 + 310 + 150 +120 ┤
  50    50    20  -170   980  -190   -40   -20   -30  -20
├1070 +1120 +1140 +1030 + 310 + 200 + 220 + 190 + 180 +160 ┤
  30    10   -20   -80  -200  -100   -30   -20   -10  -10
└ 540 ┴ 540 ┴ 530 ┴ 440 ┴ 230 ┴ 140 ┴ 120 ┴ 100 ┴ 90 ┴ 70 ┘
```

(b)   Flow through 1-km^2 blocks of aquifer (m^3/d)

Figure 6.6. Groundwater potentials and flow; $Q_1 = -1500\,\text{m}^3/\text{d}$, $Q_2 = -4500\,\text{m}^3/\text{d}$

well would be 1.37 m. However, with the major abstraction at Q_2 the water level of 4.26 m would be 0.74 m below the base of this small well. This difficulty could be overcome simply by deepening the well. Nevertheless, this simple example does illustrate that changing an abstraction pattern can have a number of side effects.

Actual well-water levels

The groundwater potentials quoted at the well nodes are not the levels that will occur in the abstraction wells. The representation of the well radius is considered in Section 9.2 but for the present the results of correcting for the actual well radius are quoted without explanation. If both wells have a radius of 0.2 m, then for the initial arrangement the groundwater potentials in the wells would be:

$$\text{well } Q_1, \text{ discharge } 4500 \text{ m}^3/\text{d} \qquad h_{\text{well}} = -6.58 \text{ m}$$

$$\text{well } Q_2, \text{ discharge } 1500 \text{ m}^3/\text{d} \qquad h_{\text{well}} = +1.68 \text{ m}$$

In Fig. 6.5 the actual groundwater potential in well Q_1 is plotted.

Alternative condition on recharge boundary

Instead of representing a recharge boundary as a known flow, some investigators adopt an alternative approach and represent it as a fixed groundwater potential. Without wells Q_1 and Q_2 in operation, the potential on the recharge boundary would be 12.04 m. Suppose that in the model, the groundwater potential on the recharge boundary is maintained at 12.0 m as wells 1 and 2 are brought into operation. The flow from this recharge boundary then becomes 8300 m^3/d, an increase of 1800 m^3/d. A consequence of this incorrect representation of the recharge boundary is that the model indicates that no water will be drawn from the sea into well Q_1 and the danger of saline contamination is thereby masked.

Indeed, the importance of allowing the potentials on the recharge boundaries to reach their correct values is emphasized by the results of Figs. 6.4(a) and 6.6(a). In these two examples the magnitude of the recharge, the combined discharge from the wells, and the conditions on all the other boundaries are identical, yet the groundwater potentials on the recharge boundary and throughout the aquifer are significantly different. This difference occurs because of the different positions of the wells.

This example has indicated the valuable information that can be gained from the simplest mathematical model of an aquifer.

CHAPTER 7

Time-variant Flow—Analog Technique

7.1 Introduction

The resistance–capacitance electrical network has been used for many years in the analysis of time-variant groundwater flow. One of the earliest detailed references to its use is by Skibitzke (1960) and a valuable review of the use of analogs for groundwater problems is presented by Prickett (1975). With the advent of present-day digital computers, the analog approach has tended to lose its popularity but this is largely because many workers are not aware of the developments that have been made in electronic components and equipment. When a small digital computer is used to control the resistance–capacitance network, the resulting system is competitive with the purely digital system in cost and convenience.

In this chapter the theory of the resistance–capacitance network will be discussed and reference will be made to the practical details of construction and operation of the network. Initial conditions are considered in detail; the method of achieving satisfactory initial conditions is important for both analog and digital solutions. The final section is concerned with the control of the resistance–capacitance network by a minicomputer.

Reference is also made to the resistance–capacitance technique in Section 9.1 where a one-dimensional problem is considered. Detailed calculations of the component values are presented and typical results are also listed. Special techniques, such as the representation of well radius and the change between confined and unconfined states, are discussed in Section 9.2. Finally, results from analog models of the Lincolnshire Limestone aquifer are included in Chapter 10.

7.2 Theory

In many of the earliest papers, the design of the resistance–capacitance network was argued from the analogy between the flow of water through a discrete portion of an aquifer and the flow of electricity through a resistor. However, the fundamental aspect of the resistance–capacitance network is that it solves the finite-difference form of the governing equation. Therefore the finite-difference approach is used below.

The equation governing regional groundwater flow as derived in Section 2.5 is

$$\frac{\partial}{\partial x}\left(T_x\frac{\partial h}{\partial x}\right) + \frac{\partial}{\partial y}\left(T_y\frac{\partial h}{\partial y}\right) = S\frac{\partial h}{\partial t} - q \tag{7.1}$$

where the symbols take the meanings as defined in Section 2.5 and the inflow q is a function of x, y, and t. In finite-difference form, for the irregular mesh of Fig. 7.1, Eq. 7.1 becomes

$$\frac{2}{\Delta x_a + \Delta x_b}\left[\frac{T_{xa}}{\Delta x_a}(h_3 - h_0) + \frac{T_{xb}}{\Delta x_b}(h_1 - h_0)\right]$$

$$+ \frac{2}{\Delta y_a + \Delta y_b}\left[\frac{T_{ya}}{\Delta y_a}(h_2 - h_0) + \frac{T_{yb}}{\Delta y_b}(h_4 - h_0)\right] = S_0\frac{\partial h}{\partial t} - q$$

where S_0 is the storage coefficient at node 0.

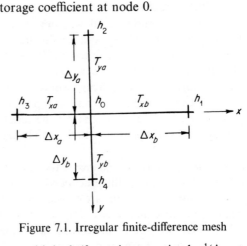

Figure 7.1. Irregular finite-difference mesh

It is convenient to multiply the foregoing equation by $\frac{1}{4}(\Delta x_a + \Delta x_b)(\Delta y_a + \Delta y_b)$

$$\frac{\Delta y_a + \Delta y_b}{2}\left[\frac{T_{xa}}{\Delta x_a}(h_3 - h_0) + \frac{T_{xb}}{\Delta x_b}(h_1 - h_0)\right]$$

$$+ \frac{\Delta x_a + \Delta x_b}{2}\left[\frac{T_{ya}}{\Delta y_a}(h_2 - h_0) + \frac{T_{yb}}{\Delta y_b}(h_4 - h_0)\right]$$

$$= \frac{1}{4}(\Delta x_a + \Delta x_b)(\Delta y_a + \Delta y_b)\left(S_0\frac{\partial h}{\partial t} - q\right) \tag{7.2}$$

Now consider the resistance–capacitance network of Fig. 7.2. Applying Kirchhoff's current law that the sum of the electrical currents entering node 0 must equal zero,

$$\frac{V_3 - V_0}{R_{xa}} + \frac{V_1 - V_0}{R_{xb}} + \frac{V_2 - V_0}{R_{ya}} + \frac{V_4 - V_0}{R_{yb}} = C_0\frac{\partial V_0}{\partial t_e} - I_0 \tag{7.3}$$

where t_e is the *electrical time*.

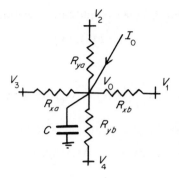

Figure 7.2. Resistance–capacitance network

Clearly Eqs. 7.2 and 7.3 are of the same form. Scaling factors are introduced so that the equations become analogous. Various schemes can be devised for scaling factors. Let the relationship between the electrical potential and the groundwater potential be

$$V = F(1)h \qquad (7.4)$$

where $F(1)$ is a convenient scaling factor. Next a relationship is required between the electrical resistance and the transmissivity, which is a function of the permeability. Let the electrical resistance be proportional to the distance between the nodal points and inversely proportional both to the cross-sectional area and the transmissivity (see Section 3.4 and Fig. 3.4(b)):

$$R_{xa} = \frac{F(2)2\Delta x_a}{(\Delta y_a + \Delta y_b)T_{xa}} \qquad R_{xb} = \frac{F(2)2\Delta x_b}{(\Delta y_a + \Delta y_b)T_{xb}}$$

$$R_{ya} = \frac{F(2)2\Delta y_a}{(\Delta x_a + \Delta x_b)T_{ya}} \qquad R_{yb} = \frac{F(2)2\Delta y_b}{(\Delta x_a + \Delta x_b)T_{yb}} \qquad (7.5)$$

Also, a third independent relationship can be defined between the electrical capacitance, which is a measure of the ability to store electricity, and the storage coefficient, namely

$$C_0 = 0.25\, F(3)(\Delta x_a + \Delta x_b)(\Delta y_a + \Delta y_b)S_0 \qquad (7.6)$$

From these three independent relationships (Eqs. 7.4, 7.5, and 7.6), two further relationships can be derived. Substituting the relationships of Eqs. 7.4, 7.5, and 7.6 into the first term on the left-hand side and all the terms on the right-hand side of the Eq. 7.3, then multiplying throughout by $F(2)/F(1)$, leads to the equation

$$\frac{\Delta y_a + \Delta y_b}{2}\frac{T_{xa}}{\Delta x_a}(h_3 - h_0) + \cdots = 0.25(\Delta x_a + \Delta x_b)(\Delta y_a + \Delta y_b)S_0\frac{\partial h}{\partial t_e}F(2)F(3)$$

$$- F(2)I_0/F(1) \qquad (7.7)$$

Comparing Eq. 7.2 and Eq. 7.7, the two additional relationships can be derived as

$$t_e = F(2)F(3)t \tag{7.8}$$

and

$$I_0 = 0.25\frac{F(1)}{F(2)}(\Delta x_a + \Delta x_b)(\Delta y_a + \Delta y_b)q \tag{7.9}$$

If Q_0 is defined as the total recharge at node 0, that is,

$$Q_0 = 0.25(\Delta x_a + \Delta x_b)(\Delta y_a + \Delta y_b)q \tag{7.10}$$

then Eq. 7.9 becomes

$$I_0 = \frac{F(1)}{F(2)}Q_0 \tag{7.11}$$

7.3 Fast-Time or Slow-Time Analogs

When resistance–capacitance networks were first introduced, the quality of capacitors was poor and therefore, to minimize the leakage currents, the networks were operated at electronic speeds. Now, however, low-leakage capacitors are available and far slower speeds may be used.

The electrical time for the influence of an effect, such as an abstraction from a well, to spread through an aquifer is given, approximately, by

$$t_e = RCN^{\frac{1}{2}}$$

where R is the average resistance value, C the average capacitance value, and N is the number of nodal points.

For *the fast-time analog* typical values are,

$$R = 1\,\text{k}\Omega, \qquad C = 0.001\,\mu\text{F}, \qquad N^{\frac{1}{2}} = 40$$

thus $\qquad t_e = 40\,\mu\text{sec}.$

At speeds of this order, an oscilloscope is needed to record the electrical potentials and expensive, complex switching equipment is required to represent variable inflows and outflows.

For *the slow-time analog* typical values are

$$R = 200\,\text{k}\Omega, \qquad C = 10\,\mu\text{F}, \qquad N^{\frac{1}{2}} = 40$$

thus $\qquad t_e = 80\,\text{secs}$

With these speeds electromechanical devices can be used both to input currents and to record electrical potentials.

Since the slow-time technique requires relatively inexpensive equipment and can be controlled by a minicomputer, the remainder of this section will be concerned only with the slow-time technique. There has been little development of the fast-time technique in recent years.

7.4 Design of Network

Scaling factors

The selection of scaling factors and component values will be described with reference to an example, similar to that of Section 6.4 but with time-variant effects added. This is just a typical example but it does indicate the order of typical component values. The aquifer parameters for an unconfined aquifer are $T = 900 \, \text{m}^2/\text{d}$; $S = 0.09$; $\Delta x_a = \Delta x_b = 1 \, \text{km}$; $\Delta y_a = \Delta y_b = 1 \, \text{km}$; recharge per nodal point on the western boundary is $1083 \, \text{m}^3/\text{d}$; and abstraction from wells $Q_1 = -4500 \, \text{m}^3/\text{d}$, $Q_2 = -1500 \, \text{m}^3/\text{d}$.

It is advisable, first of all, to select the scaling factor between physical and electrical time. A convenient factor is: *1 second electrical time represents 50 days physical time*. Then, there is often a restriction on the size of capacitors available. At present, capacitors of $10 \, \mu\text{F}$ are available, hence *10 F capacitance represents a storage coefficient of 0.09*. From Eqs. 7.6 and 7.8 it is now possible to calculate $F(3) = 1.111 \times 10^{-10}$, $F(2) = 1.8 \times 10^8$. Knowing the value of $F(2)$, it follows that *200 kΩ resistance represents a transmissivity of 900 m^2/d*.

The final factor relating electrical and groundwater potentials is chosen to suit the measuring equipment, thus $F(1) = 0.01$ and *1 volt electrical potential represents 100 m groundwater potential*. Now that all the scaling factors have been chosen it is possible to calculate from Eq. 7.11 that $-0.25 \, \mu\text{A}$ electrical current represents a flow of $-4500 \, \text{m}^3/\text{d}$. Note that a negative sign indicates an abstraction. This network design is summarized in Table 7.1.

Component types

A wide variety of different types of electrical components are available at present. When selecting the type of resistors to be incorporated in a network, there are many ranges of miniature resistors which are suitable. Carbon-film resistors are cheap and reliable. A tolerance of ± 5 per cent is usually adequate for groundwater flow models, with a rating of 0.5 W.

Great care has to be taken in the selection of the type of capacitor. For three different types of capacitor, the leakage current due to a potential difference of 1 volt across the capacitor are as follows.

Type of capacitor	Leakage current
Electrolytic	$0.01 \, \mu\text{A}$
Silver mica	$0.001 \, \mu\text{A}$
Metallized plastic film	$0.000\,01 \, \mu\text{A}$

F

Table 7.1 Analogous relationships with typical values

Physical	Typical value	Relationship	Analog	Typical value
Potential h	73.2 m	$V = F(1)h$	Voltage V	0.732 V
Transmissivity T_x	900 m^{2}d	$R_{xa} = 2F(2)\Delta x_a/[(\Delta y_a + \Delta y_b)T_{xa}]$	Resistance R_x	200 kΩ
Storage coefficient S	0.09	$C = 0.25F(3)(\Delta x_a + \Delta x_b)(\Delta y_a + \Delta y_b)S$	Capacitance C	10 μF
Time t	7 days	$t_e = F(2)F(3)t$	Electrical time t_e	0.07 sec
Quantity Q	4500 m^3/d	$I = [F(1)/F(2)]Q$	Current I	0.25 μA

For the total leakage current, these values must be multiplied by the number of capacitors in the network which *is likely to be several hundred*. Since, as was shown in the example above, a current of $0.25\,\mu A$ can represent a major abstraction site, then the only acceptable type of capacitor is one with a leakage current less than $0.000\,25\,\mu A$. The metallized plastic film and other similar low-leakage capacitors are clearly acceptable.

Specified current

Many techniques are available to provide a specified current. However, for currents of less than $1\,\mu A$, there is much to recommend the simple technique of a potential difference across a large resistance. Take, for example, the current of $-0.025\,\mu A$ which represents the abstraction rate of $-4500\,m^3/d$ in well Q_1. If a large resistance of $33.3\,M\Omega$ is chosen, then the potential difference required is $-0.25 \times 33.3 = -8.33\,V$. For well Q_2, the equivalent current of $-0.0833\,\mu A$ can be obtained by the same voltage of $-8.33\,V$ across a resistance of $100\,M\Omega$.

Though many techniques are available for producing a constant current, experience has shown that standard electronic circuits are often incapable of supplying and maintaining constant currents as small as $0.0833\,\mu A$. Even if this current can be delivered, the device must also have the property of zero current accurate to $0.0001\,\mu A$ when the pump is not operating.

7.5 Measuring Devices

An important feature of a successful resistance–capacitance model is the choice of a suitable device for measuring the electrical potentials. Certain measuring devices draw a significant current from the measuring point and therefore tend to act as an abstraction well. The standard method of defining the current demand of a device is by its input impedance. When a device of input impedance Z ohms measures a potential of 1 volt, the leakage current through the device is $1/Z$ amps. From the typical network currents quoted above, it is clear that a leakage current greater than $0.001\,\mu A$ is unacceptable and therefore an input impedance of at least $1000\,M\Omega$ is desirable.

Table 7.2 contains a list of measuring devices available at the time of writing, and their properties; further comments about certain of the instruments are given below.

Digital Voltmeters display potentials in digital form. Though some voltmeters can take a reading in 0.02 second, the human eye is unable to distinguish between such rapid readings. However, if the reading rate is reduced to two per second it is possible to follow by eye the changes of voltage with time.

Data Loggers sample the potentials at a number of points; these potentials are then converted to digital form using a digital voltmeter. When a high-speed paper tape punch is used as a means of keeping a permanent record, the maximum

Table 7.2 Types of measuring instruments

Instrument	Input impedance	Accuracy	Speed of reading	Type of output
Multimeter	20 kΩ	2%	—	Pointer (no record)
Digital voltmeter	100 000 MΩ	0.01%	50 per second	Digital display (no record)
Data logger with digital voltmeter	100 000 MΩ	0.01%	10 per second	Output on punched paper tape
U-V recorder	30 kΩ	1%	200 cm of chart per second	Graphical record
U-V recorder with pre-amplifier	1000 MΩ	1%	200 cm of chart per second	Graphical record
X–T plotter	1 MΩ	2%	10 cm on chart per sec: 0.5 sec for full-scale deflect	Plotted graph (further graphs can be superimposed)
X–T plotter with pre-amplifier	1000 MΩ	2%	as above	as above
Long-persistence oscilloscope	10 MΩ	5%	fully variable	Trace which can be photographed
Digital minicomputer with ADC*	100 000 MΩ	0.01%	10 000 per second	Stored in computer memory. Output using VDU†, teletype, etc.

* ADC = analog-to-digital converter
† VDU = visual display unit

reading rate is about ten per second. Data can be stored more rapidly using a digital computer memory.

Ultra-violet Recorders contain a number of small mirror galvanometers which reflect light beams on to photographic charts. The galvanometers are sensitive and quick to react to changes in potential. Up to ten different records can be obtained simultaneously. Since each galvanometer requires a considerable current to drive it, there must be a pre-amplifier between each galvanometer and the network. The pre-amplifier is a device which samples the potential, withdraws only a very small current from the source, and provides an output which is sufficient to drive a galvanometer. It may have unit or a higher gain depending on the sensitivity of the galvanometer.

X–Y Plotters translate electrical potentials into the *x*- and *y*-ordinates of a graph. When a time base is used for the *y*-ordinate then the device can be used to plot variations of potential with time. The response time of an X–Y plotter is not very rapid since it usually takes about one second to move from the bottom to the top of the chart. One advantage is that several results can be superimposed on a single chart. The input impedance can be improved by means of a pre-amplifier.

Long-persistence Oscilloscopes are very useful as an additional device to give an immediate graphical display of certain potentials.

Digital Minicomputers are considered in detail in Section 7.8.

7.6 Initial Conditions

With time-variant groundwater flow problems, the correct representation of initial conditions is of critical importance. Methods of obtaining satisfactory starting conditions were investigated by Rushton and Wedderburn (1973). They considered a number of typical one-dimensional problems to determine for how long any incorrect initial potentials have a significant effect. One particular example related to an aquifer of length L, transmissivity T, and storage coefficient S with the initial condition of zero potential throughout the aquifer. Then a sinusoidal inflow

$$Q_{in} = Q(1 + \cos(2\pi t/p))$$ (7.12)

was injected at one boundary whilst the other was held at zero potential. Different values of the period p were taken; the results of two cases are shown in Fig. 7.3. The full lines of graphs (b) and (d) represent the actual potentials as they build up from zero; the dotted lines represent the cyclic pattern which is achieved after a long time period.

These results indicate that the time taken to achieve dynamic balance is independent of the period of the input function. The conclusions drawn from these and a large number of other investigations is that the calculation should

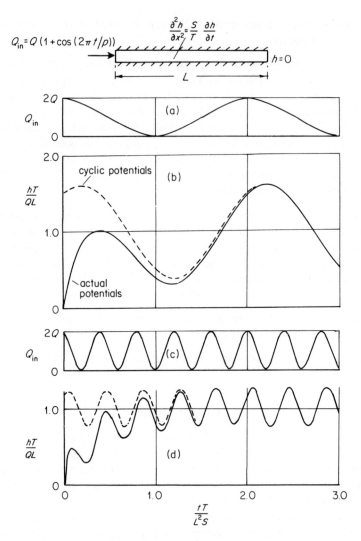

Figure 7.3. Initial conditions for one-dimensional problem. (a) and (c) are inputs, (b) and (d) are resultant groundwater potentials. Full lines represent actual potentials starting from zero; broken lines represent cyclic pattern of potentials

proceed for a period of at least

$$t = \frac{2.5\,L^2 S}{T} \qquad (7.13)$$

to achieve a dynamic balance. The dimension L should equal the length of a typical flow path within the aquifer. For aquifers which are partly confined and partly unconfined, the parameters for the unconfined region should be used.

Typical examples are as follows: Consider a confined aquifer with $T = 1200\,\text{m}^2/\text{d}$, $S = 4 \times 10^{-4}$, and $L = 25\,\text{km}$.

$$t = \frac{2.5\,L^2 S}{T} = \frac{2.5 \times 625 \times 10^6 \times 4 \times 10^{-4}}{1200 \times 365} = 1.4 \text{ years}$$

Therefore, when the calculation has been run for three years with a typical year's input data, the aquifer achieves a dynamic balance.

However, for the Lincolnshire Limestone, which is discussed in detail in Chapter 10, the parameters for the unconfined region are $T = 250\,\text{m}^2/\text{d}$, $S = 0.05$, and $L = 10\,\text{km}$. Then

$$t = \frac{2.5\,L^2 S}{T} = \frac{2.5 \times 10^2 \times 10^6 \times 0.05}{250 \times 365} = 137 \text{ years}$$

This indicates that the calculation must be run for a very long time before the dynamic-balance condition is achieved. Such a result could be anticipated when it is recalled that the groundwater velocities are so small that it takes, on average, about 100 years from infiltration until discharge from the confined region. The slow build-up, starting from zero potential, is shown in Fig. 7.4.

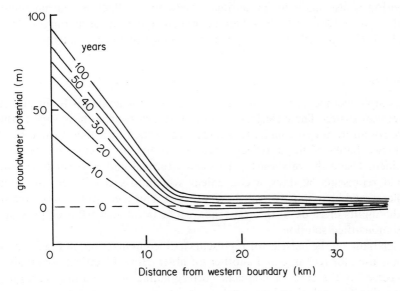

Figure 7.4. Build-up of groundwater potentials with time for Lincolnshire Limestone

For the resistance–capacitance network, the most convenient method of starting the calculation is to carry out the simulation for a number of years repeatedly using a typical historic year as input. Then, when a dynamic balance has been achieved, the years to be studied can be modelled. This, and other alternative techniques, can be used as initial conditions for digital computer simulations. Digital computer simulations of initial conditions are discussed in Section 8.9.

7.7 Apparatus for a Simple Network

The aim of this section is to provide a description of a relatively simple resistance–capacitance network and its ancillary equipment for regional groundwater flow problems. Particular reference is made to practical details. Figure 7.5 contains a diagram of the layout of the equipment plus a detailed drawing of part of the actual network.

Construction of network

The actual resistance–capacitance network is built on a sheet of 6-mm perspex, which is a good electrical insulator. A typical nodal arrangement is drawn to scale in Fig. 7.5(b): the rectangular layout is chosen for economy of space. All the resistors are mounted in slotted turrets. The resistors, marked R, are the network resistors which model the transmissivity; they are usually in the range 100 kΩ to 1 MΩ. The resistor RR is used to model either recharge or abstraction from a well. Provision is also made for mounting a capacitor C which, for an unconfined aquifer, will be 1.0 to 10.0 μF and for a confined aquifer the usual range is 0.01 to 1.0 μF. The other end of each capacitor is connected to zero potential.

Time base

A simple method of providing a time base for the analog is by means of a uniselector switch. These electro-mechanical rotary switches have wipers which move round from one contact to another, controlled by a time switch. In Fig. 7.5(a) two banks of the uniselector are drawn, but up to eight banks can be provided. Using the relationship between electrical and physical time that 1 second represents 50 days, a convenient switching speed for an unconfined aquifer would be to move forward every 0.3 sec. This is equivalent to a time step for the inputs of 15 days; the resistance–capacitance network itself provides a continuous-time solution.

Each bank on the uniselector represents a different function; in Fig. 7.5(a) one bank is used for recharge and another for abstraction. Electrical potentials are connected to the ring of contacts corresponding to the total recharge or abstraction for each time increment.

As an alternative to the uniselector switch, certain electronic switching circuits can perform the same function. A relatively straightforward system is illustrated in Fig. 7.6. The purpose of the circuit is to output a sequence of voltages on a number of channels at regular time intervals. The magnitudes of the inflows are stored in a random-access memory in an eight-bit format. It is convenient to address the memory by means of an eight-bit counter, and thus the maximum number of time steps is 256. A quartz clock is used to provide a preset

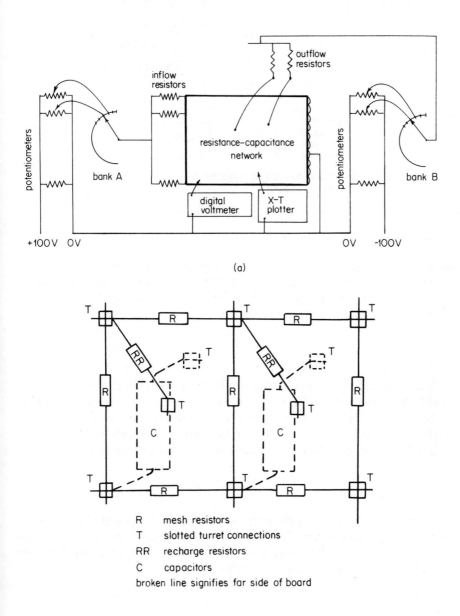

R   mesh resistors
T   slotted turret connections
RR   recharge resistors
C   capacitors
broken line signifies far side of board

(b)

Figure 7.5. Simple resistance–capacitance network: (a) network and ancillary
equipment; (b) details of a unit of the network

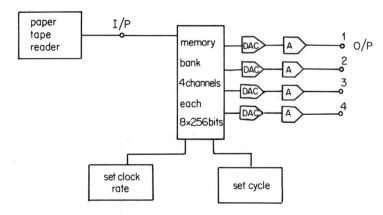

DAC   digital to analogue converter
A   output amplifiers

Figure 7.6. Electronic circuit for inflows and outflows

pulse in the range 0.0001 to 1.0 sec. The outputs from the memory are fed to inexpensive digital-to-analog converters (DAC) and then amplified ten times to provide outputs in the range ± 25.6 V. It is possible either to cycle on a specified number of outputs or to scan the full 256.

Power supplies

Two stabilized power supplies are required to provide the electrical potentials which correspond to the inflows and outflows. It is preferable to use power supplies of the order of 100 V, 0.3 A. When voltages of this order are used, the network voltages are in the range ± 5 volts, and the spurious voltages that the network might pick up due to the other equipment are small enough to be neglected.

Potentiometers

The voltages representing monthly variations in recharge and abstraction are provided by potentiometers. Multi-turn helical potentiometers are the best type of device, but a cheap alternative is to select two resistances which will divide the power supply voltage in the correct ratio.

Measuring devices

In Fig. 7.5(a) two measuring devices are shown, a digital voltmeter and a X–T plotter. The time base on the X–T recorder can be initiated automatically. Note that both of the instruments are connected to the zero-voltage line. All measuring devices and power supplies must use the common zero-voltage line as their neutral.

7.8 Control by Minicomputer

Though the resistance–capacitance network and ancillary equipment as described above is suitable for modelling periods of up to ten years, yet when the aquifer is part of a conjunctive use scheme, it is necessary to consider periods extending for hundreds of years. A technique which allows the analysis of such long time periods involves the use of a resistance–capacitance analog controlled by a digital minicomputer.

This analog–digital technique is different from the usual hybrid computer technique (Vemuri and Karplus, 1969; Hefez et al., 1975) which uses analog devices to integrate in the space dimensions but divides time into discrete steps. In the system presented here the advantage of using the resistance–capacitance network to model time as a continuous function is retained, whilst the digital computer is used both to control the complex pattern of inputs and outputs and to record the voltages which represent the changing groundwater potentials.

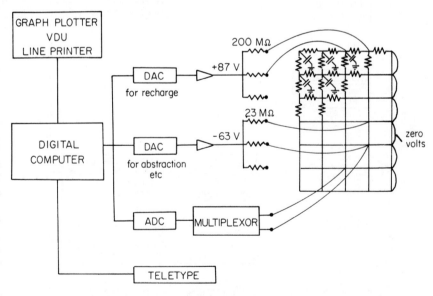

Figure 7.7. Simplified schematic diagram of connections between analog and digital computers

In a typical arrangement the digital computer is a 20K, 16-bit machine with eight 12-bit digital-to-analog converters (DAC) and also an analog-to-digital converter (ADC) multiplexed to read sixteen channels. Timing during an experiment is controlled by a real-time clock within the computer. Figure 7.7 shows how these devices are connected to the resistance–capacitance network.

The major tasks of the digital computer in servicing the resistance network are as follows:

(a) The calculation of the recharge to the aquifer from the rainfall, runoff, and evaporation and the conversion of these quantities to the appropriate binary patterns for later conversion to voltages. These voltages form the input data to the network and are changed every month (280–310 msec when 1 sec of electrical time represent 100 days).

(b) The simulation of monthly changes in abstraction, flow to rivers, and any other inflows or outflows.

(c) The recording of voltages (groundwater potentials) as required. This involves either measurement at frequent time intervals such as every 10 msec (once per day) or scanning a number of nodal points at a particular time. The computer can also be used to convert this information into a form that can be assessed easily, such as identifying the largest and smallest groundwater potentials. Data can be stored and later displayed on an X–Y plotter.

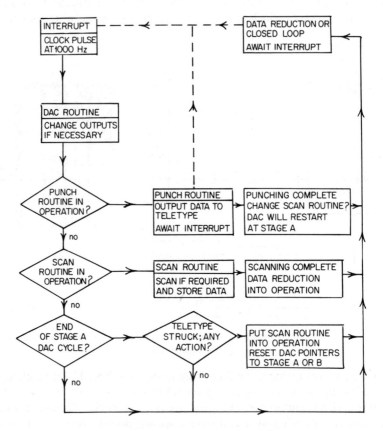

Figure 7.8. Flow chart of computer operations

(d) As the groundwater potentials change, provision can be made for the modification of the abstraction rates according to the algorithm for controlling the levels within the aquifer. This involves measuring the voltages, noting the changes and altering the abstraction rates as appropriate.

A simplified computer flow chart is shown in Fig. 7.8. Initially, the network must achieve a state of dynamic balance using inflows and outflow for a typical year, STAGE A. Once this dynamic balance has been achieved, the inflow and outflow data are changed to a sequence of years and the nodal points are scanned with abstraction rates manipulated as appropriate, STAGE B. Output of the results takes place after a run has been completed; whilst this is occurring the computer manipulates data to achieve the dynamic balance once again. This programme uses about 4K of computer store, leaving the remainder for data storage. Particular care was necessary when developing the program to ensure that timings were strictly kept when computer peripherals were in use during the same period.

Further details of the techniques and results obtained using the interactive analog–digital system can be found in Rushton and Ash (1974, 1975). This technique is becoming more versatile as the software of minicomputers develops.

7.9 Advantages of Technique

Certain of the advantages of the resistance–capacitance analog model are summarized below.

The main advantage is that time remains a continuous function. This avoids many of the difficulties and errors associated with the discrete-time digital computer solutions described in Chapter 8.

Another advantage is the simplicity of the pure resistance analog for steady-state investigations. It takes a few hours to assemble the resistors in the network and to set up the experiment. Then any inadequacies in the initial data quickly become apparent.

With both the steady-state and time-variant analogs, results are obtained virtually instantaneously. Any changes, such as a rearrangement of abstraction wells, can quickly be incorporated in the network and the results of these changes can be observed immediately by further runs of the analog.

This facility to observe the results of changes straight away is of great advantage when performing a sensitivity analysis. Often information about trends is all that is required and a series of graphs superimposed upon each other provide a convenient means of assessing the significance of such changes.

Once a resistance–capacitance analog has been built and tested, it is advisable to prepare data so that a digital computer version of the aquifer model is also available. When extensive tabulated results are required, the digital computer technique is certainly more convenient.

CHAPTER 8

Time-Variant Flow—Digital Computer Techniques

8.1 Introduction

There are many finite-difference approximations used in digital computer solutions of the governing equations for regional groundwater flow. Certain methods have been devised for economy of computing effort (Trescott *et al.*, 1976) but they tend to require considerable programming skill. In this book, attention will be restricted to methods which are comparatively easy to understand and use.

The basic difference between the various finite-difference approximations is the representation of the time-dependent term, $S\partial h/\partial t$. Each of the methods will be discussed with particular reference to the following considerations:

(a) simplicity of computer program;
(b) computational effort required;
(c) stability and convergence;
(d) representation of internal and external boundaries.

The governing equation for flow in an aquifer when the vertical components of flow are sufficiently small to be neglected (Eqs. 2.20 and 7.1) is

$$\frac{\partial}{\partial x}\left(T_x\frac{\partial h}{\partial x}\right) + \frac{\partial}{\partial y}\left(T_y\frac{\partial h}{\partial y}\right) = S\frac{\partial h}{\partial t} - q$$

where x and y are the coordinate directions [L]
  h is the groundwater potential [L]
  T_x and T_y are the transmissivities in the x- and y-directions [L^2/T]
  S is the appropriate storage coefficient
  t is the time [T]
  q is the inflow to the aquifer per unit area [L/T]

An abstraction from a well is expressed as a negative inflow and equals the actual abstraction rate divided by the appropriate mesh area.

For each of the methods the object is to calculate the groundwater potentials at time $t + \Delta t$, which is the $(n + 1)$th time increment, from the known potentials at time t, which is the nth iteration.

The commonly used finite-difference approximations can be summarized by the equation (see Section 3.8)

$$L\left[\frac{\partial}{\partial x}\left(T_x\frac{\partial h_{n+1}}{\partial x}\right) + \frac{\partial}{\partial y}\left(T_y\frac{\partial h_{n+1}}{\partial y}\right)\right] + M\left[\frac{\partial}{\partial x}\left(T_x\frac{\partial h_n}{\partial x}\right) + \frac{\partial}{\partial y}\left(T_y\frac{\partial h_n}{\partial y}\right)\right]$$

$$= S\frac{h_{n+1} - h_n}{\Delta t} - q \tag{8.1}$$

where $L + M = 1$. There are three commonly used values of L and M:

(a) $L = 0, M = 1$: this is a forward-difference approximation which leads to an explicit formulation.
(b). $L = 1, M = 0$: this is a backward-difference approximation; successive over-relaxation (SOR) and the modified alternating-direction implicit procedure (MADIP) use this approximation.
(c) $L = 0.5, M = 0.5$: this is a central-difference approximation; both the alternating-direction implicit (ADI) and the alternating-direction explicit (ADE) methods use a central-difference formulation.

The convention for the numbering of the nodes is illustrated in Fig. 8.1. Capital letters, with the nodal positions given in brackets, will be used so that expressions are then in a form suitable for the preparation of computer programs. The y-coordinate is taken as positive downwards to conform with the printing of results on a computer. Though each potential $H(\ )$ is defined in terms of three coefficients, the nodal position in the x- and y-directions and the number of the time increment, only two coefficients need to be used in the computer programs since the potentials at earlier times need not be stored. The transmissivities in the x- and y-directions $TX(\ )$ and $TY(\ )$ are defined at the mid-points between the nodes. The term $RS(\ )$ corresponds to the symbol q, the inflow to the aquifer.

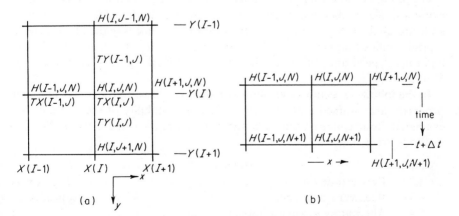

Figure 8.1. Convention for the numbering of nodes: (a) space mesh (x, y);
(b) space–time mesh (x, t)

164

For each of the numerical methods, results will be presented for a standard problem of a square aquifer (Fig. 8.2), with a regular mesh spacing and constant values of transmissivity and storage coefficient. One set of opposite boundaries are impermeable, the others are held at zero potential. This combination of boundary conditions provides a thorough test of any method. Initially all the groundwater potentials are zero and then a well at the centre suddenly starts to discharge at a constant rate, $-Q$. Groundwater drawdowns are quoted at an observation well positioned at $x = 0.1b$, $y = 0.1b$. Each side is divided into 20 mesh intervals.

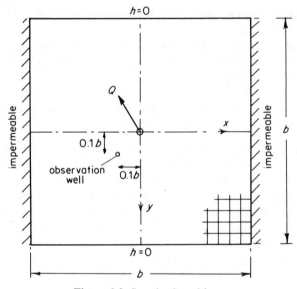

Figure 8.2. Standard problem

Graphs of groundwater potential against time are plotted with the time scale expressed in a number of different ways. Two non-dimensional forms of the time scale are used, $tT/\Delta x^2 S$ and $tT/b^2 S$. In addition, the time is quoted in days for a typical confined aquifer with $T = 1000 \, \text{m}^2/\text{d}$, $S = 2.5 \times 10^{-5}$, and $b = 20 \, \text{km}$, and for a typical unconfined aquifer with T and b taking the same values, but $S = 2.5 \times 10^{-2}$.

In the following sections five popular methods will be considered. Computer programs are outlined and the convergence and stability of the methods are examined. Restrictions on the use of the method are discussed in detail. The five methods are as follows:

Section	Method	Program
8.2	Forward difference	Fig. 8.3
8.3	Backward difference	Appendix 2
8.4	Alternating direction implicit	—
8.5	Alternating direction explicit	Fig. 8.12
8.6	Modified iterative alternating direction implicit.	—

8.2 Forward-difference Formulation

One of the earliest methods to be used for regional groundwater flow used a forward-difference approximation. If $L = 0$ and $M = 1$, then Eq. 8.1 becomes

$$\frac{\partial}{\partial x}\left(T_x \frac{\partial h_n}{\partial x}\right) + \frac{\partial}{\partial y}\left(T_y \frac{\partial h_n}{\partial y}\right) = S\frac{h_{n+1} - h_n}{\Delta t} - q \tag{8.2}$$

In finite-difference form, using the notation of Fig. 8.1 this becomes

$$\frac{2}{X(I+1) - X(I-1)}\left\{\frac{TX(I)}{X(I+1) - X(I)}[H(I+1, J, N) - H(I, J, N)]\right.$$

$$+ \frac{TX(I-1)}{X(I) - X(I-1)}[H(I-1, J, N) - H(I, J, N)]\Big\}$$

$$+ \frac{2}{Y(J+1) - Y(J-1)}\left\{\frac{TY(J)}{Y(J+1) - Y(J)}[H(I, J+1, N) - H(I, J, N)]\right.$$

$$+ \frac{TY(J-1)}{Y(J) - Y(J-1)}[H(I, J-1, N) - H(I, J, N)]\Big\}$$

$$= \frac{S(I, J)[H(I, J, N+1) - H(I, J, N)]}{\Delta t} - RS(I, J) \tag{8.3}$$

This expression is similar to Eq. 7.2; it allows for variations in transmissivity and storage coefficient and applies to an irregular mesh. Note in particular that transmissivities refer to the average between nodes.

It is convenient to write Eq. 8.3 as

$$AH(I+1, J, N) + BH(I, J-1, N) + CH(I-1, J, N)$$
$$+ DH(I, J+1, N) - (A + B + C + D)H(I, J, N)$$
$$= F[H(I, J, N+1) - H(I, J, N)] - RS(I, J) \tag{8.4}$$

where

$$A = \frac{2TX(I)}{[X(I+1) - X(I-1)][X(I+1) - X(I)]} \tag{8.5}$$

Expressions for B, C, and D take a similar form,

and
$$F = S(I, J)/\Delta t \tag{8.6}$$

Equation 8.4 can be rearranged as

$$H(I, J, N+1) = [AH(I+1, J, N) + BH(I, J-1, N) + CH(I-1, J, N)$$
$$+ DH(I, J+1, N) - (A + B + C + D - F)H(I, J, N)$$
$$+ RS(I, J)]/F \tag{8.7}$$

In this equation, the unknown potential $H(I, J, N+1)$ is written explicitly in terms of the potentials at the time step N and the flows $RS(\ )$, and therefore this is an *explicit* method.

One great advantage of the explicit procedure is the simplicity of the computer program. A one-dimensional form of the program is described in Section 9.1. Figure 8.3 lists the essential part of the program for the explicit formulation in two dimensions; this replaces the routine 'SOR CALCULATION' in the program of Appendix 2. Certain of the earlier statements together with the term SFAC are introduced to make this program consistent with the full SOR program listing.

```
C     EXPLICIT.PROCEDURE
C     SFAC MUST BE CHOSEN SO THAT DELT*T/DELX*DELX*S*SFAC.LT.0.25
C     NUM=2*B*B*S*SFAC/DELT*T, WHERE B IS LENGTH OF AV FLOW PATH
C
      IF(IFIRST.LT.0) SFAC=0.8
      NUM=1000
      ICYCLE=1
      IF(IFIRST.LT.0) GO TO 950
980   DO 940 I=2,MBOUND
      DO 940 J=2,NBOUND
940   HOLD(I,J)=H(I,J)
950   DO 960 I=2,MBOUND
      DO 960 J=2,NBOUND
      F=SFAC*S(I,J)/DELT
      IF(HFIX(I,J).GE.-10000.0) GO TO 970
      H(I,J)=(A(I,J)*HOLD(I+1,J)+B(I,J)*HOLD(I,J-1)+C(I,J)*HOLD
     1 (I-1,J)+D(I,J)*HOLD(I,J+1)-(A(I,J)+B(I,J)+C(I,J)+D(I,J)
     2 -SFAC*S(I,J)/DELT)*HOLD(I,J) +RS(I,J))/F
      GO TO 960
970   H(I,J)=HFIX(I,J)
960   CONTINUE
C
C     SECTION FOR INITIAL HEADS
      IF(IFIRST.GT.0) GO TO 990
      ICYCLE=ICYCLE+1
      IF(ICYCLE.LT.NUM) GO TO 980
C     OUTPUT SECTION FOR INITIAL STEADY HEADS
      IFIRST=100
      WRITE(6,1010)
1010  FORMAT(1X,27HINITIAL STEADY STATE HEADS      )
      CALL PRIN(H,7,2,MBOUND,2,NBOUND,DAYT)
      GO TO 9000
C     END OF EXPLICIT PROCEDURE
C
C
      END
```

Figure 8.3. Essential part of program for explicit formulation

The straightforward nature of the explicit method is apparent when it is recognized that Eq. 8.7 is the only expression involved in the calculation of the groundwater potentials. Boundary conditions are supplied either through the coefficients $A(\ )$, $B(\ )$,..., as described in Section 8.8, or if boundary or internal nodes are at a fixed groundwater potential they are enforced through the IF statement which jumps round the calculation of the new potentials.

The main disadvantage of this forward-difference explicit method is that instability is certain to occur if the time step exceeds a critical value (Smith, 1965). For a square mesh of sides Δx the condition for avoidance is

$$\Delta t \leqslant 0.25\Delta x^2 S/T \tag{8.8}$$

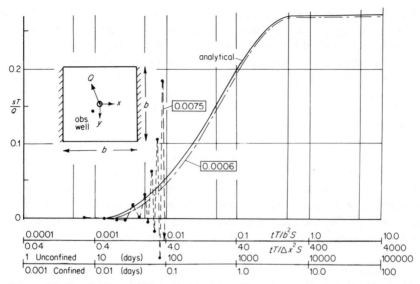

Figure 8.4. Results for explicit approximation. Numbers in the boxes are values of
$$\Delta t\, T/b^2 S$$

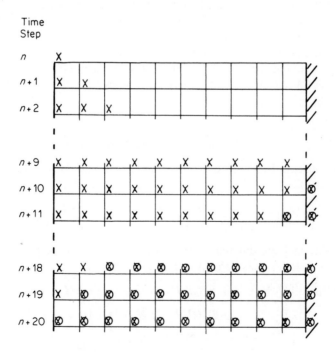

X change due to inflow

O change due to boundary

Figure 8.5. Spread of influence with explicit formulation

With variable mesh spacing, transmissivity, or storage coefficient, the smallest value of the expression anywhere within the field is the critical condition.

The explicit method has been used to analyse the standard problem of Fig. 8.2; a number of solutions are plotted in Fig. 8.4. Good results are obtained for $\Delta t\,T/\Delta x^2 S = 0.24$ (which is equivalent to $\Delta t\,T/b^2 S = 0.0006$), but if the time step is increased to $\Delta t\,T/\Delta x^2 S = 0.3$, then unstable results occur. For a typical confined aquifer, this critical time step requires that $\Delta t \leqslant 0.006\,25$ day. This is equivalent to almost 50 000 time steps each year.

An examination of the results plotted in Fig. 8.4 indicates that the differences between the full line, which plots the analytical values, and the chain-dotted line for $\Delta t\,T/\Delta x^2 S = 0.24$, are small. However, when frequent changes occur in boundary or internal conditions the *spreading of the influence* of this change throughout the aquifer is retarded. This is illustrated in Fig. 8.5, which demonstrates that the effect of a change at the left-hand boundary spreads only one mesh interval per time step, and the influence of the right-hand boundary on this change spreads equally slowly. Thus, when there are ten mesh intervals between the boundaries, it takes 20 time increments before the influence of the right-hand boundary on the change at the left-hand boundary is registered. In certain situations this delay can lead to significant errors.

Apart from the stability restriction and the delay in the spread of changes across the aquifer, the forward-difference explicit method has much to commend it, especially when first preparing a simple mathematical model of groundwater flow on a digital computer. The program presented in Fig. 8.3 allows for the calculation of initial heads. Care has to be taken in the selection of the parameters NUM and SFAC to ensure that stable initial values are obtained (see Section 8.9, method (c)).

8.3 Backward-difference Formulation

In the backward-difference formulation, $L = 1$ and $M = 0$, hence Eq. 8.1 becomes

$$\frac{\partial}{\partial x}\left(T_x\frac{\partial h_{n+1}}{\partial x}\right) + \frac{\partial}{\partial y}\left(T_y\frac{\partial h_{n+1}}{\partial y}\right) = S\frac{h_{n+1} - h_n}{\Delta t} - q \qquad (8.9)$$

Writing this in finite-difference form using the notation of Fig. 8.1 and the coefficients A, B, C..., as defined in Eqs. 8.5 and 8.6:

$$AH(I + 1, J, N + 1) + BH(I, J - 1, N + 1) + CH(I - 1, J, N + 1)$$
$$+ DH(I, J + 1, N + 1) - (A + B + C$$
$$+ D)H(I, J, N + 1)$$
$$= F[H(I, J, N + 1) - H(I, J, N)] - RS(I, J) \qquad (8.10)$$

It is not possible to write this expression in an explicit form, since this equation and the equivalent equation for other nodes form a set of simultaneous equations.

The solution of the simultaneous equations, when the complete set has to be solved at each time step, can be a time-consuming procedure, even on a large, modern digital computer. However, by adopting an iterative technique, the computational effort can become manageable. One method that has proved to be convenient for groundwater flow problems is the technique of successive over-relaxation (SOR).

In describing the SOR method it is helpful to introduce a further coefficient

$$E = A + B + C + D + S(I, J)/\Delta t = A + B + C + D + F \qquad (8.11)$$

Equation 8.10 can be rewritten in a form somewhat similar to that of the explicit procedure; thus

$$H(I, J, N + 1) = [AH(I + 1, J, N + 1) + BH(I, J - 1, N + 1)$$
$$+ CH(I - 1, J, N + 1) + DH(I, J + 1, N + 1)$$
$$+ FH(I, J, N) + RS(I, J)]/E \qquad (8.12)$$

This expression allows the direct calculation of $H(I, J, N + 1)$ but it depends on other potentials from the same time step which, in the early stages of the calculation, will not have reached their correct value. Therefore, the value of the potential given by Eq. 8.12 will be an underestimate of the true value. Over-relaxation is introduced to compensate for this. When the Mth iteration has been completed, then the *change* in potential predicted by Eq. 8.12 is

$$\Delta H(I, J, N + 1, M + 1 : M) = [\ ]/E - H(I, J, N + 1, M) \qquad (8.13)$$

where the square brackets [ ] signify the expression in square brackets on the right-hand side of Eq. 8.12. Note that each potential is now defined in terms of the two positions in space, I and J, the number of the time step, $N + 1$, and the iteration M or $M + 1$.

An over-relaxation factor ω is introduced so that a new approximation to the potential is given by

$$H(I, J, N + 1, M + 1) = H(I, J, N + 1, M) + \omega\Delta H(I, J, N + 1, M + 1 : M)$$
$$(8.14)$$

or

$$H(I, J, N + 1, M + 1) = (1 - \omega)H(I, J, N + 1, M) + \omega[\ ]/E \qquad (8.15)$$

This equation is used at each node in turn. As the iterations proceed, convergence to a steady value is achieved. Studies which have been made of the optimum over-relaxation factor indicate that a value of $\omega = 1.6$ is adequate for most regional groundwater flow problems.

Since Eq. 8.15 is a simple substitution formula similar to the explicit expression, the effect of a change at any one node spreads outwards at one mesh interval per *iteration*. Thus, the number of iterations required is roughly twice the number of nodal points in one direction if significant changes occur in the inflows or outflows.

170

To test for convergence, the error in satisfying the finite-difference form of the differential equation (Eq. 8.9) is calculated. Thus, from Eq. 8.12

$$\text{ERROR} = AH(I+1,J,N+1) + BH(I,J-1,N+1) + CH(I-1,J,N+1)$$
$$+ DH(I,J+1,N+1) - EH(I,J,N+1) + FH(I,J,N)$$
$$+ RS(I,J) \tag{8.16}$$

where the potentials for time increment $N+1$ are the values of the most recent iteration. ERROR has the same dimensions as the recharge, $[L/T]$. Thus, if the absolute value of the error is less than a certain fraction of the overall recharge (typically 0.01 per cent) at each node, then the conditions of continuity of flow are satisfied to an acceptable accuracy and the iterations can cease.

The solution of a one-dimensional problem by SOR is discussed in Section 9.1. The sections of the program in Appendix 2 entitled 'Multiplier and previous time step factors' also contain this iterative calculation. The coefficient E is calculated in the program as $TX(I,J)$; $TY(I,J)$ refers to the expression $FH(I,J,N)$ of Eq. 8.16, and AB is the expression $[\ \ ]$ as used above. Note that the arrays $TX(\ )$ and $TY(\ )$ originally contained the transmissivities. IND indicates whether the error criterion is satisfied at every node; if IND remains at zero during a complete iteration then convergence has been achieved. The statement with label 980 in the program is equivalent to Eq. 8.15.

From the above description it is apparent that this computer program is only slightly more complex than that of the explicit method. In terms of the computational effort required for calculations representing many years of aquifer behaviour, long computer runs are required.

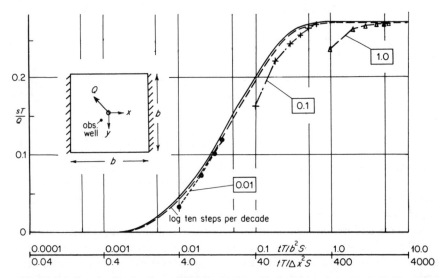

Figure 8.6. Results for backward-difference formulation. Numbers in the boxes are values of $\Delta t\,T/b^2 S$

The most important feature of the backward-difference method is that the calculation is stable, however large a time step is used. This is illustrated in Fig. 8.6, which shows solutions of the standard problem with the time step varying from $\Delta t T/\Delta x^2 S = 0.4$ to 400. Even with the largest time step the errors in the calculated groundwater potentials for the first step are approximately 10 per cent and by the fourth time step the results are effectively the same as the analytical. A time step that increases logarithmically is an economical choice when long time periods are being investigated.

Representation of internal and external boundaries is as direct as for the explicit method. Indeed, the backward-difference method has no disadvantages; the only valid reason for choosing other methods is to attempt to obtain economy in computing effort.

8.4 Alternating-direction Implicit Formulation (ADI)

The ADI method was first introduced by Peaceman and Rachford (1955) and its application to groundwater flow is described by Pinder and Bredehoeft (1968). Its attractive feature is that it is an implicit method which does not involve the costly solution of a large number of simultaneous equations. When first introduced it was claimed that, being an implicit method, the ADI technique is stable however large a time step is used. Briggs and Dixon (1968) questioned this, though the errors that they identified cause only minor irregularities in the calculated results. However, Rushton (1973, 1974a) showed that very serious errors can occur with the ADI method. For certain combinations of boundary conditions wild oscillations can occur, but with other combinations, results that look satisfactory are actually underestimating the groundwater potential by as much as 95 per cent.

In essence, ADI is a two-step procedure. Following the form of Eq. 8.1, the two steps can be written as

$$\frac{\partial}{\partial x}\left(T_x \frac{\partial h_{n+\frac{1}{2}}}{\partial x} \right) + \frac{\partial}{\partial y}\left(T_y \frac{\partial h_n}{\partial y} \right) = S\left(\frac{h_{n+\frac{1}{2}} - h_n}{0.5\Delta t} \right) - q \qquad (8.17a)$$

$$\frac{\partial}{\partial x}\left(T_x \frac{\partial h_{n+\frac{1}{2}}}{\partial x} \right) + \frac{\partial}{\partial y}\left(T_y \frac{\partial h_{n+1}}{\partial y} \right) = S\left(\frac{h_{n+1} - h_{n+\frac{1}{2}}}{0.5\Delta t} \right) - q \qquad (8.17b)$$

These equations can be combined by eliminating $h_{n+\frac{1}{2}}$; for the case where T_x and T_y are constants,

$$\frac{1}{2}\left(T_x \frac{\partial^2}{\partial x^2} + T_y \frac{\partial^2}{\partial y^2} \right)(h_{n+1} + h_n) = S\frac{h_{n+1} - h_n}{\Delta t} - q$$

$$+ \frac{\Delta t}{4} \frac{T_x T_y}{S} \frac{\partial}{\partial x^2} \frac{\partial}{\partial y^2}(h_{n+1} - h_n) \qquad (8.18)$$

The ADI approximation is therefore a central-difference formulation with a small higher-order error term.

172

Following the same convention as with Eq. 8.4, Eq. 8.17a can be written as

$$AH(I + 1, J, N + \tfrac{1}{2}) + BH(I, J - 1, N) + CH(I - 1, J, N + \tfrac{1}{2})$$
$$+ DH(I, J + 1, N)$$

$$- (A + C)H(I, J, N + \tfrac{1}{2}) - (B + D)H(I, J, N)$$

$$= 2F[H(I, J, N + \tfrac{1}{2}) - H(I, J, N)] - RS(I, J) \qquad (8.19)$$

This can be rearranged with the unknown terms at time step $N + \tfrac{1}{2}$ on the left-hand side and known terms on the right-hand side:

$$AH(I + 1, J, N + \tfrac{1}{2}) + CH(I - 1, J, N + \tfrac{1}{2}) - (A + C + 2F)H(I, J, N + \tfrac{1}{2})$$

$$= -BH(I, J - 1, N) - DH(I, J + 1, N) + (B + D - 2F)H(I, J, N)$$

$$- RS(I, J) \qquad (8.20a)$$

It is important to note that all the *unknown* terms lie along the line $Y(J)$: these are shown by the shaded circles of Fig. 8.7. The known terms from time step N are indicated on the figure by open circles. If all the unknowns for the line $Y(J)$ are written in matrix form then a tridiagonal matrix results. There will be a separate tridiagonal matrix for each of the other lines of $Y(J)$ = constant. These tridiagonal matrices can be solved by a simple, efficient Gaussian elimination routine (Smith 1965).

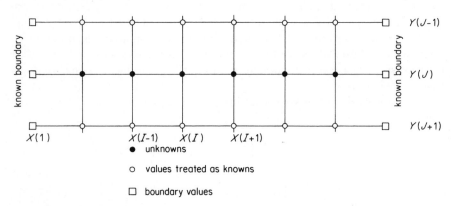

Figure 8.7. Method of working in alternate directions. Diagram shows first step

In the second stage of the calculation the appropriate equation is

$$BH(I, J - 1, N + 1) + DH(I, J + 1, N + 1) - (B + D + 2F)H(I, J, N + 1)$$

$$= -AH(I + 1, J, N + \tfrac{1}{2}) - CH(I - 1, J, N + \tfrac{1}{2})$$

$$+ (A + C - 2F)H(I, J, N + \tfrac{1}{2}) - RS(I, J) \qquad (8.20b)$$

Now all the unknowns lie along lines of constant $X(I)$. A similar procedure is used to solve these equations.

The computational effort required for a time step of the ADI method is roughly eight times that for the explicit method and twice that for one *iteration* of the SOR technique. Thus if the SOR calculation requires, on average, twenty iterations, the computational effort of ADI is only one-tenth of SOR.

Sources of error

There are a number of potential sources of error inherent in the ADI technique, as follows:

(a) Errors can occur if the time step is greater than a certain critical value. Rushton (1973) has shown that for an acceptable accuracy,

$$\Delta t T/b^2 S \leqslant 0.05 \tag{8.21}$$

where b is the shortest distance between opposite boundaries. The standard problem is used to illustrate this effect; certain results are plotted in Fig. 8.8. If $\Delta t T/b^2 S = 0.1$, then small oscillations occur, but if $\Delta t T/b^2 S = 1.0$, wildly oscillating results are obtained. With the alternative boundary conditions of all four sides at a fixed potential (Fig. 8.9), no oscillations occur and the results appear to be acceptable. However, the calculated values are very much lower than the analytical results. The safest procedure with the ADI technique is to use a time step which increases logarithmically, such that calculations are performed at

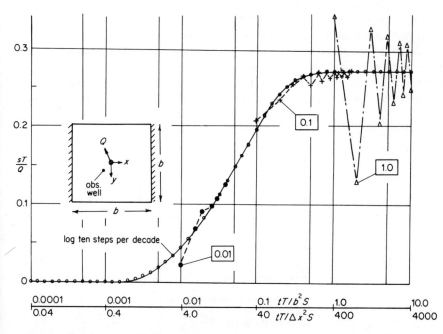

Figure 8.8. Results for ADI formulation. Numbers in the boxes are values of $\Delta t T/b^2 S$

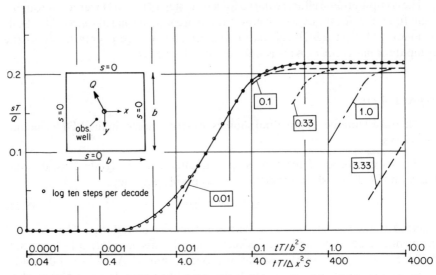

Figure 8.9. Results for ADI with all four boundaries at fixed head. Numbers in the boxes are values of $\Delta t T/b^2 S$

times of

$$t(T/\Delta x^2 S) = 10^{(0.16666n - 1.16666)} \qquad (8.22)$$

where n is the number of time steps since the last significant change, such as the *starting* or *stopping* of an abstraction well.

(b) If there are fixed groundwater potentials within the aquifer such that these potentials do not remain constant with time, then major modifications must be made to the program. An instance of the occurrence of such an internal boundary is when a lake is in contact with the aquifer. To illustrate the modifications, consider the case of a fixed potential on the line $Y(J)$ at the intermediate node

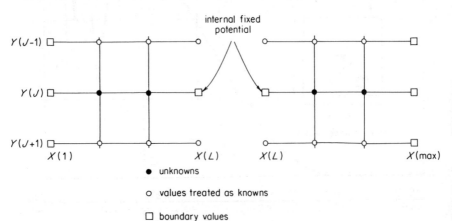

Figure 8.10. Representation of internal fixed boundary

$X(L)$ (Fig. 8.10). Because of this internal known boundary, Eq. 8.20a has to be solved by Gaussian elimination in two separate parts. In the first the limits of the calculation are $X(1)$ *and* $X(L)$, in the second the limits are $X(L)$ and $X(\text{MAX})$. This procedure disturbs the simplicity of the standard calculation. Fixed potentials which remain constant with time can be represented by a very large storage coefficient.

(c) When the external boundaries of the aquifer are such that re-entrant corners are present, the calculations will need to be subdivided in the manner described for internal fixed potentials.

(d) Care needs to be taken when representing the conditions of a leaky aquifer. A detailed discussion of leaky aquifers is to be found in Section 9.2 but the particular difficulties encountered with the ADI method are as follows.

When leakage occurs from an overlying aquifer, the simplest method is to modify Eq. 8.17(a) to become

$$\frac{\partial}{\partial x}\left(T_x \frac{\partial h_{n+\frac{1}{2}}}{\partial x}\right) + \frac{\partial}{\partial y}\left(T_y \frac{\partial h_n}{\partial y}\right) = S\left(\frac{h_{n+\frac{1}{2}} - h_n}{0.5\Delta t}\right) - q - \frac{k'}{m}(H - h_n) \quad (8.23)$$

where k' and m are the permeability and thickness of the leaky aquifer and H is its groundwater potential. If this approximation is used then unstable results may result; see Rushton (1974a, Fig. 7).

However, by adopting a slightly different approximation for the leakage term,

$$\frac{\partial}{\partial x}\left(T_x \frac{\partial h_{n+\frac{1}{2}}}{\partial x}\right) + \frac{\partial}{\partial y}\left(T_y \frac{\partial h_n}{\partial y}\right) = S\frac{(h_{n+\frac{1}{2}} - h_n)}{0.5\Delta t} - q - \frac{k'}{m}[H - 0.5(h_{n+\frac{1}{2}} + h_n)]$$

$$(8.24)$$

no instabilities occur.

Despite these limitations, the ADI method is economical in computing effort and has been used successfully in a large number of practical groundwater flow investigations.

8.5 Alternating-direction Explicit Formulation (ADE)

In this fourth finite-difference approximation a simple rearrangement of the alternating-direction technique leads to an explicit formulation which has a less stringent stability criterion than the standard explicit method. As with the ADI procedure, the time step is divided into two halves. The fundamental difference of this approach is that the unknown potentials are arranged to be at a pair of nodes inclined at 90° to each other (rather than at 180° as in the ADI method).

Thus in the notation of Fig. 8.1,

$$AH(I + 1, J, N) + BH(I, J - 1, N + \tfrac{1}{2}) + CH(I - 1, J, N + \tfrac{1}{2})$$
$$+ DH(I, J + 1, N)$$

$$- (A + D)H(I, J, N) - (B + C)H(I, J, N + \tfrac{1}{2})$$

$$= 2F[H(I, J, N + \tfrac{1}{2}) - H(I, J, N)] - RS(I, J) \quad (8.25)$$

This can be rewritten as

$$H(I, J, N + \tfrac{1}{2}) = [AH(I + 1, J, N) + DH(I, J + 1, N)$$
$$- (A + D - 2F)H(I, J, N)$$
$$+ RS(I, J) + BH(I, J - 1, N + \tfrac{1}{2}) + CH(I - 1, J, N + \tfrac{1}{2})]$$
$$/(B + C + 2F) \qquad (8.26a)$$

It will be noted that the last two terms in the square brackets of Eq. 8.26(a) are at time $N + \tfrac{1}{2}$. However, if the order of scanning the nodes is carefully planned, then the potentials at nodes $(I, J - 1)$ and $(I - 1, J)$ can be evaluated prior to the calculation at node (I, J).

This can be achieved if the rows are scanned from left to right and the columns from top to bottom. Figure 8.11 illustrates this procedure; the direction of scanning is shown by the arrows. The potentials written on the figure are the ones that were last calculated when that node was scanned. It is clear that these coincide with Eq. 8.26(a); thus the value of $H(I, J, N + \tfrac{1}{2})$ can be calculated directly.

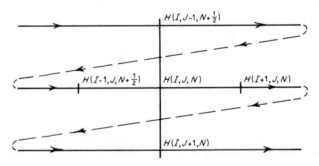

Figure 8.11. Method of scanning nodes in ADE technique.
Diagram shows first step

For the second half of the calculation, the rows are scanned from right to left and the columns from bottom to top. The appropriate equation is

$$H(I, J, N + 1) = [BH(I, J - 1, N + \tfrac{1}{2}) + CH(I - 1, J, N + \tfrac{1}{2})$$
$$- (B + C - 2F)H(I, J, N) + RS(I, J)$$
$$+ AH(I + 1, J, N + 1)$$
$$+ DH(I, J + 1, N + 1)]/(A + D + 2F) \qquad (8.26b)$$

Figure 8.12 shows the modification required to the computer program of Appendix 2. This section is inserted in the program in the same way as is the explicit procedure.

Detailed studies of the convergence of the ADE procedure are reported by Tomlinson and Rushton (1975). In Fig. 8.13, the results are presented for the test problem. The following conclusions can be deduced from this figure and from

```
C     A.D.E. PROCEDURE
C     NUM=2*B*B*S*SFAC/DELT*T, WHERE B IS LENGTH OF AV FLOW PATH
C     SFAC MUST BE CHOSEN SO THAT DELT*T/DELX*DELX*S*SFAC.LT.5.0
C
      IF(IFIRST.LT.0) SFAC=0.04
      NUM=250
      ICYCLE=1
      IF(IFIRST.LT.0) GO TO 950
980   DO 940 I=2,MBOUND
      DO 940 J=2,NBOUND
940   HOLD(I,J)=H(I,J)
C
C     SCAN FROM TOP LEFT TO BOTTOM RIGHT
950   DO 960 I=2,MBOUND
      DO 960 J=2,NBOUND
      F=SFAC*S(I,J)/DELT
      IF(HFIX(I,J).GE.-10000.0) GO TO 970
      H(I,J)=(A(I,J)*H(I+1,J)+D(I,J)*H(I,J+1)-(A(I,J)+D(I,J)-2.0*F)*
     1 H(I,J)+RS  (I,J)+B(I,J)*H(I,J-1)+C(I,J)*H(I-1,J))/
     2 (B(I,J)+C(I,J)+2.0*F)
      GO TO 960
970   H(I,J)=HFIX(I,J)
960   CONTINUE
C
C     SCAN FROM BOTTOM RIGHT TO TOP LEFT
      DO 965 II=2,MBOUND
      I=MBOUND-II+2
      DO 965 JJ=2,NBOUND
      J=NBOUND-JJ+2
      F=SFAC*S(I,J)/DELT
      IF(HFIX(I,J).GE.-10000.0) GO TO 975
      H(I,J)=(B(I,J)*H(I,J-1)+C(I,J)*H(I-1,J)-(B(I,J)+C(I,J)-2.0*F)*
     1 H(I,J)+RS  (I,J)+A(I,J)*H(I+1,J)+D(I,J)*H(I,J+1))/
     2 (A(I,J)+D(I,J)+2.0*F)
      GO TO 965
975   H(I,J)=HFIX(I,J)
965   CONTINUE
C
C     SECTION FOR INITIAL HEADS
      IF(IFIRST.GT.0) GO TO 990
      ICYCLE=ICYCLE+1
      IF(ICYCLE.LT.NUM) GO TO 980
C     OUTPUT SECTION FOR INITIAL STEADY HEADS
      IFIRST=100
      WRITE(6,1010)
1010  FORMAT(1X,27HINITIAL STEADY STATE HEADS
      CALL PRIN(H,7,2,MBOUND,2,NBOUND,TIME)
      GO TO 9000
C     END OF ADE PROCEDURE
C
```

Figure 8.12. Essential part of program for ADE technique

other results quoted in the above paper:

(a) For stable and convergent solutions, the following stability criterion applies:

$$\Delta t T / \Delta x^2 S \leqslant 5.0 \qquad (8.27)$$

When the values of S, T, Δx, or Δy vary throughout the aquifer, then the worst combination must be used. If this value is exceeded then oscillations occur only when there are three impermeable boundaries; with other boundary conditions the computed results are an underestimate.

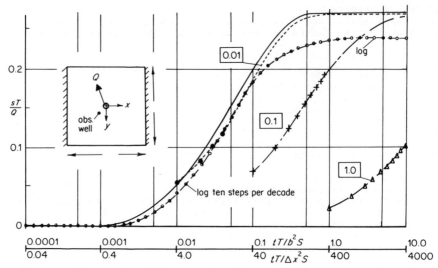

Figure 8.13. Results for ADE formulation. Numbers in the boxes are values of $\Delta t T/b^2 S$

(b) Logarithmic increases in time step are not advisable. Once the increasing time step exceeds the criterion of Eq. 8.27 then divergence from the correct results rapidly occurs.

(c) In general, the ADE method is not suitable for the analysis of aquifers which are partly confined and partly unconfined.

(d) The mesh should be arranged so that major inflows or outflows (e.g. abstraction wells) are not positioned in the upper left-hand or lower right-hand regions of the aquifer.

These conclusions do restrict the applicability of the ADE method yet it does have a permissible time step twenty times greater than the standard explicit method even though it involves roughly double the computational effort. Another advantage of the ADE method is that the influence of any changes are transmitted throughout the aquifer during a single time step.

8.6 Modified Iterative Alternating-direction Implicit Procedure (MIADI)

In an attempt to combine the advantages of many techniques, the MIADI procedure was presented by Prickett and Lonnquist (1971) in a valuable report of the Illinois State Water Survey. Later, it was shown by Rushton (1974b) that this is not a central-difference approach such as the standard ADI formulation, but that it is a backward-difference formulation.

Like the SOR procedure, the MIADI technique involves an iterative procedure. In a series of comparisons between SOR and MIADI using the same convergence criterion, it was found that there is little to choose between the two methods in terms of computational effort. Prickett and Lonnquist introduce a predictor into their program which significantly speeds the convergence. The use of a predictor is beneficial with the SOR program only when no changes occur in the inflows and outflows over a long period. When frequent changes occur, then the predictor can lead to a greater number of iterations being required.

Since it is a backward-difference method, the MIADI procedure is stable and convergent whatever the size of the time step. The report of Prickett and Lonnquist contains listings of programs with clear explanations of how different effects can be introduced into the computer program.

8.7 Summary of Finite-difference Techniques

Table 8.1 summarizes some of the more important aspects of the various finite-difference procedures and indicates the type of problem for which they are particularly suitable.

8.8 Boundary Conditions

On the boundary of an aquifer there are three possible conditions:

(a) fixed potentials;
(b) impermeable;
(c) known inflow $QIN(I, J)$ at the node (I, J).

Methods of representing a fixed potential have been discussed separately as each of the different finite-difference procedures has been presented. The boundary with a known inflow is a special case of the impermeable boundary, since it is equivalent to an impermeable boundary with a specified additional recharge.

The method of enforcing the boundary conditions in the computer solution will be described with reference to Eq. 8.7, which is copied below but with the times omitted from the left-hand side so that it can represent both explicit and implicit approaches.

$$AH(I + 1, J) + BH(I, J - 1) + CH(I - 1, J) + DH(I, J + 1)$$

$$- (A + B + C + D)H(I, J) = F(H(I, J, N + 1) - H(I, J, N))$$

$$- RS(I, J) \tag{8.28}$$

Consider the case of Fig. 8.14 where there is a vertical impermeable boundary. In effect this means that the transmissivity between nodes $(I + 1, J)$ and (I, J) is zero and node $(I + 1, J)$ becomes a fictitious node. This can be represented directly by setting coefficient $A = 0$. In addition, flow between nodes $(I, J + 1)$ to (I, J) and $(I, J - 1)$ to (I, J) will be restricted due to the removal of half of the area; therefore B and D are halved. For similar reasons F and $RS(I, J)$ are halved; this assumes that $(I - 1, J)$ and $(I + 1, J)$ are equidistant from (I, J).

When there is an inflow to the boundary of Fig. 8.14, the inflow must be expressed in the same units as the recharge, therefore it must be divided by the original nodal area. As a result, Eq. 8.28 takes the following form, where all the

Table 8.1 Summary of finite-difference procedures.

Procedure	Timestep limitation	Suitability and remarks
Explicit	$\Delta t \leqslant 0.25\Delta x^2 S/T$	Short-period analysis, particularly in unconfined aquifers provided that the boundaries do not have dominant effects
Backward difference (SOR)	None	All types of problem. A careful choice of convergence parameter to control the number of iterations is important for efficient programs.
Alternating direction implicit (ADI)	$\Delta t \leqslant 0.05b^2 S/T$	Aquifers where boundary effects are not dominant. Not suitable when frequent changes occur in the abstraction pattern.
Alternating direction explicit (ADE)	$\Delta t \leqslant 5.0\Delta x^2 S/T$	Aquifers in which boundary effects are significant. Logarithmic increases in time step can lead to errors.
Modified iterative alternating direction implicit (MIADI)	None	Identical to backward difference

Δx mesh interval (or it could be Δy) b = shortest distance between opposite boundaries

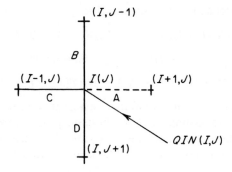

Figure 8.14. Representation of vertical
impermeable boundary

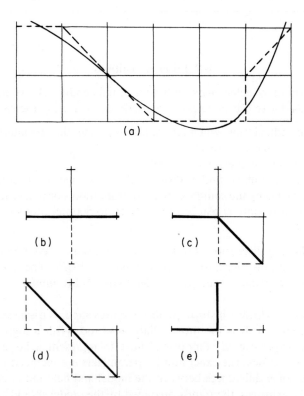

Figure 8.15. Representation of irregular boundaries

G

coefficients are as defined in Eq. 8.5 and 8.6:

$$0.5BH(I, J - 1) + CH(I - 1, J) + 0.5DH(I, J + 1) - (0.5B + C + 0.5D)H(I, J)$$

$$= 0.5F[H(I, J, N + 1) - H(I, J, N)] - 0.5RS(I, J)$$

$$-4QIN(I, J)/[(X(I + 1) - X(I - 1))(Y(J + 1)) - Y(J - 1))] \quad (8.29)$$

When there is an irregular boundary, such as that shown in Fig. 8.15, it is usually permissible to assume that the boundary passes through the nearest nodes. Four possible boundary arrangements are sketched in (b) to (e) of Fig. 8.15. The factors by which the coefficient should be multiplied are listed in Table 8.2.

Table 8.2 Factors for the various shapes of boundaries shown in Fig. 8.15

Example			Coefficients			
	A	B	C	D	F	$RECH(I, J)$
(b)	0.5	1.0	0.5	0.0	0.5	0.5
(c)	1.0	1.0	0.5	0.0	0.625	0.625
(d)	1.0	1.0	0.0	0.0	0.5	0.5
(e)	0.0	0.5	0.5	0.0	0.25	0.25

8.9 Initial Conditions

The importance of achieving satisfactory initial conditions was demonstrated in Section 7.6. There are three possible methods of starting the calculation:

(a) Perform a calculation which attempts to represent the previous history until a dynamic balance is reached
(b) Use field data as the starting condition
(c) Apply typical inflows and outflows with the storage coefficient everywhere very small. Using the standard method of solution, continue the calculation until a steady state is reached (usually 50 to 500 iterations with the SOR method).

In the following discussion results obtained from an unconfined aquifer are used as illustrations (Rushton and Tomlinson, 1975). The aquifer is the Lincolnshire Limestone; a detailed model study of this aquifer is described in Chapter 10.

Method (a), in which a historic period is represented, is discussed in greater detail in Section 7.6; it is the most reliable method. With a digital computer solution it is expensive in computing effort to represent a large number of historical years. Unless the exact aquifer parameters are included in the model there may be some differences between the model and field initial groundwater potentials. Nevertheless, the trends predicted by the model should approximate to the actual behaviour of the aquifer.

With *method (b)* where field data are used as the starting condition, the change in the groundwater potentials determined in the early stages of the calculation may even be of the wrong sign. This is illustrated in Fig. 8.16 where the potentials obtained from method (b) (broken line) diverge from those of method (a) for the first 100 days. These errors arise since the field potentials, which are often obtained from an inadequate observation well network, are not consistent with the mathematical model. When it is recognized that an inconsistency of 1 m in the initial potential in the unconfined region may be equivalent to one year's total infiltration, then it is clear that the differences between the initial potentials

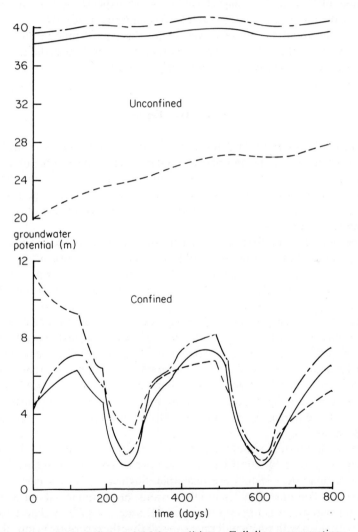

Figure 8.16. Different initial conditions. Full line, representing previous history; broken line, using field data; chain-dotted line, average inflows and outflows

estimated from the field observations and the dynamic balance can be equivalent to many years' recharge. This misplaced recharge has to be redistributed before meaningful results are obtained.

In *method* (c) the steady-state groundwater potentials may differ from those of the dynamic balance of method (a). For example, in the Lincolnshire Limestone the differences are no more than ± 2.5 m. Thus the heads during the first 800 days (the chain-dotted lines of Fig. 8.16) are reasonably close to those of method (a), though results not included here showed that after twelve years there are still discrepancies of up to 1.2 m.

Though method (a) is to be preferred, it is very expensive in computing effort. Probably the most efficient approach is to obtain the steady state of method (c) and then run about five annual cycles with the typical inflows and outflows. Starting with initial groundwater potentials estimated from field data, method (b) can lead to results which may be seriously misleading.

8.10 Mesh Design

The cost of a digital computer solution for groundwater flow problems is critically dependent on the number of nodal points and the number of time steps. Economies cannot usually be made by increasing the size of the time step, either because of convergence and stability criteria or because too large a time step would lead to a loss of information about important variations in the groundwater potentials.

An increase in the mesh spacing is an alternative means of economy. In general it is considered that an aquifer should be modelled by between 500 and 2000 nodes, though this requirement is based primarily on previous practice.

Section 3.3 shows that the approximations inherent in the finite-difference approach can be expressed as a truncation error which consists of the higher-order terms of the Taylor's expansion. However, the magnitude of these higher-order terms can only be calculated for the simplest problems. Nevertheless, the first term of the truncation error does show that the finite-difference errors are proportional to the square of the mesh interval; thus if there are errors due to the finite-difference approximations, they will increase significantly with larger mesh intervals. Further, the truncation error indicates that, if the surface to be represented is different from a smooth parabolic shape, then inaccuracies are likely to occur.

Since no formal rules about the choice of mesh spacing are available, guidance can only be gained by considering a number of specific problems for which alternative analytical results are available. Such a study was carried out by Rushton and Tomlinson (1977) and their results are summarized in Fig. 8.17. An important parameter is the non-dimensional time $t' = tT/L^2S$. These solutions were all obtained with very small time steps so that the errors are primarily due to the space discretization.

The general trends apparent from these four examples are as follows:

(a) The condition of continuity of flow is automatically satisfied by the numerical calculation so that, even if the initial results are unsatisfactory, the model may yet give adequate results at longer times (Examples 1, 3, and 4 of Fig. 8.17).

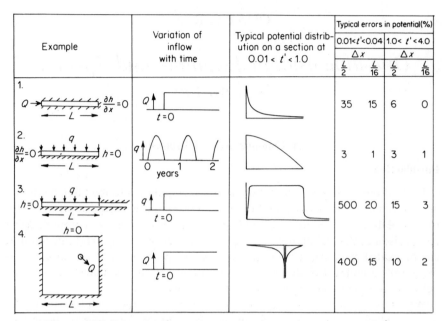

Figure 8.17. Examples illustrating choice of mesh spacing $t' = tT/L^2S$

(b) If the potential distribution shows steep gradients then fine meshes are required. Steep gradients occur with concentrated sources of recharge or discharge (Examples 1, 3, and 4).
(c) When the recharge is spread over the region, there are no steep gradients in the potential distribution and a coarse net is adequate (Example 2).
(d) If the non-dimensional time, since the recharge or the abstraction rate changed, is greater than 2.0, then a coarse net gives adequate results for most problems.

These trends were investigated further by examining two field problems. For one aquifer, an unconfined sandstone aquifer with a relatively uniform distribution of recharge and abstraction, there was little difference in the results when the mesh interval was *increased* by a factor of three which would be equivalent to a nine-fold *reduction* in the *number* of nodal points. The success of the coarse net was due to the gentle changes in the distribution of the potentials.

However, with the Lincolnshire Limestone (Chapter 10), which is partly confined, a reduction in the number of nodal points by a factor of ten produced very poor results. This indicates that the rapid changes in groundwater potential that occur in complex aquifers can only be represented adequately when a large number of nodal points are used.

CHAPTER 9

One-dimensional Example and Special Techniques

9.1 One-dimensional Problem

Introduction

Each of the analog and digital computer techniques described in Chapters 7 and 8 has particular merits, and the choice of technique will depend primarily on the availability of equipment and previous experience. A comprehensive comparison of the techniques is therefore not necessary, yet useful information can be gained by considering a one-dimensional problem as an introduction to the more complex two-dimensional situations. Both resistance–capacitance network and digital computer techniques are discussed.

With the digital computer approach, particular attention will be paid to the way in which a standard program can be adapted for different boundary conditions, parameter values or flows, and to the various finite-difference approximations. The following finite-difference procedures are considered: forward difference; alternating direction explicit; backward difference solved by successive over-relaxation; and backward difference solved by Gaussian elimination.

Often the programs contain more statements than the minimum but this is for the sake of clarity. A detailed description of the Gaussian elimination method for tridiagonal matrices is included.

A standard idealized example, for which an analytical solution exists, is chosen to illustrate the computational techniques. Consider a confined aquifer of length L, unit width, constant transmissivity T, and storage coefficient S (Fig. 9.1). On the boundary $x = 0$, no flow crosses; thus $\partial h/\partial x = 0$. Initially the groundwater potentials throughout the aquifer are zero; then the potential at the right-hand boundary is suddenly raised to H. After a time t, the groundwater potential is given by Carslaw and Jaeger (1959) as

$$\frac{h}{H} = 1 - \frac{4}{\pi} \sum_{n=0}^{\infty} \frac{(-1)^n}{2n+1} \exp(-0.25(2n+1)^2 \pi^2 t') \cos(0.5(2n+1)\pi x/L) \quad (9.1)$$

where $t' = tT/L^2 S$. This series has been evaluated using a digital computer; typically 25 terms are sufficient. The particular parameter values that will be used are $T = 2160 \, \text{m}^2/\text{d}$, $S = 0.015$, $L = 12 \, \text{km}$, $\Delta x = 1.2 \, \text{km}$, $H = 10 \, \text{m}$.

186

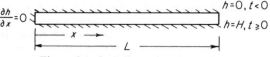

Figure 9.1. One-dimensional problem

Solution on resistance–capacitance network

The basis of the resistance–capacitance network technique was introduced in Chapter 7; this section will refer to the choice of electrical component values and the conversion of the electrical readings into groundwater potentials.

The equations relating electrical and physical parameters for a strip of aquifer of unit width, $\Delta y = 1.0$, from Eqs. 7.4 to 7.8, become

$$V = F(1)h \qquad R_x = F(2)\,\Delta x/T_x$$

$$C = F(3)S\,\Delta x \qquad t_e = F(2)\,F(3)t$$

The factor $F(1)$ is selected to be 0.01 and with resistors of $200\ \text{k}\Omega$ and capacitors of $10\ \mu\text{F}$, then $F(2) = 3.6 \times 10^5$ and $F(3) = 5.555 \times 10^{-7}$. Therefore the relationship between electrical and physical time is $t_e = 0.2t$, which means that 1 second of electrical time corresponds to 5 days.

There are eleven nodal points in the network; the components are indicated in Fig. 9.2. At the boundaries, capacitors equal to half the standard value are required; in this case $4.7\ \mu\text{F}$ capacitors were considered to be adequate. Initially all the electrical potentials are held at zero with the switch in the up position. Then, as the switch is moved down, the voltage at the right-hand boundary instantaneously increases to 0.1 V. Readings obtained by manually triggering a digital voltmeter are recorded in Table 9.1. More accurate techniques could be used but this example demonstrates what can be done with the simplest equipment.

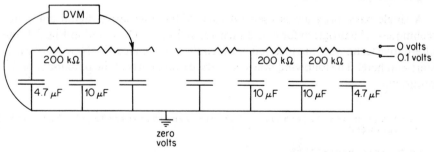

Figure 9.2. Resistance–capacitance network for one-dimensional problem

The electrical times and voltages are converted to the corresponding physical parameters and are recorded in Table 9.1. Analytical values of the groundwater potentials are also tabulated. Nowhere are the differences between analytical and experimental values greater than 2 per cent of the applied voltage. Since the

Table 9.1 Groundwater potentials obtained from electrical analog

Time		Groundwater potentials			
		$x = 0$		$x = 0.5L$	
electrical	physical	measured	analytical	measured	analytical
2 secs	10 days	0.00 m	0.00 m	0.03 m	0.01 m
4	20	0.00	0.00	0.18	0.12
6	30	0.00	0.00	0.55	0.41
8	40	0.01	0.01	0.91	0.77
12	60	0.09	0.08	1.51	1.49
16	80	0.31	0.25	2.21	2.16
20	100	0.52	0.51	2.81	2.64
30	150	1.43	1.36	3.82	3.68
40	200	2.43	2.28	4.62	4.47
60	300	4.07	3.93	5.85	5.70
80	400	5.42	5.26	6.78	6.64
120	600	7.24	7.11	8.04	7.95
160	800	8.35	8.23	8.82	8.75
200	1000	9.00	8.92	9.28	9.23

tolerance of the capacitors is ± 10 per cent and for the resistors it is ± 5 per cent, the accuracy noted above is satisfactory.

Changes to the network to allow for different boundary or recharge conditions can be made rapidly. For example, the condition of a fixed inflow can be achieved directly by supplying a current at the right-hand boundary rather than the fixed voltage as at present.

Digital computer technique

A single basic program is used for each of the alternative digital computer techniques. The program for one-dimensional flow is to be found in Fig. 9.3; the alternative techniques are inserted between the comment cards, 'New set of cards start/end here'. The following notes describe certain parts of the remainder of the program.

```
      DIMENSION X(50), A(50), C(50),H(50),BH(50),FIX(50),TV(50),ST(50)
    1   ,RECH(50)
C
C   INPUT BASIC PARAMETERS
      READ(5,100)T,S,DELX,DELT,RCH
      READ(5,110)INP,ITIME
100   FORMAT(5F10.4)
110   FORMAT(2I3)
      WRITE(6,120)T,S,DELX,DELT,RCH,INP,ITIME
120   FORMAT(1X,6HTRANS=,F8.1,6H STOR=,F8.5,6H DELX=,F8.2,6H DELT=,F8.4
     1,11H RECHARGE=  ,F8.4,11H INTERVALS=  ,I3,12H TIME STEPS= ,I5)
      WRITE(6,130)
```

```
130     FORMAT(1H0)
        NIN=INP+1
        NBOUND=INP+2
        NFICT=INP+3
        A(1) = 0.0
        C(1) = 0.0
        A(NFICT) = 0.0
        C(NFICT) = 0.0
C
        DO 20 N=1,NFICT
        TX(N)=T
        ST(N)=S
        RECH(N)=RCH
        FIX(N)=-200000.0
        H(N)=0.0
        AN=FLOAT(N)
C SET UP MESH POSITIONS
20      X(N)=(AN-2.0)*DELX
C
C   COEFFICIENTS OF FINITE DIFFERENCE EQUATIONS
        DO 30 N=2,NBOUND
        A(N)=2.0*TX(N)/((X(N+1)-X(N-1))*(X(N+1)-X(N)))
30      C(N)=2.0*TX(N-1)/((X(N+1)-X(N-1))*(X(N)-X(N-1)))
C
C   FIXED BOUNDARIES
        READ(5,140)N,FIX(N)
        H(N)=FIX(N)
140     FORMAT(I3,F10.4)
C
C   OTHER BOUNDARIES
        READ(5,150)N,AA,CC,SS,RR
150     FORMAT(I3,4F10.4)
        A(N)=AA*A(N)
        C(N)=CC*C(N)
        ST(N)=SS*ST(N)
        RECH(N)=RR*RECH(N)
        TIME=0.0
C
C NEW SET OF CARDS STARTS HERE -------------------------------------------
C FORWARD DIFFERENCE TECHNIQUE
        WRITE(6,200)
200     FORMAT(1X,6H TIME  ,28H   GROUNDWATER POTENTIALS   )
C   TIME LOOP
        DO 50 ISTEP=1,ITIME
        DO 40 N=1,NFICT
40      BH(N)=H(N)
        DO 60 N=2,NBOUND
        IF(FIX(N).GT.-100000.0)GO TO 70
        F=ST(N)/DELT
        H(N)=(A(N)*BH(N+1)+C(N)*BH(N-1)-(A(N)+C(N)-F)*BH(N)+RECH(N))/F
        GO TO 60
70      H(N)=FIX(N)
60      CONTINUE
C NEW SET OF CARDS ENDS HERE  -------------------------------------------
C
        TIME=TIME+DELT
        WRITE(6,160)TIME,(H(N),N=1,NFICT)
160     FORMAT(1X,15F8.4)
50      CONTINUE
        WRITE(6,170)
170     FORMAT(1H1)
        STOP
        END
```

Figure 9.3. Program listing including forward-difference routine

(a) Mesh positions, which are stored in array $X(N)$, where N is the number of the node, are calculated in the statements following 'Set up mesh positions'. The present program is arranged to give a constant mesh spacing, but a variable mesh spacing can be read directly into the array $X(N)$. The mesh numbering system is illustrated in Fig. 9.4(a).

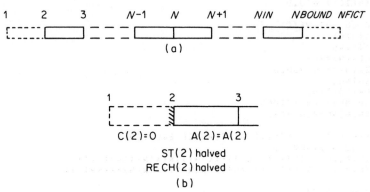

Figure 9.4. Mesh numbering system

(b) Values of transmissivity, storage coefficient and recharge are contained in arrays $TX(N)$, $ST(N)$, and $RECH(N)$. The transmissivity $TX(N)$ refers to the average value between nodes N and $N + 1$; $ST(N)$ and $RECH(N)$ refer to node N.

(c) Coefficients of the finite-difference equations for this one-dimensional problem take the same form as Eq. 8.5 though only coefficients $A(N)$ and $C(N)$ exist. Thus, for example,

$$A(N) = 2TX(N)/[(X(N + 1) - X(N - 1))(X(N + 1) - X(N))]$$

(d) Fixed potentials, which may vary with time, are read into an array FIX(N). The value of this parameter when a fixed potential does not apply is a large negative number.

(e) Boundary conditions of zero flow can be set through the statement following the comment 'Other boundaries'. For example, if the condition is that to the left of node N there is an impermeable boundary (Fig. 9.4(b)), then the coefficient $C(N)$ is set to zero and $ST(N)$ and $RECH(N)$ are halved. This is achieved by inputing coefficients AA, CC, SS, and RR which are multiplying factors for the original coefficients of the finite-difference equations.

(f) Initial groundwater potentials can be set to any value in this program. In this example the groundwater potentials are initially set to zero.

The input data corresponding to the test problem is shown below the sketch of the problem in Fig. 9.5(a). Two other problems are illustrated and the relevant data recorded. Note that in problem (b) the fixed and impermeable boundaries have been interchanged. A simplified flow chart of the common portion of this program is presented in the upper part of Fig. 9.6.

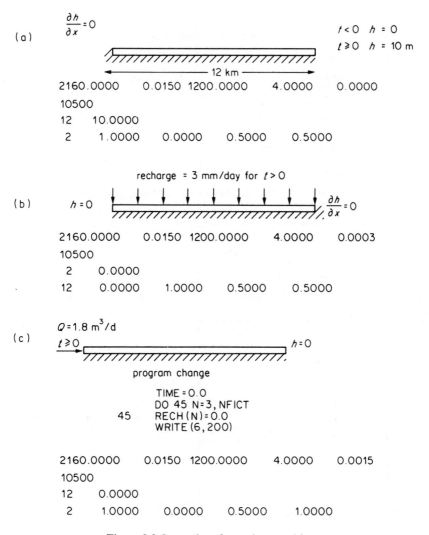

Figure 9.5. Input data for various problems

Forward-difference technique

The program for the forward-difference formulation is contained in the section 'Forward difference technique' of Fig. 9.3. The equations can be derived from the two-dimensional formulation in Section 8.2. By setting $B(N)$ and $D(N)$ equal to zero in Eq. 8.7 the main expression is obtained. The important steps of the forward-difference solution are shown in the lower part of the flow chart of Fig. 9.6.

192

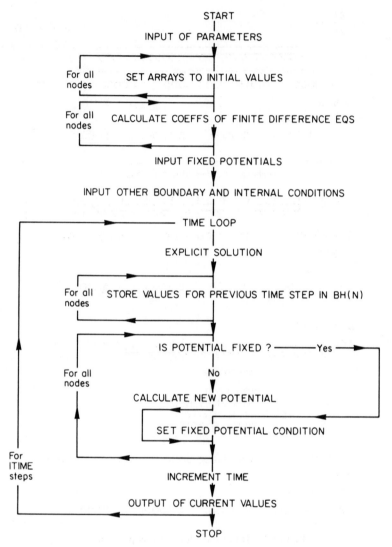

Figure 9.6. Flow chart for one-dimensional problem

For this *one-dimensional* problem, the stability criterion (Smith, 1965) requires that

$$\Delta t \leqslant 0.5\,\Delta x^2 S/T \tag{9.2}$$

When the appropriate values are substituted, the permissible time step can be calculated as 5 days. A value of 4 days is chosen since this gives smoother results. The calculation is continued for ITIME time steps.

Though the main purpose of this section is to describe the fundamental aspects of the techniques, brief comparisons of the results of different methods are made

Table 9.2 Comparison of groundwater potentials for standard example.
Time increment = 4.0 days

Time (days)	20	60	100	400
x = 0				
Analytical	0.000	0.078	0.507	5.255
Explicit	0.000	0.063	0.502	5.278
ADE	0.001	0.100	0.534	5.251
SOR	0.002	0.124	0.558	5.233
GE	0.002	0.124	0.558	5.233
x = 0.5L				
Analytical	0.124	1.489	2.643	6.644
Explicit	0.102	1.537	2.686	6.661
ADE	0.178	1.500	2.641	6.541
SOR	0.216	1.481	2.613	6.628
GE	0.216	1.481	2.613	6.628

GE signifies Gaussian elimination.

in Table 9.2. In the early stages of the calculation, the forward-difference technique underestimates due to the delay in the spreading of the effect of the sudden change; at longer times the agreement with the analytical approaches 1 per cent.

```
C   A.D.E. SOLUTION
      WRITE(6,200)
200   FORMAT(1X,6H TIME  ,28H   GROUNDWATER POTENTIALS   )
C   TIME LOOP
      DO 50 ISTEP=1,ITIME
      DO 60 N=2,NBOUND
      IF(FIX(N).GT.-100000.0)GO TO 65
      F=ST(N)/DELT
      H(N)=(A(N)*H(N+1)+C(N)*H(N-1)-(A(N)-2.0*F)*H(N) +RECH(N))/
    1  (C(N)+2.0*F)
      GO TO 60
65    H(N)=FIX(N)
60    CONTINUE
      DO 70 NN=2,NBOUND
      N=NBOUND-NN+2
      IF(FIX(N).GT.-100000.0)GO TO 75
      F=ST(N)/DELT
      H(N)=(A(N)*H(N+1)+C(N)*H(N-1)-(C(N)-2.0*F)*H(N) +RECH(N))/
    1  (A(N)+2.0*F)
      GO TO 70
75    H(N)=FIX(N)
70    CONTINUE
C
C
      END
```

Figure 9.7. Program Listing for ADE routine

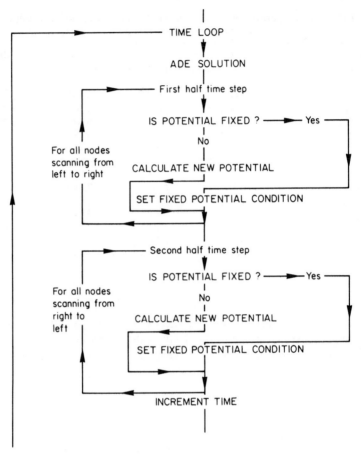

Figure 9.8. Flow chart for ADE routine

Alternating-direction explicit solution

Though the ADE method is primarily intended for two-dimensional problems, it can be adapted to one-dimensional situations. With the ADE method, one fewer array is required than for the forward-difference method since all calculations use the most up-to-data value. The appropriate equations are derived by setting $B(N)$ and $D(N)$ equal to zero in Eqs. 8.26(a) and (b). The routine, which is inserted in the section 'New set of cards' of Fig. 9.3, is listed in Fig. 9.7. No other changes are required to the standard explicit program. A simplified flow chart of this part of the program can be found in Fig. 9.8.

Results are quoted for a time step of 4.0 days in Table 9.2. Larger time steps can be used; acceptable results are obtained with a time step of 20 days, which is five times the critical value for the standard explicit solution. However, if an even larger time step is used, the results still look adequate and exhibit no oscillations, but they are significantly below the analytical values. This confirms that care must be taken in selecting the time step with the ADE method.

Successive over-relaxation solution

The SOR technique was introduced in Section 8.3. Equation 8.12 can be reduced to one-dimensional form by setting arrays $B(\ )$ and $D(\ )$ equal to zero. A further modification to the basic program is the inclusion of an extra dimension statement as shown in Fig. 9.9. A value for the over-relaxation factor of 1.5 is selected and the permissible error is set as 0.01 per cent of the typical recharge. SOR is an iterative method in which the iterations for a particular time step are stopped only when the convergence criterion is satisfied at all nodes. This is illustrated in the flow chart of Fig. 9.10.

```
       DIMENSION AMULT(50),TERMA(50)
C
C
C   SUCCESIVE OVER-RELAXATION
       WRITE(6,200)
200    FORMAT(1X,6H TIME  ,28H   GROUNDWATER POTENTIALS     )
       OFAC=1.5
       ERROR=0.0000001
C   TIME LOOP
       DO 50 ISTEP=1,ITIME
       DO 40 N=1,NFICT
       AMULT(N)=ST(N)/DELT+A(N)+C(N)
40     TERMA(N)=ST(N)*H(N)/DELT
       DO 55 I=1,100
       IND=0
       DO 60 N=2,NBOUND
       BH(N)=H(N)
       IF(FIX(N).GT.-100000.0)GO TO 70
       AB=A(N)*H(N+1)+C(N)*H(N-1)+TERMA(N)+RECH(N)
       IF(ABS(AB-AMULT(N)*BH(N)).LT.ERROR) GO TO 65
       IND=100
65     H(N)=(1.0-OFAC)*BH(N)  +OFAC*AB/AMULT(N)
       GO TO 60
70     H(N)=FIX(N)
60     CONTINUE
       IF(IND.EQ. 0)GO TO 75
55     CONTINUE
       WRITE(6,180)
180    FORMAT(45H CONVERGENCE NOT ACHIE55D IN 100 ITERATIONS  )
75     CONTINUE
C
C
C
       END
```

Figure 9.9. Program listing for SOR routine

There are two stages in the SOR routine; in the first stage values are calculated for two arrays, AMULT(N) and TERMA(N) which is a function of the head in the previous time step. In the second stage, which is the iterative portion, new approximations to the unknown potentials are obtained and the error in the finite-difference form of the differential equation due to the current estimates of

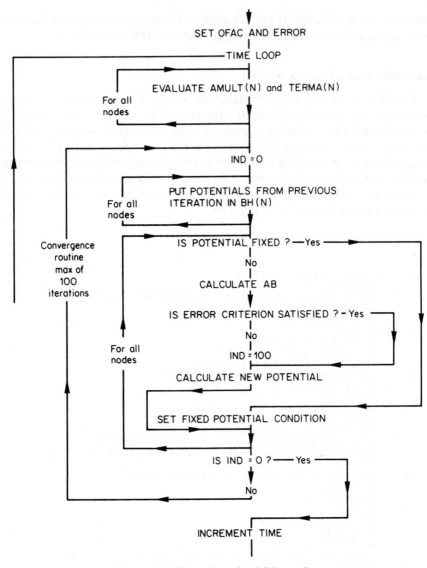

Figure 9.10. Flow chart for SOR routine

the groundwater potential is evaluated at each node. The variable IND is used as an indicator of whether the error term is less than the permissible value at each node. If IND remains at a zero value throughout an iteration, then convergence has been achieved.

With the stringent error criterion chosen for this example, about twenty iterations per time step are required in the early stages of the calculation, decreasing to about fifteen iterations in the later stages. Results for the standard problem are recorded in Table 9.2. The values are identical to those obtained by the direct solution of the same equations using Gaussian elimination as described below. This indicates that the convergence criterion is adequate. There is no restriction on the size of the time step with this procedure.

Backward-difference approximation using Gaussian elimination

It is not possible to use the ADI technique with one-dimensional problems since the technique requires a two-step procedure working first in the x- and then in the y-direction. However, useful information about the technique can be gained by considering the Gaussian elimination solution of the one-dimensional problem using the backward-difference formulation. Taking Eq. 8.10, setting the coefficients in the y-direction equal to zero, with $H(N)$ referring to the potentials at node N at time $t + \Delta t$ and $BH(N)$ referring to the potentials at time t, the resultant equation can be written as

$$-CH(N - 1) + (C + A + S/\Delta t)H(N) - AH(N + 1) = (S/\Delta t)BH(N) + RS(N)$$
$$(9.3)$$

Note that C and A refer to the coefficients of the finite-difference equations (Eq. 8.5). For the problem under consideration (Fig. 9.4(a)), the potential at the fictitious node 1 is taken as zero; hence the equation for node 2 with $N = 2$ becomes

$$(C + A + S/\Delta t)H(N) - AH(N + 1) = (S/\Delta t)BH(N) + RS(N) \qquad (9.4)$$

Then at node NMAX, which is one node to the left of the first node at which the potential is fixed, FIX(NMAX + 1), the equation is

$$-CH(\text{NMAX} - 1) + (C + A + S/\Delta t)H(\text{NMAX})$$
$$= (S/\Delta t)BH(\text{NMAX}) + RS(\text{NMAX}) + A\,\text{FIX}(\text{NMAX} + 1) \quad (9.5)$$

Equations (9.3) to (9.5) can be written in matrix form as shown in Fig. 9.11(a), where GA, GB, GC, and GD are the coefficients of the matrix. Clearly,

$$GA(N) = C$$

$$GB(N) = C + A + S/\Delta t$$

$$GC(N) = A$$

$$GD(N) = (S/\Delta t)BH(N) + RS(N)$$

$$GA(2) = 0$$

$$GC(\text{NMAX}) = 0$$

$$GD(\text{NMAX}) = (S/\Delta t)BH(\text{NMAX}) + RS(\text{NMAX}) + A\,\text{FIX}(\text{NMAX} + 1)$$

198

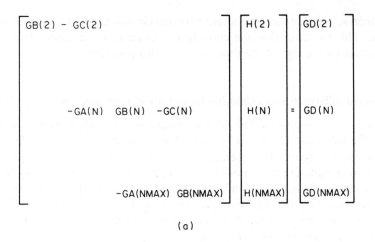

(a)

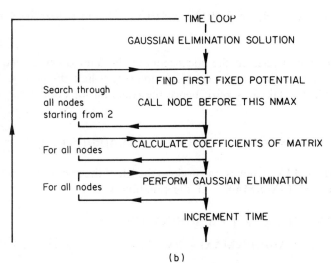

(b)

Figure 9.11. (a) Matrix for Gaussian elimination. (b) Flow chart for
Gaussian elimination

In the elimination routine, two additional variables; $U(N)$ and $V(N)$ are
introduced. The procedure is then as follows:

(a) $U(2) = GB(2)$
      $V(2) = GD(2)$

(b) $U(N) = GB(N) - (GA(N)GC(N - 1))/U(N - 1)$ For N increasing from
      $V(N) = GD(N) + (GA(N)V(N - 1))/U(N - 1)$ 3 to NMAX

(c) $H(\text{NMAX}) = V(\text{NMAX})/U(\text{NMAX})$

(d) $H(N) = (V(N) + GC(N)H(N + 1))/U(N)$ For N decreasing
                                                    from NMAX $- 1$ to 2

These equations are clearly in a suitable form for including in a computer program.

One important point to note is that if there is an intermediate node at which the potential is fixed, then the Gaussian elimination routine can only be used for the region from the left-hand boundary to this fixed node. A second elimination routine is required to include the nodes from this intermediate fixed point to the right-hand boundary. If the potential at the fixed intermediate point does not change with time, then it can be represented by introducing a very large storage coefficient at that point with a single elimination routine from the left-hand to the right-hand boundary (Prickett and Lonnquist, 1971).

A flow chart of the program is contained in Fig. 9.11(b) and the program listing in Fig. 9.12. The first stage of the solution routine is to find the position of the first fixed potential working from the left-hand boundary. For a fixed potential positioned in the middle of the field, it would be necessary to use the elimination routine again for the region to the right of this fixed potential.

The calculation of the coefficients and the elimination routine could be combined and the program made more efficient. However, for clarity the steps are

```
        DIMENSION GA(50),GB(50),GC(50),GD(50),U(50),V(50)
C
C   IMPLICIT GAUSSIAN  ELIMINATION
        WRITE(6,200)
200     FORMAT(1X,6H TIME   ,28H  GROUNDWATER POTENTIALS    )
C   TIME LOOP
        DO 50 ISTEP=1,ITIME
C
C   FIND FIRST FIXED HEAD
        DO 35 N=2,NFICT
        IF(FIX(N).LT.-100000.0) GO TO 35
        NMAX=N-1
        GO TO 37
35      CONTINUE
C
C   ELIMINATION ROUTINE
37      DO 40 N=2,NMAX
        GA(N)= C(N)
        GC(N)= A(N)
        GB(N)=A(N)+C(N)+ST(N)/DELT
40      GD(N)=ST(N)*H(N)/DELT+RECH(N)
        GD(NMAX)=GD(NMAX)+A(NMAX)*H(NMAX+1)
        U(2)=GB(2)
        V(2)=GD(2)
        DO 60 N=3,NMAX
        U(N)=GB(N)-(GA(N)*GC(N-1))/U(N-1)
60      V(N)=GD(N)+(GA(N)*V(N-1))/U(N-1)
        H(NMAX)=V(NMAX)/U(NMAX)
        DO 70 NN=3,NMAX
        N=NMAX-NN+2
70      H(N)=(V(N)+GC(N)*H(N+1))/U(N)
C
C
C
        END
```

Figure 9.12. Program listing for Gaussian elimination routine

kept separate and therefore the solution routine as presented requires six more array variables than the explicit procedure.

Results, identical to those of the SOR routine, are obtained as shown in Table 9.2. For one-dimensional problems Gaussian elimination is undoubtedly the most efficient technique, but it is not suitable for two-dimensional problems unless it is used in conjunction with the ADI procedure.

9.2 Special Techniques

Introduction

In Chapters 6 to 8 the main emphasis has been on the representation of the aquifer with transmissivity and storage coefficient which vary regionally; the simulation of boundary conditions has also been considered.

However, there are many other naturally occurring phenomena that may have a significant effect on the flow mechanism. These include the change in conditions between the confined and unconfined state, variations in transmissivity, leaky aquifers, the simulation of wells, and the representation of partially penetrating rivers. If any of these features is not included in the mathematical model it may invalidate the whole analysis. In each case the representation on analog or digital computers is described.

For each of the features to be included, the physical aspects are considered first and then the mechanism is expressed in mathematical form. Next a suitable device or technique has to be adapted to represent the mechanism. For example, a transistor switch may be used in the analog model whereas the equivalent in the digital model may be an IF statement with a modified time step. At times, the methods of representing these features will appear to be somewhat involved, but it is essential to represent the correct mechanism otherwise the mass-balance condition may be violated.

Change in storage coefficient

In many aquifers which are initially confined, the onset of heavy abstraction can cause part of the aquifer to become unconfined in the vicinity of the abstraction well. As a result, the storage coefficient changes from a low value of about 10^{-4} to a much higher value of about 10^{-1}. This can have significant effects on the flow patterns within the aquifer since large volumes of water become available in different regions.

Before discussing how this change is included in mathematical models, it is important to stress that the model must represent the correct area that becomes unconfined. For example, if the region around a well to a radius of 40 m becomes unconfined, then the area for which the unconfined coefficient applies is $0.005 \, km^2$. Clearly it is highly inaccurate to represent the whole area associated with that node (say $1.0 \, km^2$) as being unconfined. Therefore, in deciding whether the storage coefficient at a well node should be changed, the value of the groundwater potential at the well should not be used; instead the potential at the nodal point which represents the area surrounding the abstraction well should be the criterion (see discussion on 'simulation of wells' later in this section). This effect of a change in storage coefficient can be included in both analog and digital models.

The change between confined and unconfined can be represented on a *resistance–capacitance network* using a field-effect transistor switch (Fig. 9.13). The important property of this transistor is that when the voltage at the base terminal is above the switching voltage, the resistance of the transistor is effectively infinite, but when the voltage falls below the switching voltage then the resistance drops to the small value of $100\,\Omega$. This change of resistance is not instantaneous but is spread over a voltage difference of about 0.1 V. Let the voltage corresponding to the height of the top of the aquifer above datum be V_T. The reference voltage on the field effect transistor (FET) is set so that the capacitor representing the unconfined storage coefficient is isolated from the network when $V_0 > V_T$, but as V_0 falls below V_T, which is equivalent to the aquifer becoming unconfined, the large capacitance C_u is incorporated in the circuit. Note that the capacitance C_u is connected to V_T volts, not to zero as with C_c. This capacitor, C_c, which represents the confined storage, always remains in circuit. Typical practical values of the capacitances are $C_c = 0.01\,\mu\text{F}$, $C_u = 10.0\,\mu\text{F}$. FETs of type TIX42 have proved to be satisfactory; further details are given in Rushton and Wedderburn (1971).

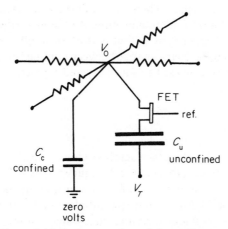

Figure 9.13. Representation of change between
confined and unconfined state

In a *digital solution*, the change between the confined and unconfined state can be achieved using an IF statement. Care must be taken to ensure that small time steps are used as the condition at a node change between the confined and unconfined states, otherwise it is possible to overshoot the top of the aquifer and violate the condition of continuity of flow.

Variation in transmissivity

The transmissivity of certain aquifers changes significantly as the saturated depth varies. Thus, when considering unconfined aquifers, the mathematical model may need to incorporate a routine which allows for a variation in transmissivity with the groundwater potential. A practical example of a chalk aquifer with a severe variation of transmissivity with depth is described in Section 12.3. The values of the transmissivity were deduced from a series of extensive pumping tests. Usually such detailed information is not available, yet regions of higher permeability can be identified by examining cores from boreholes or from velocity measurements in pumped wells. From evidence available at present, the permeability often varies by a factor of ten or more over the full depth of the aquifer.

When modelling the variation of transmissivity with depth, three different approaches are available. For certain aquifers the permeability is constant over the full depth of the aquifer and therefore the transmissivity can be calculated as the saturated depth multiplied by the permeability. In another approach, information is stored in the computer program of the variation in transmissivity with groundwater potential. If the SOR method is used, the value of transmissivity can be modified at each node as the iterations proceed. Since the relationship between the transmissivity and groundwater potential can be complicated, this procedure may demand large computer storage.

The third approach is valid when the flow in the aquifer takes place predominantly in two planes. Then the aquifer can be treated as having two systems with some form of interconnection. As the groundwater potential falls below the base of the upper system, that part of the aquifer is removed from the model in a manner similar to the change between confined and unconfined states. This technique can be extended to multi-layered aquifers.

Leaky aquifers

When an aquifer is overlain by a semi-pervious stratum, there is a possibility that water will leak through to recharge the aquifer. This leakage will occur only if the groundwater potential in the aquifer is lower than the potential in the semi-pervious medium. Should the aquifer potential be higher than the groundwater potential in the semi-pervious medium, then water will seep out of the aquifer.

The appropriate differential equation is (Walton, 1970)

$$\frac{\partial}{\partial x}\left(T_x\frac{\partial h}{\partial x}\right) + \frac{\partial}{\partial y}\left(T_y\frac{\partial h}{\partial y}\right) = S\frac{\partial h}{\partial t} - q - \frac{k'}{m}(H - h) \tag{9.6}$$

where k' is the permeability and m is the saturated thickness of the semi-pervious medium; the head in the semi-pervious medium is H. This equation assumes that the storage coefficient of the confining stratum can be neglected; the inclusion of the storage coefficient of a leaky aquifer is discussed by Trescott et al. (1976).

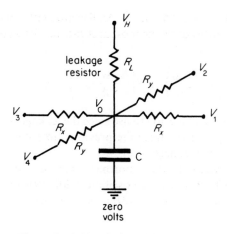

Figure 9.14. Simulation of leaky aquifer

This leakage can be included directly in a *resistance–capacitance analog solution* by means of *additional resistors*. Taking the case where T_x and T_y are constants throughout the aquifer with uniform mesh spacing Δx and Δy, the finite-difference form of Eq. 9.6 becomes

$$\frac{T_x \Delta y}{\Delta x}(h_1 - 2h_0 + h_3) + \frac{T_y \Delta x}{\Delta y}(h_2 - 2h_0 + h_4)$$

$$= \Delta x\, \Delta y\left[S\frac{\partial h}{\partial t} - q - \frac{k'}{m}(H - h_0)\right] \qquad (9.7)$$

For the electrical network of Fig. 9.14,

$$\frac{V_1 - 2V_0 + V_3}{R_x} + \frac{V_2 - 2V_0 + V_4}{R_y} = C_0\frac{\partial V_0}{\partial t_e} - I_0 - \frac{V_H - V_0}{R_L} \qquad (9.8)$$

From Eqs. 7.4 and 7.5

$$V_1 = F(1)h_1, \qquad R_x = \frac{F(2)\,\Delta x}{T_x \Delta y}$$

including these relationships in the first and last terms in Eq. 9.8,

$$\frac{F(1)(h_1 - 2h_0 + h_3)\,\Delta y T_x}{F(2)\,\Delta x} + \cdots = \cdots - \frac{F(1)(H - h_0)}{R_L} \qquad (9.10)$$

Therefore, when representing the effect of a leaky aquifer on a resistance–capacitance network, an extra resistor is introduced at node 0, with a voltage $V_H = F(1)H$ applied to the other end of the resistor, $R_L = F(2)m/(k'\,\Delta x\,\Delta y)$.

When representing a leaky aquifer in a *digital computer model*, a number of alternative finite-difference approximations are available since the leakage term can be defined at times $t, t + \frac{1}{2}\Delta t$, or $t + \Delta t$. Rushton (1974b) showed that for the backward-difference procedure, the most accurate results were obtained with the formulation

$$\frac{\partial}{\partial x}\left(T_x \frac{\partial h_{t+\Delta t}}{\partial x}\right) + \frac{\partial}{\partial y}\left(T_y \frac{\partial h_{t+\Delta t}}{\partial y}\right) = S\left(\frac{h_{t+\Delta t} - h_t}{\Delta t}\right) - q_{t+\frac{1}{2}\Delta t} - \frac{k'}{m}(H - h_t) \quad (9.11)$$

This is the simplest approximation to use since the leakage is then only a function of the groundwater potentials from the previous time step, and leakage can be included directly in the recharge term.

With the ADI method, the approximation for the leakage term of $(k'/m)(H - h_t)$ leads to results which oscillate wildly (Rushton, 1974a). The only acceptable approximation is $(k'/2m)(2H - h_t - h_{t+\Delta t})$.

Simulation of wells

In early groundwater flow solutions using finite-difference techniques, a well was modelled at the nearest mesh intersection point. Since a mesh intersection point is 'small', the mesh intersection point was considered to be an adequate representation of a well. Though it is still normal practice to represent wells as being positioned at the nearest nodal point, Prickett (1967) and Rushton and Herbert (1966) showed that corrections need to be made to model the true radius of a well.

First an analysis will be carried out to determine the effective radius r_i when the standard finite-difference equation is used without correction. Consider a square mesh of sides Δx and constant transmissivity T, with a discharge Q from node 0 (Fig. 9.15(a)). The standard finite-difference equation is

$$\frac{(h_1 + h_2 + h_3 + h_4 - 4h_0)}{\Delta x^2} = \frac{Q}{T\Delta x^2} \quad (9.12)$$

Now consider the flow pattern indicated in Fig. 9.15(b). Radial flow takes place from an outer circle, radius Δx, to an inner circle, radius r_i. The equation for radial flow is (see Eq. 3.34)

$$h - h_0 = \frac{Q}{2\pi T} \log_e\left(\frac{\Delta x}{r_i}\right)$$

Taking h as the average of h_1, h_2, h_3, and h_4, then

$$h_1 + h_2 + h_3 + h_4 - 4h_0 = \frac{2Q}{\pi T} \log_e\left(\frac{\Delta x}{r_i}\right) \quad (9.13)$$

Hence from Eqs. 9.12 and 9.13,

$$\log_e(\Delta x/r_i) = \pi/2$$

or

$$r_i = 0.208\Delta x \quad (9.14)$$

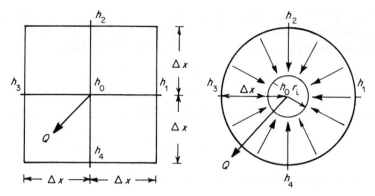

Figure 9.15. Correction for well radius

Therefore if no correction is made the effective inner radius is $0.208\Delta x$. For a mesh with $\Delta x = 1$ km the effective radius is 208 m; this is the size of a lake rather than a small-diameter well.

There are two alternative means of correcting for the localized drawdowns at a well. The first requires that the effective well radius is known and therefore the technique is only applicable for a single well. In the second approach, the correction is based on field information and can be applied to a group of wells.

For the first method *the effective well radius is known* to be r_w. Then the groundwater potential at the well, h_w, is given by

$$h_0 - h_w = \frac{Q}{2\pi T}\log_e\left(\frac{0.208\Delta x}{r_w}\right)$$

(9.15)

where h_0, as explained above, is the groundwater potential calculated at the nodal point using the standard finite-difference approximation. This correction can be inserted in the *digital computer program* as an extra statement.

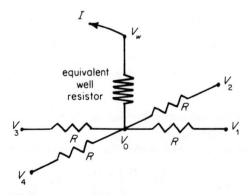

Figure 9.16. Additional resistance to represent
actual well radius

With a *resistance–capacitance analog*, the correction can be represented simply by an additional resistance. The voltages V_0 and V_w correspond to the groundwater potentials h_0 and h_w; thus for the circuit of Fig. 9.16,

$$V_0 - V_w = IR_w = \frac{F(1)}{F(2)}QR_w = \frac{F(1)QR_w}{TR} \tag{9.16}$$

This uses the relationships $I = F(1)Q/F(2)$ and $R = F(2)/T$. From Eq. 9.15

$$V_0 - V_w = F(1)(h_0 - h_w) = \frac{F(1)Q}{2\pi T}\log_e\left(\frac{0.208\Delta x}{r_w}\right) \tag{9.17}$$

Equating the right-hand sides of Eq. 9.16 and 9.17,

$$R_w = \frac{R}{2\pi}\log_e\left(\frac{0.208\Delta x}{r_w}\right) \tag{9.18}$$

With this additional resistance, the voltage V_0 represents the groundwater potential at a distance $0.208\Delta x$ from the well and the actual potential in the well is represented by V_w.

For a group of wells, or in the situation when the effective well radius is not known, *the correction must be based on field information*. The important relationship is the dependence of the abstraction rate on the additional drawdown, s_a. This additional drawdown is defined as the difference between the groundwater potential in the well and the potential at a distance of approximately $0.2\Delta x$ from the well. Typical values of additional drawdowns due to various abstraction rates are given in Table 9.3. For each increase in abstraction rate of $1000\ \mathrm{m^3/d}$ there is an additional drawdown of approximately 0.8 m.

Table 9.3 Dependence of additional drawdown on abstraction rate

Abstraction rate m/d	Additional drawdown m
2800	2.2
4600	3.7
5400	4.3

With the *digital model* this correction can be inserted as an alternative to Eq. 9.15. Thus $h_0 - h_w = 0.0008Q$.

On the resistance–capacitance model, the well resistance is given by

$$R_w = \frac{V_0 - V_w}{I} = \frac{F(1)s_a F(2)}{F(1)Q} = F(2)\frac{s_a}{Q} \tag{9.19}$$

Hence, for the numerical values given above $R_w = 0.8 \times 10^{-3}F(2)$.

Partially penetrating rivers

A river must not be represented as a known groundwater potential applied to the appropriate network nodes unless it cuts right through to the base of the aquifer and has a flow rate far greater than the quantity flowing through the aquifer. In practice, many rivers only partially penetrate the aquifer, yet the transfer of water to or from the aquifer can have a significant effect on the flow in the river. Typical relationships for the discharge to or from the aquifer and the difference between river level and groundwater potential were discussed in Section 6.2.

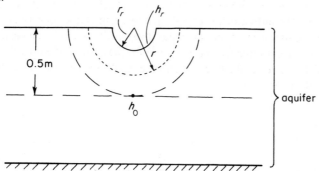

Figure 9.17. Representation of partially penetrating river

Consider the case of a partially penetrating river which is small compared with the thickness of the aquifer (Fig. 9.17). The overall flow within the aquifer is represented by the finite-difference mesh and therefore the nodal points effectively represent conditions at half the aquifer depth. Following Herbert (1970) the flow from the river perimeter to the aquifer across a semicircular arc of radius r (Fig. 9.17) is

$$Q_r = -\pi r l k \frac{dh}{dr}$$

where l is the length of the river represented by the nodal point. Integrating and noting that the groundwater potential at the river perimeter is h_r and at the nodal point in the aquifer it is h_0,

$$Q_r = \pi l k (h_r - h_0)/\log_e (0.5m/r_r) \tag{9.20}$$

or, expressing this as an inflow per unit area,

$$q \Delta x \, \Delta y = \pi l k (h_r - h_0)/\log_e (0.5m/r_r)$$

Incorporating this inflow in the general finite difference equation, Eq. 9.7, with constant transmissivities and a regular mesh interval,

$$\frac{T_x \Delta y}{\Delta x}(h_1 - 2h_0 + h_3) + \frac{T_y \Delta x}{\Delta y}(h_2 - 2h_0 + h_4)$$

$$= \pi l k (h_0 - h_r)/\log_e (0.5m/r_r) + S \Delta x \, \Delta y \frac{\partial h}{\partial t} \tag{9.21}$$

Therefore the effect of a partially penetrating river can be modelled as an applied potential with the flow from the river acting in a similar manner to leakage through an overlying stratum.

If an electrical analog is used, an additional resistance is incorporated in the circuit as shown in Fig. 9.18. The magnitude of the river resistance is

$$R_r = F(2) \log_e (0.5m/r_r)/\pi l k$$

This description assumes that the river is to be represented as a known head. In practice it is often advisable to enforce the condition as a flow, in which the magnitude and sign of the flow depends on the relative magnitudes of the river head and the groundwater potential on the perimeter of the river. If the flow entering the aquifer is Q_r, then the groundwater potential on the perimeter of the river, h_r, can be calculated from Eq. 9.20. With the analog, the flow Q_r is modelled by its equivalent electrical current I_r and the resultant potential V_r corresponds to the potential at the river.

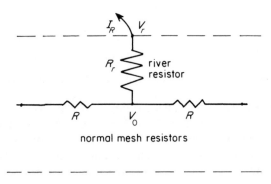

Figure 9.18. River resistance

Beds of rivers are frequently covered by silt or other less permeable materials. Corrections for such bands of lower permeability can easily be made by modifying the above analysis.

An alternative method of estimating the additional resistance was proposed by Numerov (see Aravin and Numerov, 1965; Streltsova, 1974). From detailed analyses of the sharply deformed seepage patterns, the additional seepage resistance can be superimposed on the normal flow pattern. From graphical presentations contained in the above publications, values of the additional seepage resistance for different river cross-sections can be estimated and included in a similar manner to the method of Herbert (1970). For rivers which are less than 0.2 of the aquifer thickness, the two methods lead to similar results.

The methods of Herbert and of Aravin and Numerov assume that there is a linear relationship between the flow to the aquifer and the potential difference between the aquifer and the river. However, field evidence suggests a non-linear relationship such as indicated in Fig. 6.2. Typical non-linear relationships which appear to give a fair representation are as follows:

If $\qquad h_0 \geqslant h_r$

$$Q = C_1(h_0 - h_r) + C_2\{1 - \exp\left[-C_3(h_0 - h_r)\right]\} \qquad (9.22)$$

If $\qquad h_0 \leqslant h_r$

$$Q = 0.3C_2\{\exp\left[C_3(h_0 - h_r)\right] - 1\} \qquad (9.23)$$

where C_1, C_2, and C_3 are constants depending on field conditions.

CHAPTER 10

Case Studies of Regional Groundwater Flow

10.1 Introduction

A number of case studies concerning aquifer modelling have been reported in the literature. Each aquifer has its own unique characteristics, yet much can be learnt from a study of a variety of different aquifers. Table 10.1 summarizes a number of case studies and attempts to identify the distinctive features of each.

In this section, one particular aquifer, the Lincolnshire Limestone, will be examined in detail. The discussion will illustrate that an understanding of the aquifer behaviour developed as the modelling techniques improved and as more field data became available. Indeed, at one stage during the study acceptable agreement was achieved between historic field behaviour and the model response over a period of some ten years and therefore it was anticipated that the aquifer model was satisfactory. A period of low recharge then followed and it was found that the model was unable to predict correctly the aquifer behaviour during the subsequent drought. This required further study and lead to an *improved* understanding of the flow mechanism.

This example demonstrates that the mathematical model can be used as a design tool. The flexibility of the method allows the flow mechanism to be explored and compared with the wide variety of field information. It is not possible to specify a procedure to be followed since each aquifer is unique. However, this advanced example does illustrate the major aims of a groundwater investigation.

10.2 Input Data

The aquifer to be studied in detail is the southern portion of the Lincolnshire Limestone, England. A location map, plan, and generalized section of the aquifer are presented in Fig. 10.1. The limestone in this region is a non-homogeneous aquifer exhibiting considerable facies variation in both vertical and horizontal directions. It is generally made up of shell beds, oolites, and sandy limestones (Downing and Williams, 1969; Burgess and Smith, 1978).

The presence of oolites is important since these formations are relatively porous, and groundwater flow in them is to some extent intergranular. Throughout the study area the limestone dips to the east at about 1 degree. In the east the aquifer is confined by overlying formations of sand, clay, and shale. This is shown

Table 10.1 A selection of recent regional groundwater flow studies using numerical techniques

Name of aquifer	Type	Features	Purpose of study	Type of model	References
Akrotiri, Cyprus	Gravel unconfined	Complex inflow and outflow: connection with river: saline water on three sides	Efficient development of resources	Finite difference ADE	Kitching (1975)
Burdekin Delta, Australia	Deltaic sediments, unconfined	Sediments resting on old granite surface: seepage from river	Calibration and future prediction	Finite difference ADI	Volker and Guvanasen (1977)
Cambrian-Ordovician, Illinois	Dolomites and sandstones	Leaky artesian conditions: possible change to unconfined conditions: reduction in transmissivity with depth	Predicting effect of increased abstraction	Finite difference MIADI	Prickett and Lonnquist (1971)
Columbia Basin, Washington	Unconfined overlying confined	Two aquifer system: upper unconfined clay sand and gravel: lower basalt aquifer separated by leaky confining layer	Understanding groundwater system	Finite difference iterative ADI	Tanaka et al. (1974)
Fylde, UK	Sandstone confined	Recharge from adjacent carboniferous series: some regions of hydraulic continuity with rivers: leakage to sea	Conjunctive use with surface resources	Finite difference ADI	Oakes and Skinner (1975)
Lambourn Valley, UK	Chalk, unconfined	Aquifer with parameters that vary with depth. Intermittent streams: springs	Augmentation of natural river flows during drought	Analog and finite difference	Thames Conservancy (1967, 1968, 1970, 1971, 1972); Oakes and Pontin (1976); Connorton and Hanson (1978)

Table 10.1 *Continued*

Name of aquifer	Type	Features	Purpose of study	Type of model	References
London Basin, UK	Sands, clays and chalk	Dewatering of part of aquifer has occurred: saline intrusion: aquifer-river interaction	Investigation of previous behaviour and possibility of artificial recharge	Analog and finite difference	Water Resources Board (1973)
Odessa-Lind USA	Basalt unconfined and confined	An irregular series of aquifers: recharge from infiltration, streams and canals: leakage between aquifers	Predictive model as management tool	Finite difference ADI	Luzier and Skrivan (1975)
Oued R'hir, Algeria	Limestones, clays, sands marls	Multi-layered with aquifers and aquitards: return flow from irrigation: increase in saline concentration	Determination of optimal pattern of abstraction	Finite difference SOR	De Marsily *et al.* (1978)
Parana basin, South America	Sandstone, mostly confined	950 000 km^2 in extent, methods of estimating recharge, representation of rivers	Investigation of possible exploitation	Flow balance with 52 sub-basins	Gilboa *et al.* (1976)
Rocky Mountain Arsenal, Colorado	Alluvial aquifer, unconfined	Extensive areas with zero transmissivity: recharge from canals and irrigation: connection with river	Analysis of groundwater pollution	Finite difference; various techniques	Konikow (1974); Trescott *et al.* (1976)

213

Location	Aquifer type	Description	Purpose	Method	Reference
South Downs, UK	Chalk unconfined	Aquifer parameters determined by inverse method: saline intrusion: variable transmissivity with depth	Forecast of probable consequences of different development proposals	Finite difference	Nutbrown et al. (1975)
South Humberside, UK	Chalk, confined and unconfined	Saline intrusion: overflow through blow-wells: flow mechanism dominated by gravels	Preventing future saline intrusion	Finite difference SOR	Lloyd et al. (1978)
South Platte Valley, Colorado	River gravels	Hydraulic continuity with river: pumping only during certain seasons	Studying management problems of conjunctive use	Finite difference coupled to surface water system	Young Bredehoeft (1972)
Sutter Basin, California	Multi-aquifers mainly sands and shales	Leakage between aquifers: major fault: presence of semi-confining layers	Examination of sensitivity of model to heterogeneous layers	Finite element	Gupta and Tanji (1976)
Vale of York, UK	Sandstone, unconfined and confined	Leakage to and from overlying strata dewatering to increase yield: pilot scheme results	Conjunctive use with river regulation scheme	Finite difference	Reeves et al. (1974); Birtles and Reeves (1977)

H

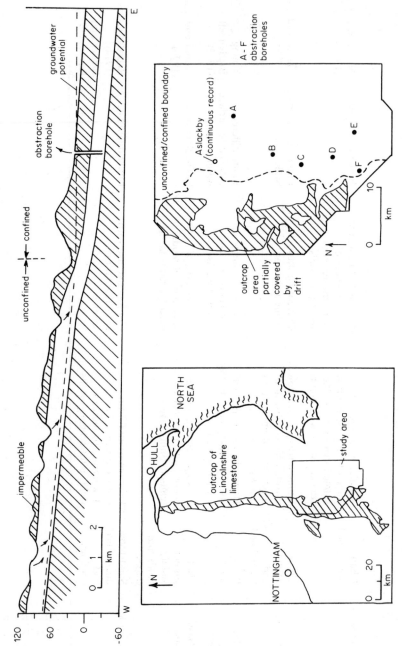

Figure 10.1. Location map, plan, and generalized section of the aquifer

diagrammatically in the section of Fig. 10.1 which also depicts the unconfined region.

When modelling the aquifer, values of the transmissivity and storage coefficient must be defined at each mesh point. From a small number of pumping tests it is deduced that the transmissivity in the confined region varies from $10\,000\,\mathrm{m^2/d}$ in the vicinity of major abstraction sites, to $1000\,\mathrm{m^2/d}$ towards the north. The confined storage coefficient is estimated to be 3.0×10^{-4}. No pumping tests have been carried out in the unconfined region, but the transmissivities are deduced from Darcy's law to vary within the range 50 to $500\,\mathrm{m^2/d}$ with an estimated storage coefficient of 0.05.

It is also necessary to specify conditions on each of the external boundaries. The region is defined on the west by a groundwater divide, whereas to the south, major faulting is thought to allow negligible flows across the boundary. To the north of the study area, the transmissivity decreases significantly and the boundary is identified by an arbitrary flow line. Even when extensive abstraction takes place it is unlikely that there will be significant flow across this northern boundary.

The boundary to the east presents the greatest uncertainty. As the limestone dips further to the east, the salinity increases, suggesting that there is little flow. Two possible boundary conditions seem appropriate. The first possibility is that the groundwater potential is maintained at a constant value at a considerable distance to the east of the area of major abstraction. The second possibility is that the boundary is totally impermeable; the true behaviour must be somewhere between these two extremes.

Accurate information about the magnitude of inflows and outflows is usually of critical importance. The outcrop area of the limestone in the study area is $342\,\mathrm{km^2}$. Recharge is calculated using the Penman–Grindley approach (Section 6.2). The calculation suggests that a soil-moisture deficit exists for most of the summer, and therefore significant recharge takes place only during the winter months. Abstraction takes place from six main sites with the total abstraction, increasing slightly each year. The fractional distribution of the total abstraction between the individual sites remains effectively constant so that abstraction patterns can be represented as a monthly total with constant distribution factors between the different sites (see the fourth column of Table 10.2).

Two small rivers flow over the outcrop area. The behaviour of these rivers is complex since there are frequent changes between influent and effluent conditions along their length. An adequate simulation can be achieved by modelling three major springs. The flow from the springs is estimated by baseflow separation of the river hydrograph, though there is now sufficient information about the springs' behaviour to represent them as a flow dependent on the difference between the groundwater potential and spring elevation.

Field data concerning variations in groundwater potential are essential if the reliability of the model is to be tested adequately. In the southern Lincolnshire Limestone, monthly observations have been made at almost 100 observation wells since 1972. From these records, contour plots of groundwater potential can be drawn (see the broken lines of Fig. 10.2). These are not a complete set of contours

Table 10.2 Daily inflows and outflows averaged over each month for the Lincolnshire Limestone during 1968.

Month	Total recharge Ml/d	Spring flow Ml/d	Abstraction Ml/d	Revised recharge distribution	
				Normal Ml/d	Fissures Ml/d
Jan	392	24.3	61.6	226	97.3
Feb	163	28.7	65.2	94	127.5
March	0	32.5	65.1	0	93.1
April	0	23.9	65.2	0	73.1
May	0	18.6	61.2	0	40.9
June	0	19.5	58.5	0	17.5
July	0	15.2	61.6	0	18.2
Aug	56	24.8	59.8	32	31.9
Sept	504	32.5	57.4	290	75.4
Oct	198	37.7	63.9	114	85.9
Nov	568	42.7	63.2	327	131.4
Dec	228	48.3	63.7	131	101.7

Distribution of abstraction
A	Jockey	12%
B	Bourne	29%
C	Wilsthorpe	41%
D	Tallington	9%
E	Etton	7%
F	Pilsgate	2%

since there is inadequate field information in certain regions. The full lines refer to results from the mathematical models and will be discussed later.

In addition, a continuous record is available since 1964 at Aslackby, an observation well in the confined region. Due to scarcity of observation points in the unconfined region and the relatively steep slopes of the water surface, the construction of a contour map in that region is difficult.

10.3 Mathematical Model

In planning a mathematical model, the first step is to select the mesh spacing. General guidance as to the desirable mesh size is given in Section 8.10. Since this aquifer is partly unconfined and partly confined, a very coarse net cannot be used. Nor is it advisable to use a graded net for there are no regions where the changes in flow pattern are so small that a coarse net would be satisfactory. Because a large number of years have to be modelled, the number of mesh intervals should be kept to a minimum and, as a compromise, a 2-km square mesh was selected with the mesh points totalling 18×22.

All the boundaries, except the eastern boundary, are impermeable. In the mathematical model they are represented by setting the transmissivity outside the

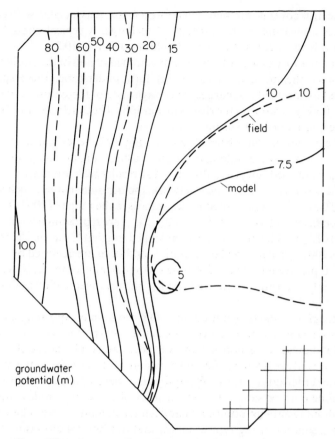

Figure 10.2. Contour plots of groundwater potential. Broken
lines, field values; full lines, model predictions

boundary to zero. There is considerable uncertainty about the behaviour of the
aquifer in the east. Two possibilities are modelled. One is that the aquifer is
impermeable so that there is no flow to the east. The other extreme condition is that
there is a fixed head on a line corresponding to the coastline.

Input data includes the magnitude of the recharge, spring flow, and abstraction.
Before describing the method of calculating these parameters, it is necessary to
decide for what time intervals they should be quoted. Each of these parameters will,
in practice, vary from day to day. However, in modelling regional groundwater
flow it is usually adequate to average them on a monthly basis. Due to the large
storage of the unconfined region, the discontinuities involved in averaging the
input data on a monthly basis tend to be smoothed out.

Recharge values are calculated in the manner described in Section 6.2. A daily
calculation is used, but the information is presented on a monthly basis. In certain
areas there is a significant boulder-clay cover over the outcrop region; in other

regions the drift cover is minimal. Therefore a factor is associated with each node which gives a measure of the intensity of recharge. A typical annual recharge distribution is recorded in the second column of Table 10.2.

A certain quantity of water leaves the unconfined region of the aquifer through springs. Since the mechanism is not yet fully understood, these springs are represented as specified discharges. The magnitude of these flows is determined from base-flow separation of the river hydrographs. The third column of Table 10.2 lists the magnitude of the spring flows.

Numerical solutions have been obtained using both the resistance–capacitance network and a backward-difference solution using a digital computer. Identical input data were used and virtually identical results were obtained.

With the *resistance–capacitance* network, the transmissivities which vary from 10 000 to 50 m^2/d are represented by resistances in the range of 100 kΩ to 20 MΩ. Storage coefficients of 5.0 $\times$ 10^{-2} and 3.0 $\times$ 10^{-4} are represented by capacitors of 2.0 and 0.012 μF. The time scale is arranged so that 1 sec of electrical time represents 100 days of physical time. Initial conditions are achieved by cycling the inputs of a typical year for about 200 times. A digital minicomputer is used to input the flows and to record the required results. The arrangement of equipment is as shown in Fig. 7.7.

For the *digital computer* solution, the SOR program of Appendix 2 is used. So that detailed results can be obtained, the calculation is performed at 1, 2, 4, 7, 10, 20 days, and at the end of the month. However, adequate results can be obtained if the calculation is only performed after ten days and on the last day of the month. The convergence criterion (see Eq. 8.16) requires that the continuity condition at each internal node must be satisfied to an accuracy of 2.5 $\times$ 10^{-6} m/day; this usually requires about twenty iterations. Dynamic initial conditions are achieved by first obtaining a solution using typical inflow and outflow conditions and with the storage coefficient set to a very low value. Then the storage coefficient is increased to its normal value and the typical year is run about five times.

10.4 Initial Results and Sensitivity Analysis

An important stage in the development of an adequate aquifer model is to check the model behaviour against field records during a historic period. It is essential to ensure that the comparison covers a sufficiently long time period for any inadequacies in the aquifer model to become apparent. The length of time over which a comparison should be made depends on two factors. Aquifers have a natural response time and for an aquifer which is totally or partially unconfined, the effect of a drought may influence the aquifer behaviour for several decades, whereas in a totally confined region the effect is dissipated within a few years. The second factor is that the historic period over which the aquifer model is tested should contain extremes of behaviour; this condition is naturally harder to satisfy.

The term 'model calibration' is used by some workers. This can be a dangerous concept if it results in certain aquifer parameters being varied until the model and field behaviour are similar. Since the aquifer response is a function of a number of different parameters (transmissivity, storage coefficient, recharge, boundary conditions, etc.) it is possible that the wrong parameters may be varied to give an apparently adequate fit over the restricted period for which reliable field data is available. The model could then give totally incorrect results when used for predictive purposes.

A *sensitivity analysis* is often a more realistic approach. Separate simulations are carried out using the mathematical model with individual parameters varied in turn over the tolerance to which they are known. For example the transmissivity may be varied by ± 40 per cent, the storage coefficient by ± 400 per cent, the recharge by ± 20 per cent, and the condition on an uncertain boundary changed from a fixed groundwater potential to a no-flow condition. In each case the sensitivity of the aquifer response in terms of changing groundwater potentials and flows is noted. This will show the sensitivity of the aquifer response to each parameter and may suggest that further field work should be carried out to define certain of the parameters more accurately. It will also indicate the accuracy that can be anticipated from the model.

A sensitivity analysis of the Lincolnshire Limestone is summarized in Table 10.3 and Figs. 10.3 to 10.7. Only certain of the results are presented; other information can be found in Rushton (1975). One conclusion from this study is that in most respects, the initial model is adequate and that it is not particularly sensitive to changes in transmissivity, storage coefficient, or conditions on the eastern boundary. However, the model response is far more sensitive to changes in the recharge patterns and therefore great care is required in estimating the recharge. The most significant conclusion from the sensitivity analysis is that satisfactory agreement cannot be obtained for the variation in groundwater potential with time in the confined region. This is discussed further in the following section.

10.5 Rapid Recharge

Field records of the groundwater potential, which show variations of up to 10 m are indicated by the full line of Fig. 10.8, whereas the model predictions, indicated by the broken line, only show annual variations of about 1 m. These results are clearly incompatible. The mathematical model was used to investigate *possible* causes of this discrepancy. Large changes in abstraction rate, reductions in the confined storage coefficient, and different conditions on the down-dip boundary were all investigated as possible causes of the difference between model and field behaviour. None of these modifications results in large variations in groundwater potential in the confined region. Therefore, it can be concluded that the model in its present form does not adequately represent the true flow mechanism.

Certain factors, observed in the field, suggested a possible modification to the flow mechanism. A large number of swallow holes can be identified in the outcrop area. Observations show that whenever significant rainfall occurs, some of the run-off enters directly into the swallow holes even if a soil-moisture deficit exists. In addition, a few days after the occurrence of this rainfall, the potentials in the confined region tend to recover; this recovery reaches a maximum after 4 to 6 weeks. Of particular importance is the fact that this recovery occurs even when there is a soil-moisture deficit in the outcrop region.

Table 10.3

Parameter varied	Results	Figure
Initial values of parameters	Adequate distribution of groundwater potentials	10.3 & 10.4
All transmissivities varied $\pm 20\%$	Groundwater potentials in unconfined region altered by $\mp 10\,\text{m}$ but in confined region still quite a good fit.	10.4
Recharge increased 15%	Groundwater potentials in confined region increased by $14\,\text{m}$ and in unconfined region by up to $25\,\text{m}$.	10.5
Storage in confined region unchanged with unconfined storage doubled (halved)	Typical annual variation decreased (increased) from $1.2\,\text{m}$ to $0.6\,\text{m}$ $(2.2\,\text{m})$. Average field variation over all unconfined boreholes during 1971–73 was $1.5\,\text{m}$.	10.6
Storage in confined region increased and decreased by 300%	Annual variation in groundwater potential as predicted by the mathematical model of about $1.0\,\text{m}$ is hardly changed, whereas the field variation is about $10\,\text{m}$.	
Eastern boundary, normally fixed, changed to impermeable	Over 10-year period, the predicted groundwater potentials diverge by no more than $1.5\,\text{m}$.	10.7

221

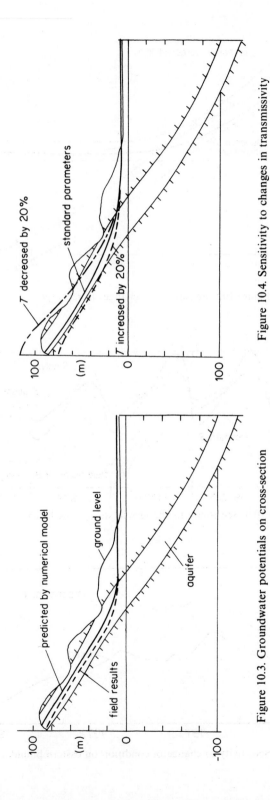

Figure 10.4. Sensitivity to changes in transmissivity

Figure 10.3. Groundwater potentials on cross-section

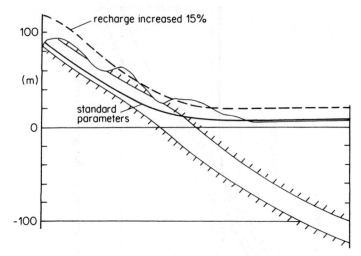

Figure 10.5. Sensitivity to changes in recharge

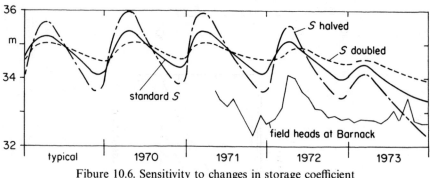

Fibure 10.6. Sensitivity to changes in storage coefficient

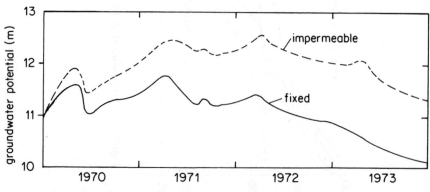

Figure 10.7. Sensitivity to change of condition on eastern boundary

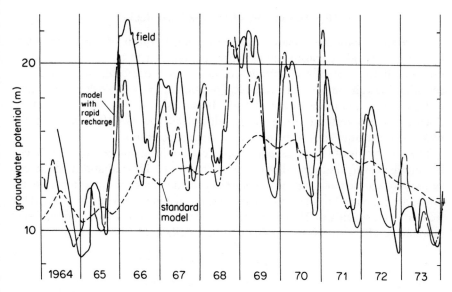

Figure 10.8. Variation of groundwater potential in confined region. Full line, field values; broken line, standard model; chain-dotted line, model incorporating rapid recharge

These observations lead to the concept of *rapid recharge*. This concept assumes that water enters swallow holes, causing a rapid transfer into the confined region. After many combinations of possible input factors had been investigated on the model, the final arrangement which gave the best overall fit between model and observed behaviour is as follows. In any one month the rapid recharge is made up from:

15 per cent of normal infiltration as recorded in the second column of Table 10.2;
10 per cent of rainfall minus (evaporation + surface runoff);
17.5 per cent of monthly rainfall in excess of 75 mm.

The rapid-recharge values for 1968 are shown in the fifth and sixth columns of Table 10.2. This quantity of water is injected directly into the line of nodes just inside the confined aquifer, with delays introduced to represent the delays observed in the field. This does not necessarily infer that the water travels from outcrop to the confined area at such a high rate, but that the recharge in the swallow holes causes a rapid transfer of water.

When this procedure is followed the mathematical model results, indicated by the chain-dotted lines of Fig. 10.8, and the field results show a satisfactory agreement. Further details can be found in Fox and Rushton (1976).

This particular study of rapid recharge illustrates the power of the mathematical model in the investigation of a wide range of flow mechanisms within the aquifer.

10.6 Dewatering of the Confined Region

Suggestions have been made that the confined region of the Lincolnshire Limestone should be considered as a storage reservoir to be used in periods of drought. If the unconfined storage coefficient is 5 per cent, then there is a considerable volume of water available. Questions such as the risk of drawing in water of poor quality from the east will be set aside for the moment and the discussion will be restricted to an investigation of the type of well field that would be required if the aquifer is to be used as a storage reservoir.

The dewatering of a homogeneous aquifer containing a square array of wells can be examined using the radial-flow model of Chapter 12. This radial-flow model can be used to represent a square array of wells provided that the outer impermeable boundary is at a radius of $0.564b$, where b is the distance between the wells. If the wells are at a spacing of 2 km then the equivalent radial-flow model to represent dewatering of the limestone aquifer can be specified as follows:

$$r_{well} = 0.2 \, m$$

$$r_{max} = 1128 \, m$$

$$\text{aquifer thickness} = 25 \, m$$

$$\text{permeability} = 100 \, m/d$$

$$\text{unconfined storage} = 0.05$$

$$\text{thickness of confining layer} = 20 \, m$$

$$\text{confined storage coefficient} = 0.0003$$

A constant discharge rate is assumed; different solutions are obtained with discharge rates within the range 1000 to 50 000 m^3/d. Initially the well discharge is provided largely by the confined storage but after an elapsed time given in the second column of Table 10.4, the aquifer becomes unconfined and is dewatered in the vicinity of the well. Abstraction continues until the level in the well falls to 20 m below the top of the confining layer; it is assumed that the water contained in the bottom 5 m cannot be exploited. The third column of Table 10.4 indicates the time between the aquifer first becoming unconfined and the maximum drawdown; the fourth column lists the volume of water abstracted expressed as a percentage of the water above the maximum permitted drawdown. It is possible to represent the decreasing yield of a pump as it has to work against an increased head, but this would only cause small modifications to the results.

The results show clearly that the higher abstraction rates quickly produce large drawdowns in the well which are aggravated by the decrease in the saturated depth in the vicinity of the well; thus the volume dewatered is comparatively small. At the lower abstraction rates, larger volumes of water can be removed, but it can take several years of steady abstraction to dewater the aquifer.

If the abstraction rate is maintained at a constant value of 10 000 m^3/d for 126 days then 42 per cent of the water could be removed. However, any increase in abstraction rate, even for a short period, would lead to a much-reduced yield.

Table 10.4 Dewatering of an aquifer with wells on a 2-km grid

Discharge rate per well m^3/d	Time to become unconfined days	Additional time to max. permissible drawdown days	% of available vol. withdrawn by dewatering
0	∞	∞	100
1 000	20	2740	89
5 000	3	382	64
10 000	1.2	125	42
15 000	0.7	49	24
20 000	0.3	14.2	8.9
30 000	0.16	0.065	0.04
50 000	0.04	0.004	0.003

There are two further important practical considerations. One is that the provision of wells on a 2-km grid may not be economical. The second is that this analysis assumes that the unconfined storage coefficient is 5 per cent over the full depth of the aquifer; this assumption had not been tested at the time at which this analysis was carried out, but has since been found to be an overestimate.

10.7 Simulation of a Historic Drought

From the above discussion it would appear that an adequate model of the aquifer is available, and predictions using the model suggest that the aquifer can provide greater resources than are currently utilized. These predictions were tested and proved unreliable in the drought period of 1975–76. During the winter 1975–76 practically no recharge occurred. When in early 1976 attempts were made to predict the probable aquifer behaviour, the mathematical model suggested that the falling piezometric head would tend to level off as indicated by the broken line of Fig. 10.9. This was not confirmed by behaviour at an observation borehole in the confined aquifer, shown by the full line.

The mathematical model suggests that during the period of very low recharge and relatively constant abstraction (May 1975 to July 1976) the groundwater levels should fall rapidly at first and then tend to level off at approximately 8 m as more and more water is drawn from the unconfined region. However, the field results showed a continuing fall during this period to a minimum level of about 3 m.

In an attempt to find the reason for this discrepancy a re-examination of the records for 1964–74 showed that there were no periods when such a low recharge occurred nor did the groundwater potentials ever fall below 8 m. Therefore there is no previous experience of the aquifer behaviour under such extreme conditions. Various attempts were made to devise a model to simulate the continuing fall in groundwater potential by reducing the storage coefficient in the unconfined region due to the fall in water levels, or decreasing the transmissivity in the confined region or changing the conditions on the eastern boundary. None of these gave the form of response observed in the field.

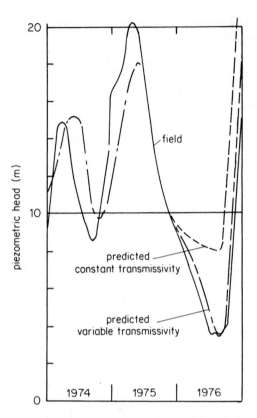

Figure 10.9. Groundwater-potential variation
during a drought

Then it was recognized that during the period of low recharge, the water levels could fall below the elevation of the main solution channels, particularly in the vicinity of the confined–unconfined interface. A consequence of the water level's falling below the main flow channels is that the flow from the unconfined region is restricted, with the result that more water is drawn from storage in the confined region. Therefore a condition was introduced in the model that the transmissivities in the x-direction in the region of the junction between the confined–unconfined regions were reduced by a factor of five as soon as the heads fell below 8.5 m. This led to the response indicated by the chain-dotted line of Fig. 10.9 which more nearly represents the field behaviour.

These results are only preliminary but they indicate the likely cause of the continuing fall in groundwater levels. Field investigations, a careful examination of all the data obtained during the period 1975–76, and additional studies using the mathematical model are all required to give a better understanding of the flow mechanism in the aquifer. The feature that the storage coefficient remained at the confined value even though the groundwater potentials were below the confining layer still remains to be explained.

One important conclusion of this study of the Lincolnshire Limestone is that an aquifer model is never complete. As new data become available the model must be re-examined, particularly if the new data are obtained during droughts or periods of high recharge. However, it is also apparent that a thorough study of an aquifer using a mathematical model should lead to a greater understanding of the aquifer behaviour.

PART IV

Numerical Method of Pumping-test Analysis

CHAPTER 11

Pumping-test Analysis Using Numerical Models

11.1 Introduction

The most widely used method of estimating the properties of an aquifer is the pumping test, which is in effect a controlled experiment to investigate the aquifer response. Known quantities of water are withdrawn from an abstraction borehole and the changes in groundwater potential both at the abstraction borehole and at other observation wells are recorded. It is assumed that the flow towards the abstraction well is radial, hence the polar form of the differential equation for groundwater flow (Eq. 2.24) can be used. Analytical solutions to this equation have been derived for a variety of conditions and, by comparing the field results with these analytical solutions using curve-matching techniques, values of the aquifer parameters can be deduced.

Since the publication by Theis (1935) of an expression which describes the time-variant drawdown due to abstraction from a well in an infinite aquifer, many additional analytical solutions and curve-fitting techniques have been developed. Two useful reference books on pumping-test analysis are Walton (1970) and Kruseman and de Ridder (1970). A wide variety of possible conditions can be represented by analytical solutions including leaky aquifers, partially penetrating wells, boundary effects, water contained within the well, aquifers which are partially confined and partially unconfined, anisotropic aquifers, delayed yield from storage, vertical components of flow, and stepped abstraction rates. However, there are combinations of these conditions that occur in practice for which no analytical solutions are available.

An alternative, versatile method of solving the same differential equation is by a discrete space–discrete time numerical model. This method has the advantage that as many conditions as necessary can be included in a single numerical solution. These conditions include: leakage; wells of finite radius; boundary effects; variable abstraction rates; recovery; change between confined and unconfined conditions; vertical components of flow; variable saturated depth; uneven distribution of well discharge with depth; water contained within the well; well losses; delayed yield from storage; variation of transmissivity and storage coefficient with depth; and multilayered aquifers.

231

With so many possible variables, matching of the field results against a set of type curves is not feasible. Instead a sensitivity approach is used. Physically realistic values of parameters are chosen for each numerical solution and the effect of changing the magnitude of these parameters in turn is investigated. It then becomes apparent which factors dominate the aquifer response and which have only minor significance.

Instead of working in terms of the groundwater potential, the variable to be used in this chapter is the *drawdown* which is the value of the groundwater potential *below* an arbitrary datum. In most cases the drawdown is assumed to be zero before the commencement of the test.

In this chapter the basis of the mathematical model will be derived first from the differential equation and then in terms of equivalent hydraulic resistances. Comparisons are then made with a number of standard analytical solutions. Next, the model is developed further to include vertical components of flow and delayed yield; the methods are validated by making comparisons with analytical solutions. In addition, certain important pumping tests, reported in the literature, are analysed using the numerical model. Parameter values which are different from those obtained from the curve-fitting methods are often deduced.

In Chapter 12 consideration is given to non-linear behaviour. This includes non-linear well losses, permeability that varies with depth, single-well tests, and the dewatering of excavations.

A summary of the topics covered in these chapters is to be found in Table 11.1. Throughout Chapters 11 and 12 the symbol T followed by brackets or a suffix refers to a 'time resistance' and not to the transmissivity.

11.2 Derivation of Discrete Model

In this section the derivation of the discrete model is restricted to cases where the *vertical components of flow are sufficiently small to be neglected*. The mathematical model can be derived in two alternative ways. Firstly, the derivation from the governing differential equation is described, then an alternative approach using a 'lumping' concept is introduced.

Derivation from differential equation

From Eq. 2.24 the differential equation describing radial flow in an aquifer can be written as

$$\frac{\partial}{\partial r}\left(mk_r\frac{\partial s}{\partial r}\right) + \frac{m}{r}k_r\frac{\partial s}{\partial r} = S\frac{\partial s}{\partial t} + q \qquad (11.1)$$

where s = drawdown below an arbitrary datum

   r = radial coordinate

   m = saturated thickness of aquifer

Table 11.1 Summary of the material in Chapters 11 and 12

Item	Section	Vertical flow	Comments	Results and figures
Derivation of discrete model	11.2	No	Derivation from differential equation or using lumping	
Computer program for discrete model	11.3	No	FORTRAN listing presented	Fig. 11.2–11.4
Change between confined and unconfined	11.3	No		
Variable saturated depth	11.3	No		
Comparison with Theis solution	11.4, 11.5	No	Idealized example; also field example (Walton, 1970)	Table 11.2 Fig. 11.5
Influence of outer boundary	11.4	No	Example with $r_{well} = 1.0, r_{max} = 10.0$	
Large-diameter well	11.4	No	Comparison with Papadopoulos and Cooper (1967)	Table 11.3
Leakage	11.4, 11.5	No	Comparison with analytical; and Kruseman and de Ridder (1970)	Table 11.4 Fig. 11.6
Vertical components of flow	11.6	Yes	Example of Huisman (1972)	Fig. 11.10
Non-uniform discharge distribution	11.6	Yes		
Delayed yield	11.7	Yes	Comparison with analytical results and field examples of Prickett (1965) and Boulton (1963)	Figs. 11.13–11.15
Non-linear well losses	12.2	Yes	Example of gravel aquifer	Figs. 12.1–12.5
Permeability varying with depth	12.3	Yes	Chalk aquifer with different rest water levels	Figs. 12.7–12.11
Single-well tests	12.4	No	Example of test in sandstone aquifer	Figs. 12.12–12.15
Dewatering of aquifers	12.5	Yes	Example with wells on a square grid	Figs. 12.17–12.18 Table 12.5

k_r = radial permeability

t = time

S = storage coefficient (confined or unconfined)

q = inflow per unit area which may be a function of the radius.

It is convenient to introduce an alternative radial variable a such that

$$a = \log_e r \tag{11.2}$$

substituting Eq. 11.2 in Eq. 11.1 as discussed in Section 3.5, and then multiplying through by r^2, leads to the equation

$$mk_r \frac{\partial^2 s}{\partial a^2} = Sr^2 \frac{\partial s}{\partial t} + qr^2 \tag{11.3}$$

Solutions to this equation with appropriate boundary and initial conditions can be obtained using finite-difference techniques. If a regular mesh Δa is used, then following the methods of Sections 3.2 and 3.8 the finite-difference form of Eq. 11.3 centred at node n is

$$\frac{mk_r}{\Delta a^2}(s_{n-1} - 2s_n + s_{n+1})_{t+\Delta t} = \frac{Sr_n^2}{\Delta t}(s_{n,t+\Delta t} - s_{n,t}) + q_{t+\frac{1}{2}\Delta t}r_n^2 \tag{11.4}$$

With a regular mesh interval Δa the nodal points are closely spaced in the vicinity of the abstraction well and become more widely spaced at greater distances. This is a convenient arrangement for the modelling of pumping tests since rapid changes in drawdown occur in the vicinity of the abstraction well. It is also advisable to use a time increment that increases logarithmically.

Equation 11.4 is a backward-difference formulation. Other finite-difference approximations for time can be used but there is always the possibility of instability (see Section 3.8).

It is convenient to visualize Eq. 11.4 in terms of equivalent hydraulic resistance as shown in Fig. 11.1. If each of the equivalent hydraulic resistances is defined as

$$H_n = \frac{\Delta a^2}{mk_r} \tag{11.5}$$

and

$$T_n = \frac{\Delta t}{Sr_n^2} \tag{11.6}$$

with a flow from each node equal to qr_n^2 (note that a positive recharge leads to a *reduction* in drawdown, hence the outward direction of the arrow), then the equation for node n becomes

$$\frac{(s_{n-1} - 2s_n + s_{n+1})_{t+\Delta t}}{H_n} = \frac{s_{n,t+\Delta t} - s_{n,t}}{T_n} + qr_n^2 \tag{11.7}$$

With the substitutions of Eqs. 11.5 and 11.6 in 11.7, Eq. 11.4 is obtained.

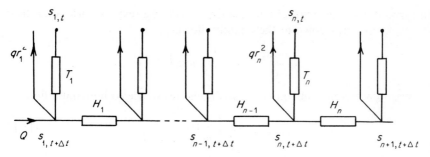

Figure 11.1. Equivalent hydraulic resistances

This concept of equivalent hydraulic resistance is introduced because it is of value when considering the representation of boundary conditions. For example, if node 1 represents the region from the edge of the well to a distance $0.5\,\Delta a$ into the aquifer, the quantity of water entering the aquifer due to a change in storage is roughly half the value that would apply if the node represented a region extending $0.5\,\Delta a$ on either side. Thus the value of the 'time resistance' T_1 is double the magnitude calculated from Eq. 11.6.

The abstraction rate Q is represented as a negative recharge at node 1. Some manipulation is required to convert the abstraction rate, $[L^3/T]$, to an equivalent recharge, $[L/T]$, at a nodal point. The recharge at a node is given by

$$q = \frac{-Q}{A_n} \tag{11.8}$$

where A_n = area represented by a node

$$= 2\pi r\,\Delta r$$

Since $a = \log_e r$, $\Delta r = r\,\Delta a$. Hence

$$A_n = 2\pi r^2\,\Delta a \tag{11.9}$$

thus

$$q = \frac{-Q}{2\pi r^2\,\Delta a} \tag{11.10}$$

Derivation using lumping concept

In this alternative approach (Rushton and Booth, 1976), the aquifer is considered as a series of rings having radii $r_{n-1}, r_n, r_{n+1}, \ldots$.

Radial-flow component

The equation for the quantity flowing towards the origin through an annulus of width dr at a radius r due to a change in drawdown ds is, from Darcy's law,

$$Q_r = -2\pi rmk_r\frac{ds}{dr} \tag{11.11}$$

Integrating and inserting the limits r_n and r_{n+1} leads to an expression for the ratio of the change in drawdown to the quantity flowing:

$$\frac{s_n - s_{n+1}}{Q_r} = \frac{1}{2\pi k_r m} \log_e \left(\frac{r_{n+1}}{r_n} \right)$$

This ratio represents an equivalent hydraulic resistance to horizontal flow,

$$\phi_{r,n,n+1} = \frac{1}{2\pi k_r m} \log_e \left(\frac{r_{n+1}}{r_n} \right) \qquad (11.12)$$

Water released from storage

During a time period Δt the ratio of the increment in drawdown, Δs, to the quantity of water released from storage, Q_s, is

$$Q_s = \text{plan area} \times S \frac{\Delta s}{\Delta t} = 0.25\pi \left[(r_{n+1} + r_n)^2 - (r_n + r_{n-1})^2 \right] S \frac{\Delta s}{\Delta t}$$

This can be expressed as an equivalent 'time resistance'.

$$\phi_{t,n} = \frac{\Delta s}{Q_s} = \frac{\Delta t}{0.25\pi \left[(r_{n+1} + r_n)^2 - (r_n + r_{n-1})^2 \right] S} \qquad (11.13)$$

Recharge

Due to recharge, the quantity entering node n equals

$$R_n = 0.25\pi \left[(r_{n+1} + r_n)^2 - (r_n + r_{n-1})^2 \right] q \qquad (11.14)$$

From Eqs. 11.11 to 11.14, the flow-balance equation at node n can be written as

$$\left[\frac{s_{n+1} - s_n}{\phi_{r,n,n+1}} + \frac{s_{n-1} - s_n}{\phi_{r,n-1,n}} \right]_{t+\Delta t} = \frac{s_{n,t+\Delta t} - s_{n,t}}{\phi_{t,n}} + R_n \qquad (11.15)$$

Comparison of approaches

The derivation from the differential equation, Eq. 11.4, and the lumping approach, Eq. 11.15, lead to identical numerical models. Since the mesh increases logarithmically with $a = \log_e r$, the increment $\Delta a = \log_e (r_{n+1}/r_n)$. Also the expression $0.25\pi \left[(r_{n+1} + r_n)^2 - (r_n + r_{n-1})^2 \right]$, the area represented by the node, was shown in Eq. 11.9 to equal $2\pi r^2 \Delta a$. From these relationships Eqs. 11.12, 11.13, and 11.14 can be rewritten as

$$\phi_{r,n,n+1} = \frac{\Delta a}{2\pi k_r m}$$

$$\phi_{t,n} = \frac{\Delta t}{2\pi r_n^2 \Delta a S}$$

$$R_n = 2\pi r_n^2 \Delta a q$$

When these factors are substituted in Eq. 11.15 and each term is multiplied through by $1/2\pi \Delta a$, the following equation results:

$$\left[\frac{s_{n+1} - s_n}{\Delta a^2 / mk_r} + \frac{s_{n-1} - s_n}{\Delta a^2 / mk_r}\right]_{t + \Delta t} = \frac{s_{n,t+\Delta t} - s_{n,t}}{\Delta t / Sr_n^2} + qr_n^2$$

This is identical to Eq. 11.7, which was derived directly from the differential equation.

11.3 Numerical Solution

This section describes the method of obtaining a numerical solution for the mathematical model of horizontal radial flow through an aquifer to an abstraction well using a digital computer. Factors that are represented in the solutions include a well of finite radius with allowance for the free water initially contained within the well, various conditions on the outer boundary, a change between the confined and unconfined states, and variations in the saturated depth of an unconfined aquifer. A flow chart of the numerical techniques is contained in Fig. 11.3 and the program listing is given in Fig. 11.4.

Numbering of nodes and mesh spacing

Figure 11.2 shows the system of numbering the nodes. Note that node 1 now represents the *region within the well*. The well radius is $R(2)$ and the outer boundary $R(NMAX)$. The increment, $\Delta a = 0.383\,76$, is selected so that there are six mesh intervals for a tenfold increase in radius, with the final interval ending at $R(NMAX)$. The notation and equations are written in a form helpful to the preparation of the computer program.

Equations

At a typical node N the appropriate hydraulic resistors $H(N)$ and $H(N-1)$ are calculated from Eq. 11.5, and $T(N)$ is calculated from Eq. 11.6. The unknown

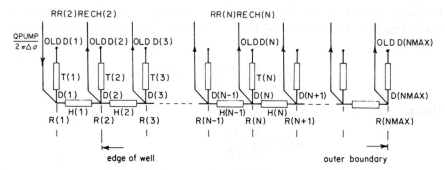

Figure 11.2. Discrete model with allowance for water in well

drawdown is $D(N)$ and the drawdown at the previous time step is $OLDD(N)$. Thus the standard equation, Eq. 11.7, at node N is written as

$$-\frac{D(N-1)}{H(N-1)} + D(N)\left(\frac{1}{H(N-1)} + \frac{1}{H(N)} + \frac{1}{T(N)}\right) - \frac{D(N+1)}{H(N)}$$

$$= \frac{OLDD(N)}{T(N)} - RR(N)RECH(N) \tag{11.16}$$

where $RR(N) = R(N)R(N)$.

Special consideration must be given to the well nodes. In the derivation of the discrete model, the analysis considered only the flow in the aquifer. However, the water actually contained within the well is also significant. This can be simulated simply by assuming that the aquifer extends within the well but with different properties. If the properties of the 'aquifer' within the well are such that the transmissivity is very high and the storage coefficient is unity, then this will correspond to free water within the well.

Node 1 represents the whole of the region within the well. Since the water surface within the well is constant, the horizontal hydraulic resistance $H(1)$ is set to a low value. A special 'time resistance' $T(1)$ represents the free water within the well. Some manipulation is required to calculate the 'time resistance'.

From Eqs. 11.6 and 11.9, the expressions for the 'time resistance' and the area represented by node 1 are

$$T(1) = \frac{\Delta t}{Sr_1^2}$$

$$A_1 = 2\pi r_1^2 \Delta a$$

Since node 1 must now represent the plan area of the well which has a radius r_2,

$$A_{well} = \pi r_2^2$$

Thus the ratio of the area of the well to the unmodified area of node 1 is

$$\frac{A_1}{A_{well}} = \frac{2\pi r_1^2 \Delta a}{\pi r_2^2}$$

In addition the free water within the well is equivalent to a storage coefficient of unity. Therefore, to represent the water in the well, the 'time resistance' must be modified to

$$T(1) = \frac{\Delta t}{r_1^2} \frac{2\pi r_1^2 \Delta a}{\pi r_2^2} = \frac{2\Delta t \Delta a}{r_2^2}$$

Hence

$$T(1) = \frac{2\Delta t \Delta a}{RR(2)}$$

where $RR(2)$ is the square of the well radius.

Node 2 represents the region from the edge of the well to a distance half a mesh interval into the aquifer. Hence the 'time resistance' $T(2)$ is set to double the value given by Eq. 11.6.

At node 1, the quantity abstracted by the pump, $QPUMP$, must be modelled. $QPUMP$ represents the total volume of water abstracted from the well in unit time. In the early stages of the test, part of this total is contributed by the free water contained within the well; the remainder is drawn from the aquifer.

The discharge $QPUMP$ is represented as an equivalent negative recharge. From Eq. 11.10

$$q = \frac{-QPUMP}{2\pi RR(1)\,\Delta a}$$

Thus the equivalent flow-balance equation for node 1 which is derived by modifying Eqs. 11.7 and 11.16 becomes

$$D(1)\left(\frac{1}{H(1)} + \frac{1}{T(1)}\right) - \frac{D(2)}{H(1)} = \frac{OLDD(1)}{T(1)} + \frac{QPUMP}{2\pi\,\Delta a} \tag{11.17}$$

At the outer boundary radius $R(NMAX)$, the model is terminated. This is represented by arranging for $H(NMAX)$ to have a very large value.

Thus the equation at node NMAX becomes

$$-\frac{D(NMAX-1)}{H(NMAX-1)} + D(NMAX)\left[\frac{1}{H(NMAX-1)} + \frac{1}{T(NMAX)}\right] =$$

$$\frac{OLDD(NMAX)}{T(NMAX)} - 0.5\,RR(NMAX)\,RECH(NMAX) \tag{11.18}$$

When this outer boundary is impermeable, the time resistance is set to be double the appropriate value. If the drawdown at the outer boundary remains at a constant value, the time resistance is set to a small value, which is equivalent to a very large storage coefficient. With such a large storage, the drawdown at the outer boundary remains virtually unchanged, thereby satisfying the condition of a constant drawdown. Another modification at this outer boundary is that the non-standard spacing of the last mesh interval requires a special equation in the calculation of $H(NMAX-1)$ and $T(NMAX-1)$.

Confined–unconfined

Whether the confined or unconfined storage coefficient should be used depends on the drawdown relative to the top of the aquifer; these storage coefficients are defined in Section 2.4. If a node has a drawdown less than TOP, where TOP is the position of the top of the aquifer below datum level, then the confined storage coefficient applies. When the drawdown is greater than TOP, the unconfined coefficient should be used. As the calculation proceeds, the groundwater potential may cross the top of the aquifer. Thus the storage coefficient and therefore the value of $T(N)$ will change. When this is likely to happen it is advisable to use smaller, constant time steps during that period rather than the usual logarithmic increase.

Variable saturated depth

The horizontal hydraulic resistance is a function of the saturated depth m (Eq. 11.5). If the aquifer is confined, then the saturated depth will be equal to the aquifer thickness, BASE $-$ TOP. If, however, the aquifer is unconfined, then the aquifer depth can be calculated as $SD = $ BASE $- D(N)$. Note that the saturated depth is a function of the unknown drawdown, $D(N)$, which necessitates the use of an iterative scheme each time step, in which the calculation is repeated four times with successively improved approximations to the saturated depth.

Time step

The initial time step must be small enough for an accurate representation at the smallest radius. This can be achieved by setting the initial value of the time step as

$$\Delta t = \frac{0.025 r_w^2 S_c}{T} \tag{11.19}$$

where S_c is the confined storage coefficient and r_w is the radius of the abstraction well. Equation 11.19 is equivalent to a value of the non-dimensional aquifer parameter

$$u = \frac{r^2 S}{4Tt} = 10.0$$

where $r = r_w$. With this large value of u, the drawdown at the first and all other nodes will be small (Kruseman and de Ridder, 1970, p. 54).

Thereafter, the time increases by a factor of $10^{0.1}$ for each calculation. This provides ten time steps for a tenfold increase in time. Each time the abstraction rate changes, including the case when it is reduced to zero, the time step must return to the small value given by Eq. 11.19. In certain situations, such as a change between the confined and unconfined state or when delayed yield occurs, the increase in Δt should be smaller or Δt should be constant for a period.

Position of observation wells

In this simple program with a regular logarithmic mesh spacing, the observation wells are positioned at nodal points. The node numbers of the observation positions are read in as data. If more exact observation-well positions are required the mesh can be redesigned to coincide with the desired positions. In general, however, the variation of the drawdown with radius is sufficiently smooth for the use of a regular mesh spacing to be adequate.

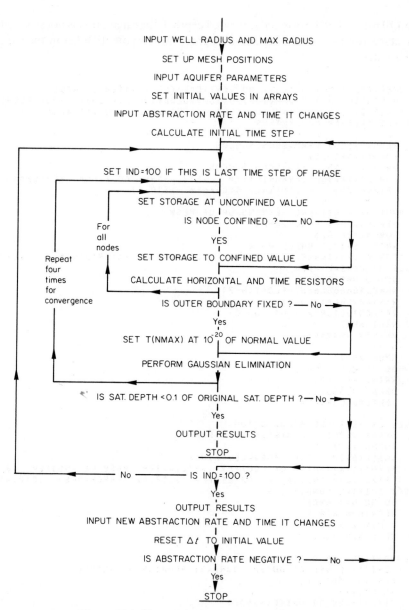

Figure 11.3. Flow chart for pumping-test program

Excessive drawdown

After the new drawdowns have been calculated, a check is made to see whether the drawdowns are excessive. Once the saturated depth becomes less than half of the original saturated thickness, rapid changes in the drawdowns occur. The calculation is only stopped when the depth of water at the abstraction well is less than 10 per cent of the original saturated depth. If the calculation is not stopped at this stage, results may be obtained with groundwater potentials below the base of the aquifer.

```
C     NUMERICAL PUMPING TEST, NO VERTICAL FLOW, WATER IN WELL
      DIMENSION R(100),RR(100),D(100),OLDD(100),T(100),H(100),RECH(100)
     1 A(100),B(100),C(100),E(100),U(100),V(100),NOB(6),ARRAY(6)
C     INPUT AQUIFER PARAMETERS
C
      READ(5,510) PERM,SCON,SUNCON
510   FORMAT(4F10.5)
      WRITE(6,515) PERM,SCON,SUNCON
515   FORMAT(1X,14HPERMEABILITY=   ,E12.4,19H  CONFINED STORAGE=
     1 E12.4,22H   UNCONFINED STORAGE=  ,F10.5)
C
C     INPUT OF RADII AND SETTING UP OF MESH
      READ(5,500) RWELL,RMAX
500   FORMAT(2F10.3)
      WRITE(6,505) RWELL,RMAX
505   FORMAT(1X,14HWELL RADIUS =    ,E12.4,13H  MAX RADIUS=     ,E12.4)
C     SET UP RADIAL MESH
      DO 10 N=1,100
      AN=0.16666666667*FLOAT(N-2)
      R(N)=RWELL*10.0**AN
      IF(R(N).LT.RMAX) GO TO 10
      R(N)=RMAX
      RR(N)= RMAX*RMAX
      NMAX=N
      NMONE=N-1
      GO TO 20
10    RR(N)=R(N)*R(N)
20    DELA= 0.383765
      DELA2=DELA*DELA
C
C     LEVELS MEASURED BELOW DATUM
      READ(5,520) TOP,BASE,WLEVEL,RCH
520   FORMAT(4F10.5)
      WRITE(6,525) TOP,BASE,WLEVEL,RCH
525   FORMAT(1X,16HTOP OF AQUIFER  ,E12.4,17H  BASE OF AQUIFER    ,
     1 E12.4,22H INITIAL WATER LEVEL=  ,E12.4,11HRECHARGE=   ,E12.4)
C     SET INITIAL CONDITIONS
      DO 30 N=1,NMAX
      RECH(N)= RCH
      D(N)= WLEVEL
30    OLDD(N)= WLEVEL
C
C     CONDITIONS ON OUTER BOUNDARY
C     IF JFIX=1 LEVEL AT OUTER BOUNDARY REMAINS AT WLEVEL
      READ(5,530) JFIX
530   FORMAT(I1)
      IF(JFIX.EQ.1) WRITE(6,532)
      IF(JFIX.NE.1) WRITE(6,534)
532   FORMAT(22H  ** FIXED BOUNDARY**   )
534   FORMAT(22H ** FREE BOUNDARY **   )
C
C     GIVE NODE NUMBER OF SIX OBS WELLS
      READ(5,540) (NOB(J),J=1,6)
540   FORMAT(6I3)
C
C     INPUT ABSTRACTION RATE
      TSTART=0.0
      READ(5,550)QPUMP,TSTOP
```

```
550    FORMAT(2F10.3)
       WRITE(6,555) QPUMP,TSTOP
555    FORMAT(1X,15HPUMPING RATE=      ,E12.4,14H TILL TIME OF    ,E12.4,
      1 5H DAYS   )
C
C    CONVERTING ABSTRACTION TO QABST
       PI=4.0*ATAN(1.0)
       QABST=0.5*QPUMP/(PI*DELA)
       IND=0
C
C    INITIAL TIME AND DELT
       TIME=0.0
       DELI=0.025*RR(1)*SCON/(PERM*(BASE-TOP))
       DELT=DELI
       DO 35 I = 1,6
       I1= NOB(I)
35     ARRAY(I)= R(I1)
       WRITE(6,545) R(1),(ARRAY(J),J=1,6),R(NMAX)
545    FORMAT(1X,24HTIME  (DAYS)     (PHASE)      ,8F12.4)
C
C    CALCULATION FOR A SPECIFIC TIME, IND=0 FOR LAST STEP
40     TIME=TIME + DELT
       IF(TIME.LT.TSTOP) GO TO 50
       DELT=TSTOP-TIME+DELT
       TIME=TSTOP
       IND=100
C
50     CONTINUE
C
C    CALCULATION REPEATED FOUR TIMES FOR CONVERGENCE
       DO 60 NUM=1,4
       DO 70 N=1,NMONE
C
C    TAKE AVERAGE SATURATED DEPTH BETWEEN N AND N+1
       SD=BASE- 0.5*(D(N)+D(N+1))
       STOR=SUNCON
       IF(SD.LT.(BASE-TOP)) GO TO 80
       SD=BASE - TOP
       STOR=SCON
80     H(N) = DELA2/(SD*PERM)
70     T(N)=DELT/(STOR*RR(N))
C
C    TO REPRESENT WATER IN WELL
       H(1)=0.0001*H(1)
       T(1)=2.0*DELT*DELA/RR(2)
       T(2)=2.0*T(2)
       H(NMAX-1)=(ALOG(R(NMAX))-ALOG(R(NMAX-1)))*(ALOG(R(NMAX))-ALOG
      1  (R(NMAX-1)))/(SD*PERM)
       H(NMAX)=1.0E+10
       T(NMONE)=2.0*DELT*DELA/((R(NMAX)-R(NMONE-1))*STOR*R(NMONE))
       T(NMAX )=1.0*DELT*DELA/((R(NMAX)-R(NMONE))*STOR*R(NMAX))
       IF(JFIX.EQ.1) T(NMAX)=1.0E-10*T(NMAX)
C
C    GAUSSIAN ELIMINATION
C
C    CALCULATION OF COEFFICIENTS
C    EQUN IS   -A(N)*D(N-1) + B(N)*D(N) - C(N)*D(N+1) = E(N)
       B(1)=1.0/H(1) + 1.0/T(1)
       C(1)= 1.0/H(1)
       E(1)= OLDD(1)/T(1) + QABST
       DO 90 N=2,NMONE
       A(N)= 1.0/H(N-1)
       B(N)= 1.0/H(N-1) + 1.0/H(N) + 1.0/T(N)
       C(N)= 1.0/H(N)
```

```
90       E(N)= OLDD(N)/T(N) - RR(N)*RECH(N)
         A(NMAX)= 1.0/H(NMONE)
         B(NMAX)= 1.0/H(NMONE) + 0.5/T(NMAX)
         E(NMAX)= 0.5*OLDD(NMAX)/T(NMAX) - 0.5*RR(NMAX)*RECH(NMAX)
C
C   ELIMINATION
         U(1)= B(1)
         V(1)= E(1)
         DO 100 N=2,NMAX
         U(N)= B(N) - (A(N)*C(N-1))/U(N-1)
100      V(N)= E(N) + (A(N)*V(N-1))/U(N-1)
         D(NMAX)= V(NMAX)/U(NMAX)
         DO 110 NN=1,NMONE
         N=NMONE-NN+1
110      D(N)=(V(N) + C(N)*D(N+1))/U(N)
C   TEST FOR EXCESSIVE DRAWDOWNS
         DRAWMX= 0.9*BASE + 0.1*TOP
         IF(D(1).LT.DRAWMX) GO TO 60
         WRITE(6,580)
580      FORMAT(1X,20H EXCESSIVE DRAWDOWN)
         WRITE(6,565)
         DO 105 N=1,NMAX
105      WRITE(6,570)N,R(N),RR(N),H(N),T(N),D(N)
         STOP
60       CONTINUE
C
C   OUTPUT AND CHANGE PARAMETERS
         TIMIN=TIME - TSTART
         DO 120 I = 1,6
         I1 = NOB(I)
120      ARRAY(I) = D(I1)
         WRITE(6,560) TIME,TIMIN,D(1),(ARRAY(J),J=1,6),D(NMAX)
560      FORMAT(1X,10E12.4)
         DO 130 N=1,NMAX
130      OLDD(N)=D(N)
         DELT= TIMIN*0.25892
         IF(IND.EQ.0) GO TO 40
C   END OF CALCULATION FOR A SPECIFIC TIME
C
         WRITE(6,565)
565      FORMAT(1X,8HNODE NO.,4X,6HRADIUS,14X,14HRADIUS SQUARED,6X,
         1 13HHORIZ HYD RES,7X,15HTIME RESISTANCE,5X,8HDRAWDOWN)
         DO 140 N=1,NMAX
140      WRITE(6,570) N,R(N),RR(N),H(N),T(N),OLDD(N)
570      FORMAT(1X,I4,5E20.6)
C
C   NEW PUMPING PHASE
C   RESET PARAMETERS
         DELT=DELI
         IND=0
         TSTART=TIME
         READ(5,550)QPUMP,TSTOP
         WRITE(6,555) QPUMP,TSTOP
         QABST=0.5*QPUMP/(PI*DELA)
         IF(QPUMP.GE.0.0) GO TO 40
         STOP
         END
```

Figure 11.4. Computer program for discrete model of pumping test

Solution of equations

When written in matrix form, Eqs. 11.16, 11.17, and 11.18 form a tridiagonal matrix. Therefore an efficient solution can be obtained using Gaussian elimination in the manner described in Section 9.1. Instead of using the notation GA, GB, GC, and GD, this program uses coefficients A, B, C, and E. A further difference from the previous Gaussian elmination routine is that, for this problem, node 1 is the first unknown. Since the relative magnitudes of the quantities vary greatly, the elimination must be carried out to a high precision; for some computers it is necessary to resort to double precision to achieve an acceptable accuracy.

11.4 Standard Examples

This section considers the use of the numerical technique to investigate certain standard examples. These include the Theis (1935) solution for homogeneous aquifers, an aquifer with a finite impermeable outer boundary, the effect of water contained in large diameter wells, and an aquifer fed by leakage from a semi-permeable stratum.

Theis solution

In the Theis solution, the well has an infinitely small radius and the aquifer extends to a sufficient distance for the outer boundary to have no influence. At times less than zero, the discharge rate is zero, then at $t = 0$ the quantity withdrawn from the aquifer instantaneously increases to Q and is maintained at that value. Two non-dimensional parameters used in the solution are

$$u = \frac{r^2 S}{4Tt} \tag{11.20}$$

and

$$W(u) = \frac{4\pi Ts}{Q} \tag{11.21}$$

The relationship between these parameters is

$$W(u) = \int_u^\infty \frac{e^{-x}}{x} \, dx \tag{11.22}$$

Extensive tabulations of this function are available (Walton, 1970, p. 136).

When testing the numerical model against the Theis solution, it is convenient to use the following parameter values which is equivalent to working in non-dimensional terms.

$$r_{well} = 10^{-5} \, m \qquad r_{max} = 10^5 \, m$$

$$k = 10^{-8} \, m/d \qquad m = 2.5 \times 10^5 \, m$$

hence $T = km = 2.5 \times 10^{-3}\,\text{m}^2/\text{d}$

$$S = 10^{-2} \qquad\qquad Q = \pi \times 10^{-2}\,\text{m}^3/\text{d for } 10^6 \text{ days}$$

Consequently

$$u = \frac{r^2}{t} \quad \text{and} \quad W(u) = s$$

An aquifer of thickness 2.5×10^5 m is clearly very large but it does ensure that the drawdown is small compared with the aquifer thickness. Likewise the maximum radius of 10^5 m is sufficiently large to be beyond the influence of the abstraction well. Results from the numerical model using these parameters can be compared directly with the Theis values. Table 11.2 contains typical values, selected from the extensive computer results. The agreement between numerical and analytical results is good.

Anisotropic conditions can be included following the approach of Hantush (1966) in which the anisotropic properties are transformed to an equivalent isotropic case.

Outer boundary

The ability of the numerical model to represent an outer impermeable boundary can be tested by analysing the following problem:

$$r_{\text{well}} = 1.0\,\text{m} \qquad\qquad r_{\text{max}} = 10.0\,\text{m}$$

$$T = 2 \times 10^5\,\text{m}^2/\text{d} \qquad S = 0.1$$

$$Q = 1.0\,\text{m}^3/\text{d for a period of 10 days}$$

With such a high value for the transmissivity, the drawdown is effectively constant with radius. At the end of the ten-day period, the drawdown should equal

$$s = \frac{Q \times t}{\text{plan area} \times S} = \frac{10}{\pi(10^2 - 1^2) \times 10^{-1}}$$

$$= 0.321\,\text{m}$$

The actual value calculated from the program is 0.309 m, an error of nearly 4 per cent. This error arises from an inexact representation of the outermost 'time resistance', $T(\text{NMAX})$. In the program it is set to be double the value for an internal node; it should strictly be set to a value slightly larger than this because of the logarithmic mesh spacing.

Large-diameter well

The program has been designed to include the effect of the water contained within the abstraction well. In the analytical solution of Papadopoulos and Cooper (1967) only the drawdown at the well face is calculated, so that no comparisons are possible at points within the aquifer.

Table 11.2 Selected results for standard problem

Time (days)	$\frac{4tT}{Sr^2}$	r = 0.001 m Drawdown (m)		$\frac{4tT}{Sr^2}$	r = 1.0 m Drawdown (m)	
		Numerical	Analytical		Numerical	Analytical
2.54×10^{-7}	2.54×10^{-1}	0.0047	0.0043	2.54×10^{-7}	0.0000	0.0000
6.74×10^{-6}	6.74	1.43	1.47	6.74×10^{-6}	0.0000	0.0000
6.57×10^{-1}	6.57×10^{5}	12.78	12.83	6.57×10^{-1}	0.0099	0.0095
3.39	3.39×10^{6}	14.42	14.47	3.39	0.894	0.907
6.71×10^{2}	6.71×10^{8}	19.71	19.75	6.71×10^{2}	5.90	5.93
1.03×10^{4}	1.03×10^{10}	22.45	22.47	1.03×10^{4}	8.63	8.66

Input data for the numerical solutions is as follows:

$$r_{well} = 1.0\,m \qquad r_{max} = 10^5\,m$$

$$k = 10^{-8}\,m/d \qquad m = 2.5 \times 10^5\,m$$

$$S = 10^{-2} \qquad Q = \pi \times 10^{-2}\,m^3/d \text{ for } 10^6 \text{ days}$$

The analytical expression of Papadopoulos and Cooper for the drawdown in the well is

$$\frac{4\pi s_w T}{Q} = \frac{32S^2}{\pi^2} \int_0^\infty \frac{1 - \exp(-\beta^2/4u_w)}{\beta^3 \,\Delta\beta}\,d\beta \qquad (11.23)$$

where $u_w = r_{well}^2 S/4Tt$

β = a dummy variable

$\Delta\beta = [\beta J_0(\beta) - 2SJ_0(\beta)]^2 + [\beta Y_0(\beta) - 2SY_1(\beta)]^2$ in which $J_0(\beta)$, $Y_0(\beta)$, and $Y_1(\beta)$ are Bessel functions.

Table 11.3 Selected results for large-diameter well, $S = 10^{-2}$

Time (days)	$\dfrac{tT}{Sr_w^2}$	Drawdown (m) at well	
		Numerical	Analytical
0.6	0.6	0.0059	0.006
4.92	4.92	0.0481	0.048
5.47×10	5.47×10	0.491	0.492
1.02×10^3	1.02×10^3	4.49	4.54
1.0×10^6	1.0×10^6	13.2	13.2

Comparisons between certain of the numerical results and analytical values, which are recorded in Table 11.3, show excellent agreement. This set of results applies only to a storage coefficient of 10^{-2}. Values of the storage coefficient ranging from 10^{-1} to 10^{-4} have been investigated and consistently good agreement is achieved.

Leakage

Leakage to or from an overlying aquifer is a phenomenon that is important in pumping tests. It can be incorporated as a recharge which depends on the current value of the drawdown. For a leaky aquifer, the recharge per unit area (Walton, 1970, p. 144) can be expressed as

$$q = k_r m(s_{j,t+\Delta t} - s_{j,w})/B^2 \qquad (11.24)$$

where $s_{j,t+\Delta t}$ is the current value of the drawdown

$s_{j,w}$ is the position, below datum, of the water surface in the overlying stratum

B is the leakage coefficient $[L]$.

Inclusion of the leakage in the computer program (Fig. 11.4) requires just one extra statement between labels 80 and 70. The statement is

$$\text{RECH(N)} = \text{PERM*SD*(OLDD(N)} - \text{WLEVEL)/BL2}, \qquad (11.25)$$

where BL2 signifies B^2. In this expression it is assumed that the water level in the overlying leaky stratum is WLEVEL.

The particular test problem used data identical to the Theis problem, with the additional parameter $B = 1.25$ m. Selected results which are contained in Table 11.4 show good agreement between the numerical model and the analytical solution.

11.5 Field Examples

Two field examples from the literature are considered in this section. One refers to a pumping test in a confined aquifer, the second is concerned with a leaky aquifer.

Confined aquifer

Walton (1970, p. 236) describes a pumping test in a confined gravel aquifer. The aquifer, which is approximately 5 m thick, lies at a depth of 85 m below ground level; the initial piezometric level is 20 m below ground level. The aquifer is underlain by impermeable shales and overlain by clays. During the test the discharge rate from the well was 1390 m^3/d for a period of 500 mins.

Drawdown data were obtained at an observation well positioned 250 m from the abstraction well. This data has been analysed using conventional curve-matching techniques and the aquifer parameters are deduced to be

$$T = 125 \, \text{m}^2/\text{d}, \qquad S = 2.0 \times 10^{-5}$$

The test has also been analysed using the numerical model. No information is available about the abstraction-well radius; for this discussion it is assumed to be 0.25 m. After a few trials with various values of the transmissivity and storage coefficient, the parameters were deduced to be

$$T = 155 \, \text{m}^2/\text{d}, \qquad S = 2.0 \times 10^{-5}$$

A comparison between the numerical and field results is given in Fig. 11.5.

The difference between Walton's results and those deduced from the numerical model occurs principally because of the significance of the water contained within the well which was ignored in the curve matching solution. The significance of the water contained in the well is demonstrated in Table 11.5. The abstraction rate of 1390 m^3/d is equivalent to 0.97 m^3 per minute. However, in the first minute the volume of water actually removed from the well is roughly 0.6 m^3, leaving only 0.37 m^3 abstracted from the aquifer. After ten minutes, 0.95 m^3/min is abstracted from the aquifer.

Time–drawdown data were also obtained for the abstraction well (denoted by + in Fig. 11.5); hence the numerical model can be used to investigate the aquifer

Table 11.4 Selected results for leaky aquifer, $B = 1.25$

Time (days)	$\dfrac{4tT}{Sr^2}$	$r = 0.1\,\text{m}$ Drawdown (m)		$\dfrac{4tT}{Sr^2}$	$r = 1.0\,\text{m}$ Drawdown (m)	
		Numerical	Analytical		Numerical	Analytical
5.03×10^{-3}	5.03×10^{-1}	0.051	0.050	5.03×10^{-3}	0.0000	0.0000
2.50×10^{-1}	2.50×10^{1}	2.64	2.65	2.50×10^{-1}	0.006	0.004
1.67	1.67×10^{2}	4.30	4.30	1.67	0.380	0.387
10.2	1.02×10^{3}	5.21	5.21	10.2	1.05	1.05
100.0	1.00×10^{4}	5.30	5.29	100.0	1.13	1.13

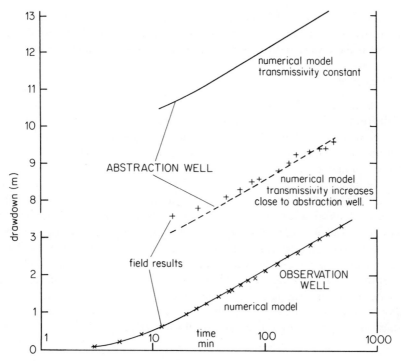

Figure 11.5. Pumping test in confined aquifer and discrete-model simulation

Table 11.5 Relative volumes of water drawn from the aquifer and from the well

Time period	Volume from well	Volume from aquifer
0–1 min	0.6 m³	0.37 m³
1–2 min	0.27 m³	0.70 m³
2–3 min	0.18 m³	0.79 m³
9–10 min	0.02 m³	0.95 m³

behaviour in the vicinity of the abstraction well. With a well radius of 0.25 m, the drawdowns in the abstraction well are as shown by the full line of Fig. 11.5. Clearly the assumption of a homogeneous aquifer does not represent the behaviour in the field. The inclusion of well losses, as suggested by Walton, would lead to even greater drawdowns and therefore *increased* values of transmissivity must be introduced. After a number of attempts, the aquifer was represented as having a transmissivity at the face of the well equal to 8 times the value at the observation well, decreasing to 3 times the standard value at 1.0 m, and 1.5 times the standard value at 10 m. For distances greater than 50 m from the abstraction borehole, the standard transmissivity value of 155 m²/d is used. Since this analysis was carried out, further information has become available (Bruin and Hudson, 1955) which confirms the postulated transmissivities. The broken line of Fig. 11.5 indicates that this variable transmissivity predicts drawdowns that are close to the field values. An improved understanding of the flow processes around the well would only be possible if there were more observation wells close to the abstraction well.

Leaky aquifer

A well-documented pumping test in a leaky aquifer is described by Kruseman and de Ridder (1970). The pumping test was conducted at a site about 1500 m from the river Waal in the Netherlands. Geological sections show an aquifer of mainly medium-coarse and medium-fine sand having a depth of approximately 35 m. The aquifer is underlain by an impervious stratum and overlain by a semi-pervious cover of sandy clays and peat. A constant discharge rate of 761 m³/d was maintained from the abstraction borehole for a period of 8 hours. Drawdown data, obtained at observation wells positioned at 30, 60, 90, and 120 m from the abstraction well, are plotted in Fig. 11.6. The sensitivity analysis used to obtain these results is summarized in Table 11.6. Reference is only made to the observation well at 30 m though in practice account is taken of all the observation wells. Field values are quoted to the left of the table and the remaining columns from left to right indicate how each of the aquifer parameters was altered in turn until an improved approximation was obtained. The final values in the right-hand column were deduced by examining the response at all four wells.

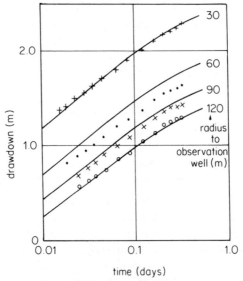

Figure 11.6. Pumping test in leaky aquifer

Figure 11.6 shows the best compromise result. Parameter values are deduced to be:

$$\begin{aligned}
\text{transmissivity} &= 1680 \pm 50 \text{ m}^2/\text{d} \\
\text{storage coefficient} &= 1.5 \pm 0.2 \times 10^{-3} \\
\text{leakage factor} &= 850 \pm 100 \text{ m}
\end{aligned}$$

Table 11.6 Sensitivity analysis, leaky aquifer. Results for observation well at 30 m

T (m²/d)	F	1000	3000	300	2000	2000	2000	2000	1500	1500	1680
S	I E L	1.0×10^{-3}	1.0×10^{-3}	1.0×10^{-3}	1.0×10^{-3}	1.0×10^{-3}	1.0×10^{-2}	1.0×10^{-4}	1.0×10^{-3}	1.5×10^{-3}	1.5×10^{-3}
B (m)	D	100	100	100	100	1000	1000	1000	1000	1000	850
Time (day)						Drawdown (m)					
0.0167	0.140	0.160	0.055	0.436	0.083	0.132	0.065	0.193	0.164	0.149	0.137
0.0497	0.173	0.166	0.055	0.530	0.083	0.163	0.097	0.212	0.207	0.191	0.174
0.0859	0.189	0.166	0.055	0.548	0.083	0.177	0.113	0.216	0.227	0.212	0.191
0.178	0.218	0.166	0.055	0.554	0.083	0.195	0.134	0.219	0.251	0.238	0.214
0.369	0.231	0.166	0.055	0.554	0.083	0.208	0.155	0.219	0.271	0.260	0.231

It was not possible to obtain an improved fit for the observation well at 60 m even though localized changes in transmissivity and leakage factor were modelled. The strata logs indicate that the piezometer is positioned near to a clay bed; this may modify the flow mechanism in the vicinity of this observation well.

11.6 Vertical Components of Flow

The development of the numerical technique to include vertical components of flow is now considered. Significant modifications to the flow mechanism result from these vertical components. Certain examples are included to illustrate the application of the technique.

Derivation

In the derivation of the aquifer model in Section 11.2 the assumption was made that vertical components of flow could be neglected. In many practical situations this condition does not hold. Vertical components of flow can significantly modify the potential distribution, particularly in the vicinity of a well in an unconfined aquifer.

It is possible to represent the vertical components of flow by dividing the aquifer depth into a number of subdivisions, thereby introducing a mesh in the vertical as well as the radial direction (Boulton, 1951). However, the significant increase in unknown potentials leads to a very large increase in the computing time, making the technique too expensive to be used for pumping-test analysis.

Instead, the simplest possible idealization is introduced to model approximately the vertical flow components. The drawdown is defined at two levels; at the free surface with drawdown s_f, and at one-quarter of the saturated depth above the base of the aquifer, where the drawdown is written as s_b. These drawdowns are illustrated in Fig. 11.7(a). For confined aquifers s_f represents the drawdown at the top of the aquifer.

The differential equation which allows for vertical components of flow, Eq. 2.25, when multiplied through by the saturated depth m, becomes

$$\frac{\partial}{\partial r}\left(mk_r\frac{\partial s}{\partial r}\right) + \frac{m}{r}k_r\frac{\partial s}{\partial r} + mk_z\frac{\partial^2 s}{\partial z^2} = S\frac{\partial s}{\partial t} \tag{11.26}$$

where the storage coefficient $S = S_s m$.

Release of water from storage can occur in two ways. Firstly, a change in groundwater potential results in a change in pressure in the aquifer, and therefore the quantity released for each unit volume of aquifer depends on the specific storage S_s. However, for an unconfined aquifer the release of water from storage depends primarily on the change in the free-surface drawdown, s_f, and equals $S_y \partial s_f/\partial t$, where S_y is the specific yield. This quantity is combined with any recharge or flows into the aquifer at the water-table elevation. Consequently the release of water at the free surface is a boundary condition at the upper surface.

Having specified the differential equation and the boundary conditions, the derivation of the numerical model can now be considered.

The first two terms of Eq. 11.26 are rewritten as in Eqs. 11.1 and 11.3, using the relationship $a = \log_e r$; thus

$$\frac{\partial}{\partial r}\left(mk_r\frac{\partial s}{\partial r}\right) + \frac{m}{r}k_r\frac{\partial s}{\partial r} = \frac{mk_r}{r^2}\frac{\partial^2 s}{\partial a^2}$$

Using a constant mesh interval Δa, these terms are represented by a finite-difference approximation as described in Section 11.2.

However, the flow is represented as taking place in an upper and a lower region and therefore the equivalent horizontal hydraulic resistances are double the value given by Eq. 11.5. Thus

$$HU_n = HL_n = \frac{2\Delta a^2}{mk_r} \tag{11.27}$$

where U and L signify upper and lower.

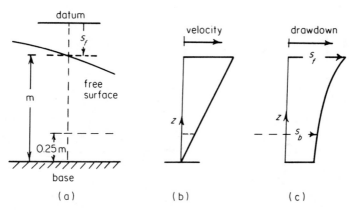

Figure 11.7. Definitions of drawdowns s_f and s_b

Next consider the vertical flow term $mk_z\partial^2 s/\partial z^2$; this can be expressed in terms of the two drawdowns s_f and s_b. Assuming that the vertical flow component decreases uniformly from a maximum at the free surface to zero at the impermeable base of the aquifer (Fig. 11.7(b)),

$$v_z = cz$$

where c is an unknown constant. As a consequence the drawdown will take a parabolic form

$$s = 0.5\,cz^2 + d \tag{11.28}$$

Since the drawdowns s_f and s_b are defined at $z = m$ and $z = 0.25m$ respectively, then

$$c = \frac{s_f - s_b}{0.468\,75m^2}$$

256

Therefore at node n,

$$mk_z \frac{\partial^2 s}{\partial z^2} = \frac{k_z(s_{fn} - s_{bn})}{0.468\,75m} \tag{11.29}$$

Then multiplying by r^2 as in the derivation of Eq. 11.3, a vertical hydraulic resistance can be written as

$$V_n = \frac{0.468\,75m}{k_z r_n^2} \tag{11.30}$$

Consequently the discrete model takes the form shown in Fig. 11.8. It is important to note that for an unconfined aquifer the 'time resistances' which are defined in Eq. 11.6 are connected to the upper line of nodes, thereby representing the water released from storage entering the top of the aquifer. For a confined aquifer the storage coefficient equals the specific storage multiplied by the aquifer thickness; this is represented by 'time resistances' of double the value of Eq. 11.6 connected to both the upper and lower lines of nodes.

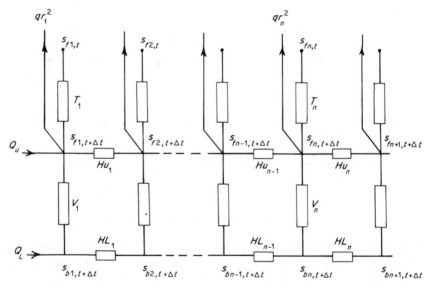

Figure 11.8. Discrete model allowing for vertical flow components

Distribution of discharge

Since the vertical coordinate is now represented in the numerical model, it is possible to represent very approximately the distribution of discharge on the well face. The total quantity withdrawn from the well is divided into two parts, a portion from the upper region of the aquifer Q_U and another portion from the lower region, Q_L.

Features of the aquifer such as fissures and regions of higher permeability together with the form of the well construction have a significant effect on the distribution of the discharge on the well face. Little is known about this distribution, but in a homogeneous unconfined aquifer with no screening in the well, detailed three-dimensional studies indicate that between 80 and 90 per cent of the total discharge is withdrawn from the lower part of the aquifer. This distribution occurs because the upper region of the well face acts as a seepage face and therefore the quantity flowing from this region is small.

As suggested above, the form of screening and casing in the well has a significant effect on the flow distribution on the well face. Frequently, solid casing is provided on the seepage face and perforations are only found below the well-water level. In these circumstances the entire abstraction may take place from the lower region of the aquifer.

Discrete model

The equivalent hydraulic network for an unconfined aquifer is sketched in Fig. 11.8. For each radius there are two unknowns, $s_{fn,t+\Delta t}$ and $s_{bn,t+\Delta t}$. Equations for continuity of flow at the upper and lower nodes are as follows:

$$\left[\frac{s_{fn+1} - s_{fn}}{HU_n} + \frac{s_{fn-1} - s_{fn}}{HU_{n-1}} + \frac{s_{bn} - s_{fn}}{V_n}\right]_{t+\Delta t} = \frac{s_{fn,t+\Delta t} - s_{fn,t}}{T_n} + qr_n^2 \tag{11.31}$$

$$\left[\frac{s_{bn+1} - s_{bn}}{HL_n} + \frac{s_{bn-1} - s_{bn}}{HL_{n-1}} + \frac{s_{fn} - s_{bn}}{V_n}\right]_{t+\Delta t} = 0 \tag{11.32}$$

Solutions of these equations for particular problems can be obtained using a specially written elimination routine. Alternatively, standard library matrix routines can be used; the bandwidth of the matrix is five if the unknowns are written in the order $s_{bn-1}, s_{fn-1}, s_{bn}, s_{fn}, s_{bn+1}, s_{fn+1}$, etc. Apart from the solution routine, the computer program is similar to that of Section 11.3. The greater complexity of the routine for solving the simultaneous equations and the fact that there are twice the number of unknowns lead to computing times which are roughly four times greater than when vertical components of flow are not included.

Example

At present there are no exact analytical solutions for time-variant unconfined radial-flow problems. Thus different facets of the numerical solution must be tested separately. As a first example consider the situation illustrated in Fig. 11.9. Radial flow takes place through an aquifer of constant radial permeability k_r from a fixed head H at a radius r_{max}, towards a well of radius r_w discharging at a constant rate Q. If the vertical permeability is so small that the vertical components of flow can be neglected then the *steady-state* drawdown is shown by Huisman (1972, p. 95) to equal

$$s = \frac{Q(\log_e r_{max} - \log_e r_w)}{\pi k(2H - s)} \tag{11.33}$$

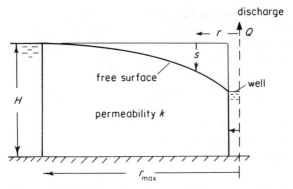

Figure 11.9. Problem involving variable saturated depth

Since the unknown drawdown s occurs on both sides of the expression, a computer program using a searching technique is required to calculate analytical results for particular problems.

Consider the example where $H = 10\,\text{m}$, $r_{max} = 100\,\text{m}$, $k_r = 20\,\text{m/d}$, $k_z = 0.0$, $r_{well} = 0.2\,\text{m}$, and $Q = 800\,\text{m}^3/\text{d}$. Numerical results obtained using Eq. 11.33 are plotted in Fig. 11.10 as a broken line. This problem has also been solved using the numerical model. The condition of no vertical components of flow was enforced by setting k_z to a very low value. The well discharge is distributed uniformly, thus $Q_U = Q_L = 400\,\text{m}^3/\text{d}$. With the storage coefficient set to a low value and the calculation continued for a sufficient time period, a steady-state condition was achieved: this is plotted as the full line in Fig. 11.10. Good agreement between the analytical and numerical results is apparent.

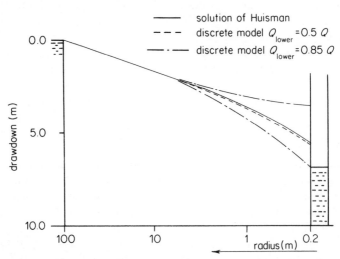

Figure 11.10. Discrete-model results for steady-state flow with variable saturated depth. Broken line indicates uniform discharge distribution on well face; chain-dotted line indicates most of discharge from lower part of the aquifer

The solution described above ignores vertical components of flow which are significant in the vicinity of the abstraction well. A solution which is a better representation of the true behaviour can be obtained by assuming that $Q_U = 0.15Q$, $Q_L = 0.85Q$, and $k_z = k_r$. Results for this case are shown in Fig. 11.10 by the chain-dotted lines. The upper line represents the free-surface profile and corresponds to values of s_f, whilst the lower line plots the average groundwater potential in the lower portion of the aquifer, s_b. This lower line intercepts the face of the well at the position of the water surface in the well.

From this calculation, with an unequal distribution of Q_U and Q_L, the depth of water in the well is calculated to be 3.3 m and the height of the top of the seepage face above the base of the aquifer is 6.5 m. Yet, when vertical flow components are not included, the seepage face is ignored and the well-water level is calculated to be 4.4 m. Therefore, expressions such as Eq. 11.33 which ignore the vertical components of flow give an inadequate picture of the flow mechanism in the vicinity of the abstraction well.

11.7 Delayed Yield

Concept

The concept of delayed yield was introduced by Boulton (1963). He suggested that water is not released from storage instantaneously, but instead it takes a considerable length of time before the total quantity of released water enters the aquifer. Recently there has been discussion on the true mechanism of delayed yield and a number of alternative expressions have been devised. In this discussion the mechanism suggested by Boulton will be used to show how the discrete model can be modified to represent complex mechanisms. The representation of other mechanisms is also possible.

Following Boulton, a change in drawdown does not lead to an instantaneous release of the total quantity of water from storage. During a time period from t to $t + \delta t$ an increase in drawdown δs occurs. This leads to an instantaneous release of water per unit horizontal area $S \delta s$ (where S is the instantaneous storage coefficient) plus a delayed yield to the aquifer which at a later time t' is equal to

$$\delta s \alpha S_y \exp\left[-\alpha(t' - t)\right] \tag{11.34}$$

where S_y is the specific yield of the aquifer and α is the reciprocal of the delay index. The validity of Eq. 11.34 can be checked by integrating with respect to t' between the limits t and infinity; the resultant volume is $S_y \delta s$.

Representation in discrete model

To devise a technique of simulating the delayed-yield concept on the discrete model, the quantity of water entering a typical node during the nth time step will be examined. During the nth time step the time increases from $t_n - \Delta t_n$ to t_n and the

260

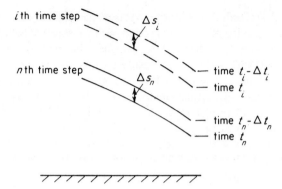

Figure 11.11. Free-surface positions during delayed-
yield calculation

drawdown increases by an amount Δs_n (see Fig. 11.11). Apart from infiltration, there are the following inflow components which all have the dimension [L/T]:

(a) A quantity arising from the *instantaneous storage coefficient*,

$$S\frac{\Delta s_n}{\Delta t_n}$$

(b) The inflow arising from delayed yield due to *previous drawdowns*. During the ith time step $(i < n)$ a drawdown Δs_i occurred. This produces a contribution to the inflow at the later time t_n which, from Eq. 11.34, equals

$$\alpha\, \Delta s_i S_y \exp\left[-\alpha(t_n - t_i)\right]$$

and the total inflow due to all the previous drawdown increments will equal

$$\alpha S_y \sum_{i=1}^{n-1} \Delta s_i \exp\left[-\alpha(t_n - t_i)\right] \qquad (11.35)$$

(c) During the current time step a contribution occurs due to the delayed yield. From Eq. 11.34 the inflow during the current time increment Δt_n is

$$\Delta Q = \int_0^{\Delta t_n} ds\alpha S_y \exp\left[-\alpha(\Delta t_n - t_s)\right] \qquad (11.36)$$

where t_s is the time since the start of the time increment. Assuming a linear increase in head during the time step, then

$$ds = (\Delta s_n/\Delta t_n)\, dt_s$$

Therefore Eq. 11.36 becomes

$$\Delta Q = (\Delta s_n/\Delta t_n)\alpha S_y \int_0^{\Delta t_n} \exp\left[-\alpha(\Delta t_n - t_s)\right] dt_s$$

$$= (\Delta s_n/\Delta t_n)S_y\left[1 - \exp(-\alpha\Delta t_n)\right] \qquad (11.37)$$

Hence the contribution during the current time step can be represented by an equivalent storage coefficient $S_y[1 - \exp(-\alpha\Delta t)]$.

Gathering together these three components, (a), (b), and (c), the right-hand side of Eq. 11.26 becomes

$$\{S + S_y[1 - \exp(-\alpha\Delta t_n)]\}\frac{\Delta s_n}{\Delta t_n} + \alpha S_y \sum_{i=1}^{n-1} \Delta s_i \exp[-\alpha(t_n - t_i)] + q$$

$$(11.38)$$

The effect of delayed yield can therefore be included in the discrete digital model by introducing an effective storage coefficient

$$S' = S + S_y[1 - \exp(-\alpha\Delta t_n)] \qquad (11.39)$$

and an effective recharge

$$q' = \alpha S_y \sum_{i=1}^{n-1} \Delta s_i \exp[-\alpha(t_n - t_i)] + q$$

Following Ehlig and Halepaska (1976) this can be written as

$$q' = \alpha S_y \left[\sum_{i=1}^{n-2} \Delta s_i \exp[-\alpha(t_{n-1} - t_i)] + \frac{\Delta s_{n-1}}{\alpha\Delta t_{n-1}}[1 - \exp(-\alpha\Delta t_{n-1})] \right]$$

$$\exp(-\alpha\Delta t_n) + q \qquad (11.40)$$

Equation 11.40 is convenient to evaluate, since the first term in square brackets is the *summation* from the previous time step, whilst the second term depends on the drawdown of the previous time step.

Therefore delayed yield can be represented by a differential equation with the right-hand side written as

$$S'\frac{\Delta s_n}{\Delta t_n} + q'$$

with S' and q' as defined by Eqs. 11.39 and 11.40.

When delayed yield occurs in an aquifer, the initial response is dominated by the instantaneous storage coefficient, whereas at very long times the response is that of an aquifer with storage coefficient equal to the specific yield. That the model follows this behaviour is apparent from Eq. 11.38. For small times, Δt is small, thus the effective storage coefficient tends to the instantaneous storage coefficient S. However, for long times when Δt becomes large, the effective storage coefficient approaches S_y.

Program and test results

The effect of delayed yield can be included directly in the discrete model by the additional statements indicated in Fig. 11.12. This figure is concerned with the

```
       DIMENSION X(100),Y(100)
C
C    INPUT AQUIFER PARAMETERS
       READ(5,510) PERM,SCON,SUNCON ,ALPHA
510    FORMAT(4F10.5)
       WRITE(6,515)PERM,SCON,SUNCON,ALPHA
515    FORMAT(1X,14HPERMEABILITY=   ,E12.4,19H  CONFINED STORAGE=
      1 E12.4,22H  UNCONFINED STORAGE=     ,E12.4,8H  ALPHA=,F10.5)
       DO 25 N=1,NMAX
25     Y(N)=0.0
C
C
C
C
C    ADDIIIONAL STATEMENTS FOR DELAYED YIELD
50     F=ALPHA*DELT
       IF(F.GT.100.0) GO TO 45
       FACA=EXP(-F)
45     FACB=1.0 - FACA
       FACC=FACB/(ALPHA*DELT)
       DO 55 N=1,NMAX
       X(N) = FACA*Y(N)
55     RECH(N)= ALPHA*SUNCON*X(N) + RCH
       DO 60 NUM=1,4
       DO 70 N=1,NMONE
C
C    TAKE AVERAGE SATURATED DEPTH BETWEEN N AND N+1
       SD=BASE- 0.5*(D(N)+D(N+1))
       STOR=SCON+FACB*SUNCON
       IF(SD.LT.(BASE-TOP)) GO TO 80
       SD=BASE - TOP
       STOR=SCON
80     H(N) = DELA2/(SD*PERM)
70     T(N)=DELT/(STOR*RR(N))
C
C
C
C
C    ADDITIONAL 2 STATEMENTS FOR DELAYED YIELD
       DO 115 N=1,NMAX
115    Y(N)=X(N)+FACC*(D(N)-QLDD(N))
C
C
C
C
       END
```

Figure 11.12. Additional computer statements to represent delayed yield

modification of the model which included only horizontal flow components, but identical changes are made for the case where vertical flow components are also included. The term in square brackets of Eq. 11.40 is stored in array $Y(N)$ and is calculated from the value of $Y(N)$ at the previous time step and the change in drawdown during the previous time step.

Since there are only a few extra statements involved in the calculation of the effective recharge due to the delayed yield, the run time of this program is only slightly longer than that of a program without delayed yield.

This program can be tested by using the following parameters:

$$r_{well} = 0.0001 \text{ m}$$

$$r_{max} = 10000.0 \text{ m}$$

$$T = kd = 10^{-7} \times 250000.0 = 0.025 \text{ m}^2/\text{d}$$

$$S = 0.0001$$

$$S_y = 0.1 \text{ (or } 9 \times 10^{-4})$$

$$Q = 0.1\pi \text{ m}^3/\text{d}$$

$$\alpha = 0.25 \text{ d}^{-1} \quad \text{thus} \quad B = (T/\alpha S_y)^{\frac{1}{2}} = 1.0 \text{ m}$$

$$1/u = 4Tt/S_y r^2 = t/r^2 \text{ (or } 1000 \, t/9r^2)$$

Certain results, with $(S + S_y)/S = 1000$ and 10, are plotted in Fig. 11.13. The lines (both full and broken) refer to analytical results calculated from Boulton's integral expression (Boulton, 1963). For comparison the discrete-model results were obtained with a constant saturated depth and no vertical components of flow. With six mesh intervals and twelve time steps per tenfold increase in radius and time respectively, the discrete-model results are indicated by the various symbols in Fig. 11.13. On the whole, the differences between the analytical and discrete-model results are small, though at large times with $r/B = 2.0$ the differences are greater. If many more time steps are used then the differences can be made smaller.

Field examples

Two field studies will be considered. They each demonstrate that the discrete model can lead to an improved understanding of the aquifer behaviour. The first, a

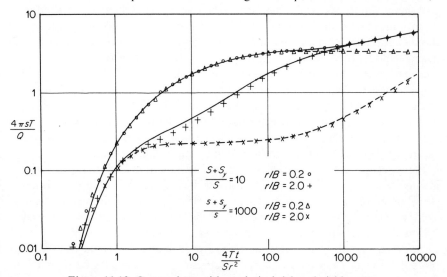

Figure 11.13. Comparison with analytical delayed-yield curves

valuable example of the occurrence of delayed yield in an aquifer, is described by Prickett (1965). The glacial drift aquifer at Lawrenceville (USA) consists of a mixture of sands and gravels and it is assumed to have a constant saturated depth of 30 m. A constant discharge of 5450 m^3/d was withdrawn from a well of 0.4 m diameter for a period of one day. In Fig. 11.14 the drawdowns at a radius of 61 m from the abstraction well are indicated by circles.

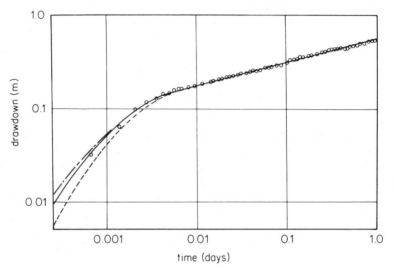

Figure 11.14. Study of delayed yield in glacial drift aquifer. Chain dotted line, $r_w = 0.0001$ m; full line $r_w = 0.2$ m; broken line, $r_w = 0.5$ m

When a solution was obtained using the discrete model with a well radius of 0.2 m, and a decreasing saturated depth, slightly different aquifer parameters from those deduced by Prickett are required for a good fit between field results and the discrete model prediction, which is shown by the full line. The discrete model parameters, with Prickett's values shown in brackets, are as follows: permeability 110 (110) m/d; instantaneous storage coefficient 1.8×10^{-3} (2.33×10^{-3}); specific yield 0.023 (0.0317); and delay index 13.8 (13.8) day^{-1}. These slight differences between the discrete model and those deduced from the classical-type curve methods are due to the effects of water contained in the well and variable saturated depth; the differences are of only minor significance.

In the second example, the classical-type curve method leads to results that are inconsistent. A practical example of delayed yield was introduced by Boulton (1963) in his well-known introductory paper. The aquifer, which consists of coarse sands and gravels, has a saturated depth of about 27 m. Abstraction from a well of 0.45 m diameter continued for a period of 50 hours at a rate of 5880 m^3/d. Drawdowns in two observation wells, one at 22 m and the second at 168 m, are plotted in Fig. 11.15. Separate analyses were carried out using the curve matching techniques. Table 11.7 summarises the parameters deduced by Boulton which

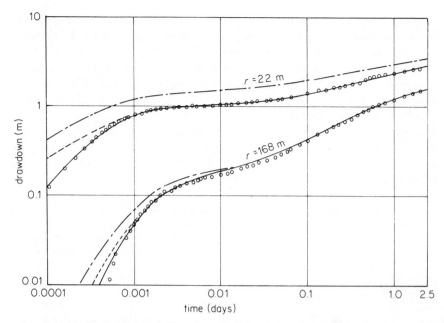

Figure 11.15. Comparison of discrete model with delayed-yield results presented by Boulton (1963). Chain-dotted line, constant parameters; broken lines, variable permeability; full lines, increased abstraction-well diameter

include a common value for the permeability but different values for the instantaneous storage coefficient, specific yield and delay index at two observation wells.

The important question to be resolved by the numerical model is whether Boulton's parameter values are consistent. The steps in obtaining discrete model solution are described below. Certain of the model results are plotted in Fig. 11.15; the field results are indicated by the circles.

(a) For the first trial solution, parameters deduced by Boulton at the two observation wells were incorporated in a single discrete model. The value of the permeability was everywhere 130 m/d, the instantaneous storage coefficient decreases from 27.3×10^{-4} at a radius of 22 m to 6.3×10^{-4} at 168 m, whilst the specific yield shows a similar reduction from 8.8×10^{-2} to 2.0×10^{-2}. An average value, 0.067, was taken for the reciprocal of the delay index. These parameters lead to results on the discrete model that are so unsatisfactory that they are not included on Fig. 11.15.

(b) The next approach was to assume that the parameters deduced by Boulton at the distant observation well apply to all radii. The predicted drawdowns at each of the observation wells are indicated by the chain-dotted line of Fig. 11.15. The fit at the distant observation well is acceptable apart from short times, whilst the drawdowns at the nearer observation well show a roughly constant deviation from the field results.

(c) Since the nearer observation well shows a constant deviation, it seems probable that the agreement could be improved by introducing a permeability which increases towards the well. After a number of trials, the results indicated by the broken lines in Fig. 11.15 were obtained with permeability 1000 m/d at the abstraction well reducing to 130 m/d at distances greater than 100 m. The other parameters which applied to the whole aquifer were $S = 6 \times 10^{-4}$, $S_y = 1.6 \times 10^{-2}$, $\alpha = 0.05 \, \text{day}^{-1}$.

(d) Further improvements at short times can be achieved by setting the well radius equal to 0.9 m, thereby increasing the effect of water within the well. The final results are shown by the full lines in Fig. 11.15, and the parameter values are listed in the right-hand columns of Table 11.7.

Table 11.7 Factors used in delayed-yield example

	Boulton		Discrete model	
	$r = 22\,\text{m}$	$r = 168\,\text{m}$	$r = 22\,\text{m}$	$r = 168\,\text{m}$
$k(\text{m/d})$	129	129	250	130
S	0.002 7	0.000 63	0.000 6	0.000 6
S_y	0.088	0.019 8	0.016	0.016
$\alpha(\text{d}^{-1})$	0.077	0.057	0.05	0.05
effective well radius (m)	(0.0)		(0.9)	

This example has shown that the classical curve-fitting analysis often leads to different values at the two observation wells. When these two sets of results are used in a single solution the results are totally unsatisfactory. This demonstrates one of the serious disadvantages of curve-matching techniques. However, by using the discrete model with a larger effective diameter of the abstraction well and higher permeabilities close to this well, satisfactory agreement has been obtained at both observation wells simultaneously. Further observation-well information, particularly in the vicinity of the abstraction borehole, would be required to confirm that parameters deduced from the discrete model are the correct values.

CHAPTER 12

Non-linear Behaviour in Pumping-test Analysis

12.1 Introduction

The discussion of Chapter 11 has shown that pumping tests can be analysed satisfactorily using a discrete model. In particular, the possibility of including a number of features in a single model has lead to an improved interpretation of certain field examples.

There are other features that occur in pumping tests which cannot be included in conventional analyses. These include non-linear well losses, permeability which varies with depth, tests in which drawdown data are only available at the abstraction well, and the dewatering of aquifers. These will be considered in the following sections.

12.2 Non-linear Well Losses

In all the problems discussed thus far it has been assumed that there are no losses at the well. However, the complex flow patterns in the vicinity of a well often result in head losses which must be added to the changes in head described by Darcy's law.

For example, the conventional approach by Cooper and Jacob (1946) is that the well loss is proportional to the square of the discharge, hence

$$s_w = CQ^2 \tag{12.1}$$

where s_w is the additional drawdown in the well and Q is the discharge. The constant C can be estimated from a step drawdown test using the formula

$$C = \frac{(\Delta s_i/\Delta Q_i) - (\Delta s_{i-1}/\Delta Q_{i-1})}{\Delta Q_{i-1} + \Delta Q_i} \tag{12.2}$$

where Δs represents an increase in drawdown due to an increase in discharge ΔQ for two successive steps, $i - 1$ and i. This approach has proved to be acceptable for certain pumping tests in confined aquifers but is rarely suitable for unconfined-aquifer tests, particularly when the abstraction-well drawdown is a significant proportion of the saturated depth.

The approach adopted in the discrete model is to simulate the non-linear behaviour in the vicinity of the abstraction well by modifying the horizontal hydraulic resistances adjacent to the well. The non-linear resistance takes the value

$$H' = H[1 + F(v/v_0)^n] \qquad (12.3)$$

where H is the normal value of the equivalent horizontal hydraulic resistance (Eqs. 11.5 or 11.27)

F is an unknown well factor

v is the average horizontal velocity at the abstraction well

v_0 is the initial horizontal velocity at the abstraction well

n is a coefficient which usually equals 2.

Important aspects of this expression are that the well losses depend on the velocity, not on the quantity, thereby reflecting the increased well resistance as the saturated depth decreases. Also, coefficients other than the power of two suggested by Cooper and Jacob can be used.

The inclusion of these non-linear hydraulic resistances is straightforward. Only one additional statement is inserted in the program to modify the horizontal hydraulic resistance adjacent to the well, $H(2)$ (or $HU(2)$ and $HL(2)$ in the case where vertical components of flow occur). The statement, inserted immediately after label 70, (Fig. 11.4), takes the form

$$H(2) = H(2)*(1.0 + F*(QABST/(SD*VD))**CN) \qquad (12.4)$$

where VD is a term which represents the initial velocity and equals the initial discharge rate divided by (BASE—WLEVEL), SD is the saturated depth, and CN is the coefficient n. The factors F and CN are deduced by a trial-and-error procedure; in many cases the coefficient CN takes a value around 2.0.

The expression of Eq. 12.4 assumes that all the losses occur very close to the well. However, if the field data indicate that the effect spreads to a certain radius, it is possible to modify all the horizontal hydraulic resistances to that radius. The factor F needs to be reduced in proportion to the number of horizontal hydraulic resistances that are modified.

Example of gravel aquifer

As an example of the importance of non-linear well losses, a pumping test carried out in a gravel aquifer near Hereford, England, is considered (Rushton and Booth, 1976). The Yazor Brook gravels are part of a glacial drift deposit; the sands and gravels are composed dominantly from the local Old Red Sandstone which underlies the area but they contain rock fragments and pebbles. A borehole log for the pumping-test site is to be found in Fig. 12.1; the saturated thickness of the gravels is not constant; at the abstraction well the saturated thickness is 5 m, but 1 km to the east the saturated thickness decreases to 2.7 m, whilst at 1 km to the west it increases to 6.6 m.

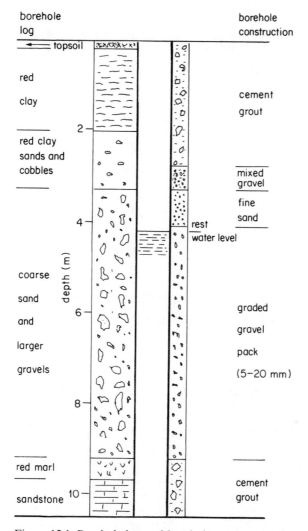

Figure 12.1. Borehole log and borehole construction in
gravel aquifer

A pumping test was carried out with a stepped abstraction rate. The construction of the borehole is shown in Fig. 12.1; the well screen was 350 mm in diameter with slot openings of 1.5 mm. The thickness of the gravel pack was intended to be 75 mm, but due to difficulties during the construction it often extended to 200 mm or more. Four fully penetrating observation boreholes were provided, one in the gravel pack and three at distances of up to 100 m from the abstraction well, as shown in Fig. 12.2. Abstraction well water levels were recorded and found to be virtually identical to the levels in the gravel pack.

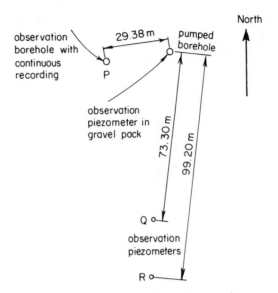

Figure 12.2. Layout of pumping test site

Durations of each step and the quantities abstracted are listed below:

Elapsed time (mins)	Abstraction rate (m³/day)
0– 5 760	1 814
5 760– 6 120	2 056
6 120– 7 140	2 142
7 140–12 975	0

Difficulties were encountered in achieving and maintaining the abstraction rate from the borehole. Therefore the readings from the piezometers at the beginning of each stage tend to be scattered. The abstraction rates were measured at a 'V' notch and from an examination of the field records it seems probable that variations of up to 15 per cent did occur in the discharge rate.

Time–drawdown results for the pumped well and observation well P are included in Fig. 12.3, observation wells R and Q gave similar but smaller readings than P. Note that in this example drawdown is plotted as positive downwards.

Discrete-model solutions for this aquifer are described below. The pumped well, which fully penetrated the aquifer, has a radius of 0.175 m. It is assumed that 85 per cent of the inflow to the well takes place through the lower half of the saturated depth, a value that has been deduced from a number of theoretical analyses. For the initial attempts the gravel aquifer was assumed to be homogeneous, of constant initial saturated depth of 5.0 m, and to extend to a radial distance of 3.1 km. Figure 12.3 contains results for $k = 288$ m/day and $S = 0.3$; the results are a reasonable fit for the first two stages at the pumped well, but otherwise they are unsatisfactory.

On the other hand, Fig. 12.4 indicates that results with $k = 800$ m/day and $S = 0.05$ give a good approximation at the observation well but are

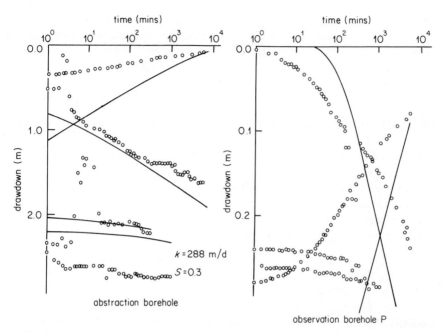

Figure 12.3. Time–drawdown curves at pumped and observation well. Full lines indicate first attempt at modelling field behaviour

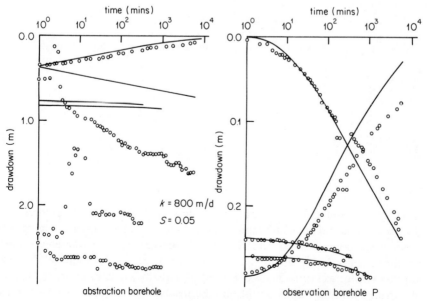

Figure 12.4. Second attempt at modelling field behaviour

unsatisfactory at the abstraction well. Yet a careful examination of the results at the abstraction well for the recovery phase show good agreement. This suggests that these parameters provide an adequate representation of the aquifer behaviour but do not represent the response close to the abstraction well during the pumping phases.

Several other attempts have been made to achieve a better agreement:

(a) The distribution of flow into the well was modified so that all the flow came from the lower portion of the aquifer.
(b) The permeability of the lower portion of the aquifer was set as one-fifth of the upper portion.
(c) A vertical coefficient of permeability, 20 per cent of the horizontal coefficient, was introduced.
(d) The depth of the aquifer was reduced to 3.0 m.
(e) Delayed yield from storage was introduced.

Cases (a) and (d) produce a negligible improvement, though case (e) in which delayed yield is included does give a better fit in the early stages.

For the next attempt a well-loss factor, described by Eq. 12.3, was included in the form of increased hydraulic resistances. In deciding over what region this additional resistance should be applied, account was taken of the construction of the gravel pack: thus the well-loss effect extended to a radius of 0.7 m. The new equivalent hydraulic resistances were:

$$H'(2) = [1 + F(v/v_0)^2]H(2)$$
$$H'(3) = [1 + 0.5F(v/v_0)^2]H(3)$$
$$H'(4) = [1 + 0.2F(v/v_0)^2]H(4)$$

The following parameters were used in the discrete-model solution:

$$\text{Depth} = 5.0 \text{ m}$$
$$k_z = k_r = 705 \text{ m/d}$$
$$S = 0.05 + 0.15(1 - e^{-17t}), \text{ where } t \text{ is the}$$
$$\text{time in days}$$
$$Q_L = 0.85 Q_{total}$$
$$F = 6.3$$

and the results of Fig. 12.5 were obtained. These show only minor differences from the field results and are a great improvement on any of the former model results.

An important question remains: are these the correct parameter values? Further studies have shown that this is not a unique solution. With other parameters, for example $F = 3.0$ and $n = 4$, equally satisfactory results can be obtained. This uncertainty could only be resolved if data were available from additional observation wells positioned close to the abstraction well.

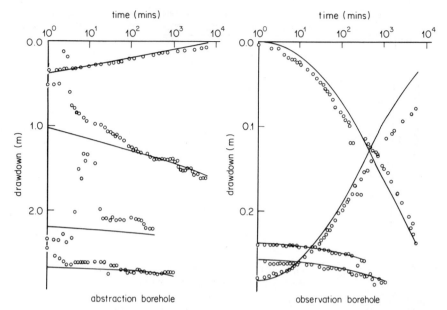

Figure 12.5. Final attempt at modelling field behaviour. Non-linear well losses are included

12.3 Parameters Varying with Depth

In many aquifers, both the permeability and the storage coefficient vary with depth. Frequently the permeability decreases with depth, so that when the water level falls below a more permeable region the borehole yield decreases rapidly. This effect can be included in the discrete model.

Principles

The consequences of the permeability varying with depth require careful examination. Consider, for example, an idealized aquifer having a local permeability variation, as shown in Fig. 12.6(a). For simplicity the aquifer is considered to be only 10 m thick. The local permeability takes a constant value of 10 m/d for 5 m from the base of the aquifer, and then increases linearly to 60 m/d at the top of the aquifer.

Having defined the permeability, it is now necessary to calculate the transmissivity with the water table at different positions. When the *drawdown* of the water table is 6 m, the transmissivity equals the area for the lowest four metres of Fig. 12.6(a), that is $40\,\text{m}^2/\text{d}$. For a drawdown of 2 m, the transmissivity is $125\,\text{m}^2/\text{d}$, and when the aquifer is fully saturated the transmissivity reaches $225\,\text{m}^2/\text{d}$. These values are shown in the third column of Table 12.1 and are plotted in Fig. 12.6(b). By dividing the transmissivity by the saturated depth, the average permeability can be calculated; this is recorded in the fourth column of Table 12.1 and plotted in Fig. 12.6(c).

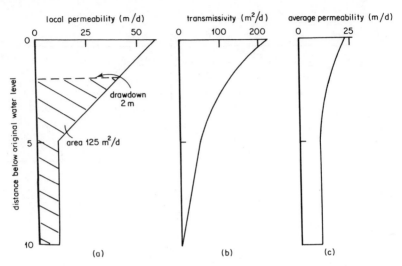

Figure 12.6. Example of relationship between local permeability, transmissivity, and average permeability

Table 12.1 Calculation of average permeability from local permeability variations

Distance below rest-water level (m)	Local permeability (m/d)	Transmissivity (m²/d)	Average permeability (m/d)
0	60	225	22.5
1	50	170	18.9
2	40	125	15.6
3	30	90	12.9
4	20	65	10.8
5	10	50	10
6	10	40	10
7	10	30	10
8	10	20	10
9	10	10	10

If this average permeability k_r is used to calculate the equivalent hydraulic resistance as derived in Eq. 11.5,

$$H_n = \frac{\Delta a^2}{m k_r}$$

then the numerical model can be used directly to represent variations of permeability with depth. At each node, the permeability must be adjusted according to the current drawdown. Thus, in the vicinity of the abstraction well, the average permeability may be significantly smaller than the permeability at a more distant observation point. Variations in storage coefficient can be included in a similar manner.

Example of chalk aquifer

The importance of including the variation of permeability and storage coefficient with depth will be demonstrated by an example of pumping tests in a chalk aquifer; full details are given in Rushton and Chan (1976). Two pumping tests were conducted in the same abstraction well in an unconfined chalk aquifer in England. The tests were separated by a period of four months, during which the rest water level fell by 7.5 m. The abstraction borehole is over 100 m deep with a nominal well diameter of 0.76 m; an observation borehole is situated 220 m from the abstraction borehole. Datum level is defined as the rest water level of the first phase of pumping.

In April 1974 a pumping test (Phase I) was conducted with an abstraction rate of 6450 m³/d for 14 days. Field data which were obtained for drawdown and recovery in both the abstraction and observation boreholes are denoted in Fig. 12.7 by crosses.

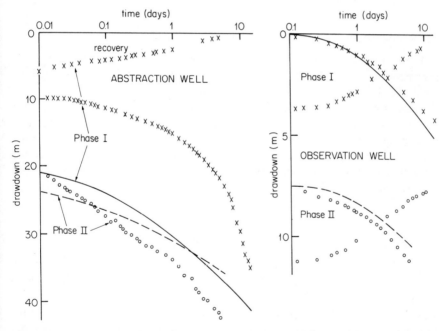

Figure 12.7. Field drawdowns for both abstraction and observation boreholes for Phases I and II. Model prediction using constant permeability and storage coefficient shown by full and broken lines

At the end of July a second test (Phase II) was carried out with a rest water level 7.5 m below that of Phase I. After 6 days, with an abstraction rate of 4650 m³/d, the drawdown in the abstraction borehole exceeded 43 m and it was not possible to maintain the discharge. The test was therefore discontinued. Drawdown and recovery data for the observation borehole and drawdown data for the abstraction well are shown by the circles in Fig. 12.7.

From borehole logs and velocity measurements, it is apparent that significant inflows occur through four major and two minor fissures situated between 10 and 65 m below datum. At depths greater than 65 m below datum there is no significant inflow; thus the effective depth of the aquifer below datum is taken to be 65 m.

Phase I and II drawdown data have been analysed by means of the conventional methods of pumping-test analysis (Walton, 1970). For Phase I the transmissivities were deduced to be in the range 265–345 m^2/d with the storage coefficient 1.0–1.05 $\times$ 10^{-2}. Corresponding values for Phase II are 200–210 m^2/d and 0.9–1.0 $\times$ 10^{-2}. An analysis of the drawdown in the abstraction borehole suggests transmissivities for Phase I and II of 190 and 120 m^2/d respectively. These differences indicate that unless an improved method of analysing the pumping test is available, a reliable interpretation of this test is not possible.

Three attempts at using the discrete model to represent the pumping test behaviour are discussed below.

Constant value of permeability and storage coefficient

The first trial used an average horizontal permeability of 4.6 m/d, a vertical permeability of 1.0 m/d, and a storage coefficient of 0.9 $\times$ 10^{-2}. Decreases in the saturated depth are represented in the solution.

Predictions of the drawdown in both the abstraction and observation wells for the pumping stage of Phase I and II are plotted as full and broken lines respectively in Fig. 12.7. An examination of the results shows that though the predicted and field drawdowns at the observation borehole show fair agreement for both phases, the modelling of the abstraction borehole is totally unacceptable. In particular it should be noted that the predicted drawdowns for Phase I at the abstraction borehole are serious overestimates.

Modification of permeability and storage coefficient to allow for fissures

Borehole logging techniques have identified major fissures at depths of 10, 22, 26, and 44 m below datum and minor fissures at depths of 55 and 65 m. These fissures have been identified at the abstraction well, yet it is unreasonable to assume that they extend at these levels through the aquifer. However, it is possible that the fissures at the abstraction borehole indicate bands of higher permeability and storage coefficient. Figure 12.8(a) has been constructed on the assumption that the major fissures coincide with bands of higher permeability of 50 units and minor fissures a permeability of 5 units, each extending for a depth of 5 metres. The normal intergranular permeability combined with the effect of the inclined joints is taken to be 5 units. Thus Fig. 12.8(a) represents a possible distribution of local permeability with depth.

However, the average permeability over the total saturated depth is the parameter which is required for the discrete model. Accordingly, Fig. 12.8(b) is constructed to relate the average permeability to the saturated depth. The storage

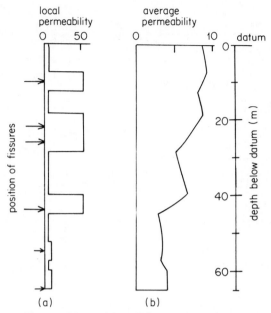

Figure 12.8. Position of fissures and resultant
average permeability

coefficient is assumed to be directly proportional to the local permeability and therefore follows the form of Fig. 12.8(a).

Trial values of permeability and storage coefficient can be obtained by multiplying the values of Fig. 12.8 by appropriate constants. For example, with the average permeability and storage coefficient for zero drawdown equal to 7 m/d and 0.4×10^{-2}, then the discrete-model predictions are as shown in Fig. 12.9. Although the results for the observation borehole are less satisfactory than with constant permeability and storage coefficient, the representation of the abstraction borehole shows some improvement.

Permeability and storage coefficient for best fit

Next an attempt was made to select values of the permeability and storage coefficient to give a best fit between field and model results. Since the model predictions are not very sensitive to changes in storage coefficient, it is assumed that the storage coefficient varies in a similar manner to the average permeability. Values of average permeability are specified at 5 m vertical intervals with intermediate values obtained by fitting cubic curves between four adjacent points.

An acceptable fit, Fig. 12.10, can be obtained with the variations in average permeability and storage coefficient shown in Fig. 12.11(a). Different distributions can be selected to give an improved agreement in certain regions, but quoted values are selected to provide an overall agreement at the abstraction and observation boreholes during pumping and recovery for both phases.

K

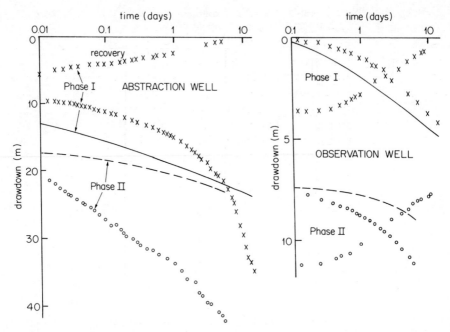

Figure 12.9. Discrete-model predictions when permeability and storage coefficient vary as shown in Figure 12.8

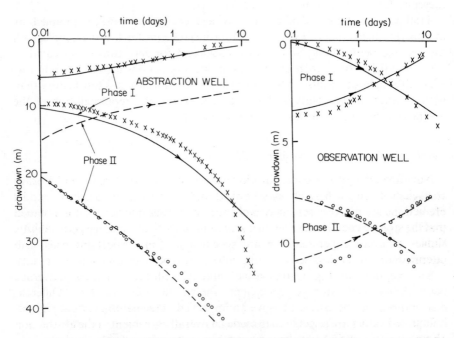

Figure 12.10. Best overall fit using discrete model

From this average permeability distribution it is possible to deduce the equivalent local permeability, Fig. 12.11(b); the calculation of the local permeability is very sensitive to the rate of change of average permeability. Nevertheless, it is clear from the model that there is a highly permeable zone between 5 and 15 m below datum. The negative values for the first 4 m may arise from the influence of the well casing, yet it is equally possible that they originate from the sensitivity of the method of calculation. Since the agreement between the field and model results is less satisfactory below 26 m, little reliance can be placed on the quoted local permeabilities below that depth.

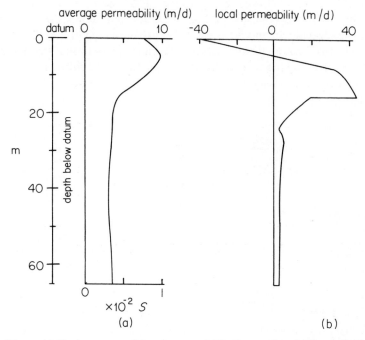

Figure 12.11. Average and local permeability for results of Figure 12.10

An assumption that has been made in this study is that the permeability and storage coefficient are functions of drawdown but are not functions of radius. However, field evidence does suggest that these parameters do vary with distance from the abstraction well, and further studies by Connorton and Reed (1978) have shown that an improved fit can be obtained when a radial variation in permeability is included.

12.4 Single-well Tests

In many situations, pumping tests are carried out without the provision of any observation wells. Therefore the only information available is the variation with time of the water level within the abstraction well. The numerical model is well suited to the examination of such a situation. This section describes preliminary studies of this topic.

First example

The approach will be illustrated by considering a particular example. An abstraction borehole of 0.1 m diameter extends through an impermeable cover 10 m thick into a sandstone aquifer. The well is cased from ground level to a depth of 12 m. The test consisted of a phase of nearly 3 days at a constant abstraction rate of 2.5 litres/sec followed by recovery of 2.5 hours. This was followed immediately by a stepped test with rates of 1.0, 1.5, and 2.25 l/s, each lasting for 2 hours. Time–drawdown and recovery curves are presented in Fig. 12.12. Broken lines are plotted through the observed values; it is apparent that there was some difficulty in achieving the correct discharge rate during the third phase. Since the maximum drawdown was only 80 cm, the measured values are probably liable to an error of about ± 2 cm.

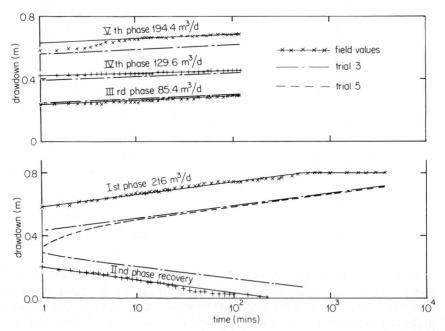

Figure 12.12. Field results for single-well test and trials 3 and 5 (first example) using discrete model

The results indicate that this is a confined aquifer, and with no evidence of vertical flow components the simpler discrete model of Section 11.3 is used. The well and maximum radius are set to be 0.05 and 10^5 m respectively. Abstraction rates are listed in Table 12.2.

Table 12.2 Abstraction rates for pumping test

Quantity m³/d	l/s	Finish time (days)
216	2.5	2.67
0	0	3.0
86	1.0	3.083
130	1.5	3.167
194	2.25	3.25

The procedure followed in obtaining an adequate representation of the pumping test is outlined in Table 12.3. In this table twelve trial solutions are considered and the drawdowns at certain specific times are tabulated. This allows comparisons to be made between the field results, which are contained in the first column, and calculated values from the numerical model.

Trials 1 to 4 are the initial attempts at obtaining a fit; one of these, trial 3, is plotted as a chain-dotted line in Fig. 12.12. None of these results exhibits the curvature shown in Phase I at larger times. In an attempt to understand the cause of this curvature, the effective well radius was increased to 0.25 m yet the behaviour for Phase I, which is indicated by the dashed line of Fig. 12.12, shows little improvement.

In trials 6 and 7 the well radius was set to its original value of 0.05 m and the outer radius on which the drawdown remains zero is reduced to 10 km then to 1 km. This last trial approximates to the field behaviour giving a drawdown which remains effectively constant when the time exceeds 100 min. In trials 8 and 9 the outer radius was set at 1600 m and the values of T and S adjusted to give a best overall fit with the field values. This was achieved with a well radius of 0.04 m in trial 9.

A reduced well radius is not physically realistic but it does suggest that there is a head loss close to the well. Therefore the correct radius of 0.05 m is used but the hydraulic resistance for the mesh length adjacent to the well face is doubled, which is equivalent to halving the transmissivity. The factor by which the hydraulic resistance is modified is recorded in the fifth row of Table 12.3. Trials 10, 11, and 12 use a factor of 2.0. Results for trials 10 and 11 are plotted in Fig. 12.13; the values of trial 12 are so close to the others that they are not included in the figure. Any of these last three trials provides an acceptable fit. Thus the aquifer parameters are deduced to be:

transmissivity, $450 \pm 50 \text{ m}^2/\text{d}$

storage coefficient, $7.0 \pm 3.0 \times 10^{-5}$

outer boundary, zero drawdown at 1.6 km

well loss factor, 2.0

Table 12.3 Various trials for single-well test

Trial No.	Field	1	2	3	4	5	6	7	8	9	10	11	12
Trans. (m²/d)		1000	500	500	500	500	500	500	500	450	460	460	460
Storage (10^{-5})		10.0	10.0	100.0	5.0	5.0	5.0	5.0	7.0	7.0	7.0	10.0	4.0
Well radius (m)		0.05	0.05	0.05	0.05	0.25	0.05	0.05	0.05	0.04	0.05	0.05	0.05
Outer radius (m)		100 000	100 000	100 000	100 000	100 000	10 000	1000	1600	1600	1600	1600	1600
Factor × $H(2)$		1.0	1.0	1.0	1.0	1.0	1.0	1.0	1.0	1.0	2.0	2.0	2.0
Phase I													
0.9 min	0.58	0.26	0.50	0.43	0.53	0.32	0.53	0.53	0.53	0.59	0.59	0.57	0.60
103	0.74	0.35	0.67	0.59	0.69	0.58	0.69	0.67	0.68	0.71	0.77	0.75	0.78
307	0.77	0.37	0.71	0.63	0.73	0.62	0.73	0.67	0.71	0.80	0.80	0.79	0.80
918	0.81	0.38	0.75	0.67	0.77	0.66	0.77	0.67	0.71	0.80	0.80	0.80	0.80
3840	0.81	0.41	0.80	0.72	0.82	0.71	0.82	0.67	0.71	0.80	0.80	0.80	0.80
Phase II													
0.9 min	0.20	0.14	0.29	0.29	0.29	0.40	0.29	0.14	0.19	0.21	0.21	0.23	0.19
9.6	0.12	0.10	0.21	0.21	0.21	0.21	0.21	0.06	0.11	0.12	0.12	0.14	0.10
103	0.03	0.06	0.13	0.13	0.13	0.12	0.13	0.00	0.03	0.03	0.03	0.04	0.02
Phase III													
0.9 min	0.23	0.14	0.28	0.25	0.29	0.20	0.29	0.21	0.21	0.24	0.24	0.23	0.24
120	0.29	0.17	0.34	0.31	0.35	0.30	0.35	0.27	0.28	0.31	0.31	0.31	0.32
Phase IV													
0.9 min	0.42	0.23	0.44	0.39	0.45	0.31	0.45	0.38	0.38	0.43	0.43	0.42	0.44
120	0.46	0.25	0.48	0.43	0.49	0.42	0.49	0.40	0.42	0.47	0.47	0.47	0.48
Phase V													
0.9 min	0.62	0.33	0.63	0.56	0.65	0.52	0.65	0.56	0.57	0.65	0.65	0.64	0.66
120	0.68	0.35	0.69	0.62	0.71	0.61	0.71	0.61	0.63	0.71	0.71	0.70	0.72

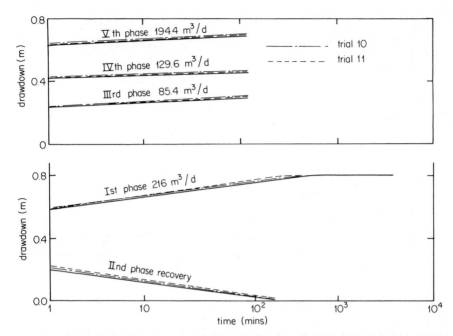

Figure 12.13. Trials 10 and 11 (first example) using discrete model

There is in fact a lake at a distance of about 5 km from the abstraction well but the effect of the fixed head at 1.6 km is more likely to be due to an increased transmissivity at that distance from the abstraction well. If observation-well data were available it would be possible to identify the nature of the boundary.

Second example

It is important to ascertain the accuracy that can be achieved by this method. The accuracy will, to a large extent, depend on the details of each test. Nevertheless, useful information can be gained by typical examples.

One such investigation is described by Rushton (1978). The discrete model was itself used to generate data of the drawdown in an abstraction well. Then an attempt was made to deduce the aquifer parameters from this abstraction-well data. In this way an assessment of the reliability of the technique is possible.

The drawdowns for the abstraction and recovery phases are shown by the full lines of the log-normal plots, Figs. 12.14 and 12.15. It is clear that in the early stages of both the abstraction and recovery there is a marked deviation from a straight line which may be due to the free water in the well, or to well losses. A further important feature is the slight upward curvature after three days of the abstraction phase.

Fifteen solutions were obtained using the numerical model covering a wide range of parameter values. Reference will only be made to the five included in Figs. 12.14 and 12.15; for further details the paper by Rushton (1978) should be

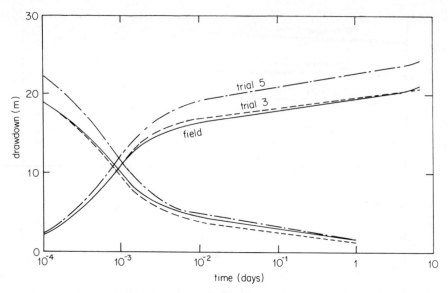

Figure 12.14. Generated field data (full lines) and trials 3 and 5 (second example)

consulted. The parameters used in these five trials are listed in Table 12.4. An examination of trial 3, Fig. 12.14, suggests that the parameters are close to their correct values. It is only the upward curvature for times greater than three days which is not reflected by the numerical model solution. This upward curvature suggests an impermeable boundary but when the condition on the outer boundary is modified to represent an impermeable barrier at 10 km, the results of

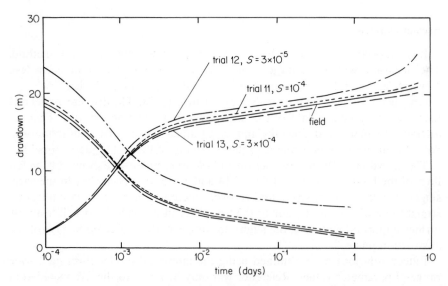

Figure 12.15. Effect of changing the storage coefficient

trial 5 are obtained. Though the upward curvature is now reproduced, there are significant differences from the required drawdowns.

Table 12.4 Aquifer parameters for the trials plotted in Figs. 12.14 and 12.15

Trial No.	3	5	11	12	13
Permeability (m/d)	10	10	9	9	9
Storage coefficient (10^{-5})	10	10	10	3	30
Outer radius (km)	100	10	6	6	6
Type of boundary	F	I	I	I	I
Factor × H(2)	20	20	16	16	16

F indicates fixed boundary
I indicates impermeable boundary

Following further adjustments, the results of trial 11 are obtained. This provides a better overall fit. Trials 11, 12, and 13 examine the sensitivity of the response to the magnitude of the storage coefficient.

Parameter values deduced from the numerical model are listed below. Values in brackets are the original parameters from which the abstraction-well data were generated.

permeability, 8.5 to 9.5 m/d (9.2 m/d)

storage coefficient, 1.3 to 3.0 $\times 10^{-4} (2.7 \times 10^{-4})$

impermeable boundary at 4.3 to 5.0 km (4.0 km)

multiplying factor for well losses, 15 to 17 (17.6)

The accuracy of these estimates is encouraging.

Perhaps the most important conclusion is that if the test had only continued for 3 days then the influence of the boundary would not have been apparent. Consequently there would be no information to question the reliability of the results of trial 3. Each of these studies suggest that a single-well test will only provide realistic estimates of the aquifer parameters when the test is continued for a sufficient time for the boundaries to influence the drawdowns.

12.5 Large Drawdowns Causing Dewatering

One method of exploiting the storage of unconfined aquifers is to regulate the abstraction rates in such a manner that a significant portion of the aquifer is dewatered. Dewatering requires that the aquifer is covered by an array of wells, preferably lying on a square grid of sides b (Fig. 12.16(a)).

As an alternative to the analysis of the complete system of a rectangular array of wells, it is acceptable to consider the equivalent problem of radial flow to a single abstraction well from an impermeable boundary at a radial distance of 0.564b, Fig. 12.16(a) (Rushton and Turner, 1974).

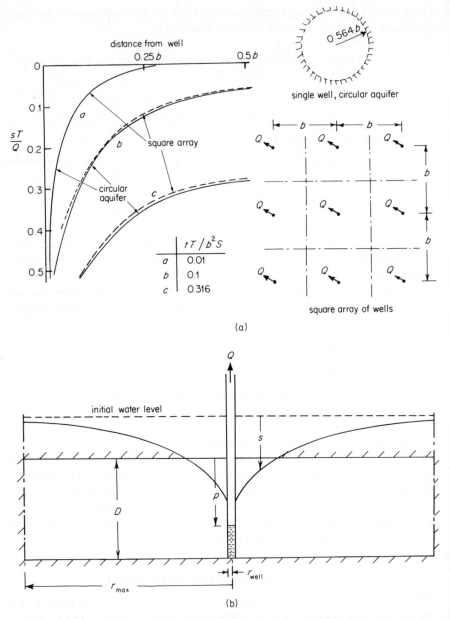

Figure 12.16.(a) Comparison of circular and square aquifers. (b) Parameters for large drawdowns

The plan area of a circular aquifer of radius $0.564b$ equals $1.0b^2$ and Fig. 12.16(a) indicates that the drawdowns for the circular and square aquifer are virtually identical. Therefore the discrete radial model as described in Section 11.6 will be used to examine the dewatering of an aquifer. In the following discussion well losses are not included, consequently a family of representative solutions can be derived.

Since the primary purpose of this analysis is the determination of the total quantity of water that can be abstracted from an aquifer, certain non-dimensional quantities are introduced for the problem sketched in Fig. 12.16(b). The non-dimensional quantities are indicated by a prime.

$$V' = V/\pi D^2 p \quad \text{volume}$$

$$Q' = Q/kDp \quad \text{quantity}$$

$$t' = tk/SD \quad \text{time}$$

$$s' = s/D \quad \text{drawdown}$$

$$p' = p/D \quad \text{maximum drawdown in abstraction well below top of aquifer}$$

$$r'_w = r_w/D \quad \text{well radius}$$

$$r'_{max} = r_{max}/D \quad \text{maximum radius}$$

Note that V is the volume of the aquifer that is dewatered (the actual volume of water removed is $V \times S$), D is the original saturated depth of the aquifer, and p is the maximum permitted lowering below the top of the aquifer in the abstraction well.

Unconfined aquifer

Considering first the case where the aquifer is always unconfined, this situation can be represented by the discrete model with an allowance made for the decrease in saturated depth. The whole of the discharge is taken from the lower region of the abstraction well (i.e. $Q_{lower} = Q_{total}$) since this represents the most severe situation. As a first example, the particular case of a well radius $r'_w = 0.01$ and a maximum radius $r'_{max} = 100.0$ will be studied. The abstraction well is pumped at a steady rate Q' until the drawdown in the well, p', is 0.2. Using the numerical model of Fig. 11.8, values are obtained for the volume of water removed from the aquifer and the time taken to reach the level $p' = 0.2$ in the abstraction well. A range of values of Q' are considered and the results are plotted as the full lines of Fig. 12.17 (a) and (b).

Two limiting cases can be identified. If the permeability is very high then the water table tends to be horizontal as indicated in insets (i) on Fig. 12.17(a) and (b). Consequently the volume dewatered can be calculated directly as

$$V' = V/\pi p D^2 = 10^4$$

Also the time taken to reach this condition is

$$t' = \frac{k}{SD} t = \frac{k\pi}{SD} \frac{(100D)^2 pS}{Q} = \frac{10^4 \pi}{Q'} \qquad (12.5)$$

These relationships are shown by the broken lines of Fig. 12.17(a) and (b).

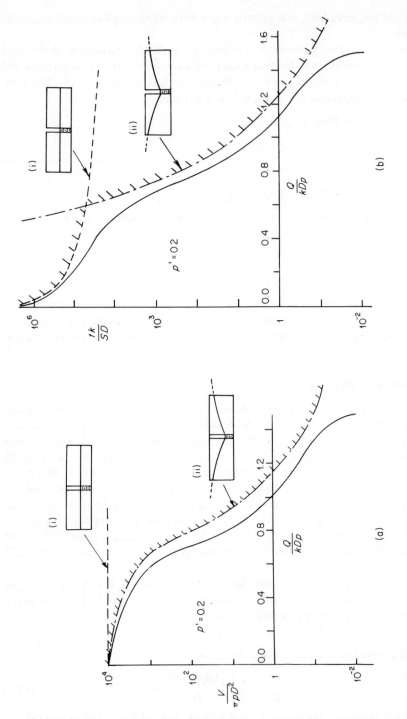

Figure 12.17. Non-dimensional curves compared to limiting cases: (a) maximum volume dewatered; (b) time when drawdown is 0.2

An alternative limiting case is given by the standard Theis analysis (insets (ii) on Fig. 12.17). If it is assumed that the decrease in saturated depth can be neglected and the outer radius extends to infinity then the drawdown at the well is given by the Theis (1935) expression

$$s = \frac{Q}{4\pi kD} \int_{r^2 S/4Tt}^{\infty} \frac{e^{-x}}{x} dx$$

Rearranging and including certain of the non-dimensional terms,

$$\frac{s}{p} = \frac{Q'}{4\pi} \int_{r'^2/4t'}^{\infty} \frac{e^{-x}}{x} dx \tag{12.6}$$

In addition, it is possible to estimate the total volume dewatered from the face of the well to the outer radius. This is given by

$$V = \int_{r_w}^{r_{max}} 2\pi s r \, dr$$

Thus

$$V' = \frac{V}{p\pi D^2} = 2 \int_{r'_w}^{r'_{max}} \frac{s}{p} r' \, dr'$$

$$= \frac{Q'}{2\pi} \int_{r'_w}^{r'_{max}} \int_{r'^2/4t'}^{\infty} \frac{e^{-x}}{x} dx \, r' \, dr' \tag{12.7}$$

The time for the drawdown in the well to reach p can be determined by setting $s/p = 1.0$ and $r' = r'_w$ in Eq. 12.6; tabulated values for the exponential integral are available. The volume of water withdrawn can then be estimated from Eq. 12.7. Numerical integration using the trapezoidal rule with the intervals increasing logarithmically, give acceptable results. These limiting cases are indicated by the chain-dotted lines of Fig. 12.17(a) and (b).

An examination of Fig. 12.17(a) and (b) shows that the actual results, indicated by the full lines, always lie within the limiting cases. For $Q' < 0.2$ the results tend towards the situation when the water table is horizontal, thereafter the curves approach the values derived from the Theis expression; Eqs. 12.6 and 12.7. However, for $Q' > 1.3$, there is a rapid fall in water level in the vicinity of the well. This becomes more severe due to the vertical components of flow and results in a small dewatered volume. In fact for $Q' > 1.5$, the fall of water in the well is so rapid that the well immediately runs dry. This will not occur in practice since the effect of water contained within the well is ignored.

A set of non-dimensional curves are presented in Fig. 12.18 for which $r'_w = 0.01$ and $r'_{max} = 100$; additional sets of curves are presented in Rushton and Turner (1974). Four values of the maximum drawdown are selected: $p' = 0.001, 0.2, 0.5,$ and 0.8. The results for $p' = 0.001$ generally are close to the Theis curve, yet at higher abstraction rates, when $Q' > 1.51$, the drawdown for all four cases increases so rapidly that the well immediately runs dry.

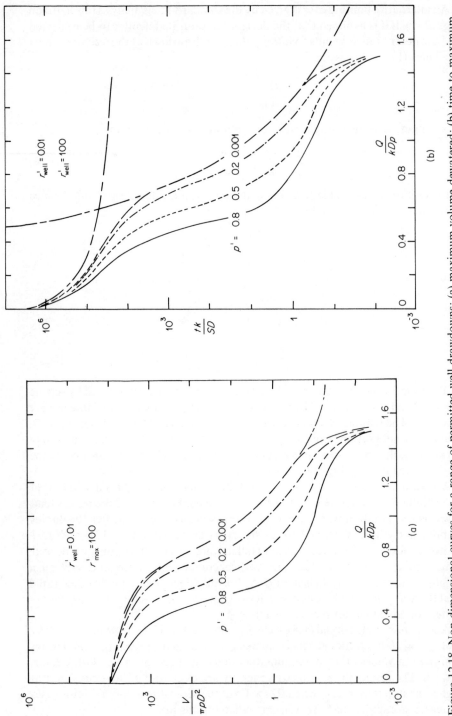

Figure 12.18. Non-dimensional curves for a range of permitted well drawdowns: (a) maximum volume dewatered; (b) time to maximum drawdown

Aquifer initially confined

The above discussion has been concerned with the situation in which the aquifer is always unconfined, and it is convenient to obtain sets of non-dimensional curves. However, when the aquifer is initially confined it is not practicable to include the additional effects in the sets of non-dimensional curves and therefore each particular example has to be considered separately.

Consider a typical borehole which is part of a large well field with wells on a square grid of sides 6 km. Further parameters are:

aquifer thickness, 25 m

permeability, 40 m/d

confined storage coefficient, 0.0003

unconfined storage coefficient, 0.05

well diameter, 0.7 m

initial water level above top of aquifer, 0.0 m or 40.0 m

maximum drawdown in well below top of aquifer, 18.75 m

Using the numerical model, a simulation is carried out in which the abstraction well is pumped until the drawdown below the top of the aquifer is 18.75 m. The time to reach this level is determined together with the quantity removed, expressed as a percentage of the volume of the aquifer above the maximum drawdown line multiplied by the unconfined storage coefficient. Abstraction rates within the range of 1000 to 50 000 m^3/d are considered. Two different initial water levels are modelled, the level coinciding with the top of the aquifer and an initial piezometric level of 40 m above the top of the aquifer.

The results for the first of these cases when the aquifer is always unconfined are recorded in the second and third columns of Table 12.5. They indicate that for low pumping rates, a large percentage of the water can be removed, but this requires a steady abstraction rate for a long period. For example, if 60 per cent of the water is to be removed, the steady abstraction must be maintained for almost ten years.

It is only when the well spacing is reduced to half a kilometre or less that dewatering of the aquifer can be achieved efficiently. These results also show that if the abstraction rate exceeds a certain critical value, in this example 20 000 m^3/d, then the drawdowns in the well increase so rapidly that only a very small proportion of the water is removed.

When the aquifer is initially confined, the behaviour as predicted by the numerical model is recorded in the last three columns of Table 12.5. The fourth column gives the time at which the aquifer first becomes unconfined and the fifth column indicates how long it takes before the maximum drawdown is reached. For the lower pumping rates there is little difference from the situation when the aquifer is initially unconfined; the percentage volume removed is the same though the times taken are longer. However, for the higher abstraction rates the

Table 12.5 Time for well drawdown to become 18.75 m and percentage of water removed. (a) Aquifer always unconfined; (b) Aquifer initially confined

Discharge rate m³/d	(a) Always unconfined		(b) Initially confined		
	Time (days) max. drawdown	% removed	Time (days) unconfined	Time (days) max. drawdown	% removed
1 000	24 630	90	340	25 190	90
5 000	3 426	63	65	3 516	63
10 000	1 085	39	30	1 120	39
15 000	371	21	18.2	404	21
20 000	62	4.7	12.6	88	4.7
25 000	2.2	0.22	9.1	16	0.17
30 000	0.18	0.022	6.9	10.6	0.011
35 000	0.033	0.005	5.3	8.2	0.002
40 000	0.013	0.002	4.1	6.5	0.0006
50 000	0.005	0.001	2.7	4.2	0.0001

volumes removed are far smaller. Initially, the drawdowns in the confined portion spread uniformly, but once the aquifer becomes unconfined in the vicinity of the well, then the drawdowns in the region close to the well increase rapidly and the maximum drawdown condition is reached.

12.6 Conclusion

In Chapters 11 and 12 an alternative method of analysing pumping tests has been introduced. This method should not replace the established curve-fitting methods which do not require a computer; instead they should be seen as being complementary. Usually the initial attempts at interpreting pumping-test results should be carried out by means of curve-fitting methods. Then, if there are some aspects that cannot be explained, the numerical method should be used to explore possible reasons.

The importance of explaining all the features of pumping-test response has been illustrated by many of the examples in these chapters. Misleading interpretations can easily result unless a very critical approach is used. Indeed, it is advisable to check all pumping-test results against a numerical solution even though the curve-matching techniques appear to give an adequate fit.

The program for this numerical method can easily be run on a minicomputer or even on a desk-top computer. It is economical in computing effort; a complete run including several changes in pumping rate takes a similar time to the evaluation of a single analytical value for one of the more sophisticated analytical solutions. Further developments are certain to be made to the numerical technique as more experience is gained in its application to practical problems.

APPENDIX 1

Practical Details of Electrical Analogs

This appendix is intended to complement material to be found throughout the text. Table A1.1 indicates where practical information is given.

Table A1.1

Item	Sections
Resistors	3.10, 7.4, 9.1, 10.3
Capacitors	7.3, 7.4, 9.1, 9.2, 10.3
Digital voltmeters	3.10, 7.5, 7.7
Other measuring equipment	7.5, 7.7
Uniselector switches	7.7
Power supplies	3.10, 7.7
Minicomputer control	7.8

Basic Principles

The basic principle of the electrical analog method, as described in this book, is that a mesh of resistors or capacitors, either two- or three-dimensional, provide by analogy a solution in electrical terms to the governing partial differential equation when expressed in finite-difference form.

Mesh

A square, rectangular, triangular, or alternatively an irregular triangular mesh which offers complete freedom in the choice of nodal positions, may be used. The authors consider that the use of an orthogonal grid has many advantages and that there is a risk of significant errors occurring when a random mesh is used.

A disadvantage of the discrete network compared with a continuous conducting medium is in the representation of irregular-shaped boundaries. Various authors have considered the problem of representing irregular boundaries, namely Redshaw (1948), MacNeal (1953), and Karplus (1958). In each method arguments are based on some form of lumping of the irregular portion of the field into a single resistance and the methods often lead to different values.

In practice it will be found that the very simple method described by Redshaw (1948), in which the resistance value is proportional to the reduced length of the mesh when an irregular boundary cuts a regular mesh, gives a very small error. This point has been discussed at some length by Herbert and Rushton (1966).

Graded Mesh

An advantage of the discrete network over a continuous conducting medium is that a graded network can be used, thus enabling the mesh separation to be increased in regions remote from locations of importance. This device effects a considerable saving in space, number of components, and assembly time.

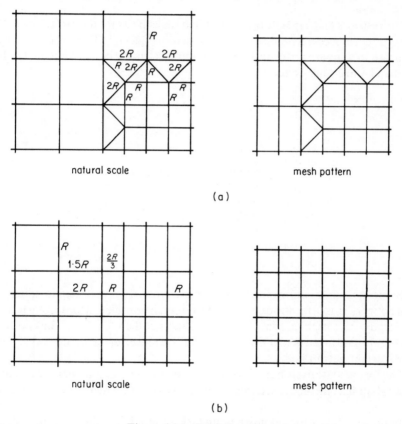

natural scale   mesh pattern

(a)

natural scale   mesh pattern

(b)

Figure A1.1. Graded meshes

There are two fundamental types of graded mesh for orthogonal grids. In one, Fig. A1.1(a), the field is divided into squares or rectangles of increasing size, with diagonal resistors introduced where the mesh changes in size. A particular advantage of this approach is that only two values of resistors are required. Details of this technique have been reported by Palmer and Redshaw (1957).

In the other approach all the mesh lines are continued to the boundary, Fig. A1.1(b). Consequently there are certain mesh points that are not really needed, though their presence does simplify the construction of the networks.

Practical Construction Details

Though there are various methods of constructing networks, the following suggestions relate to one method that has proved to be adequate. An important feature is that components are not built into the circuit by soldering; instead a screwed connection is used, allowing for greater flexibility in changing components.

The resistors and capacitors are mounted in deep square nuts which have two slots at right angles extending for half their depth (Fig. A1.2). These square nuts are mounted in a square array on a panel of non-electrically conducting material by means of valve pins (Redshaw, 1959).

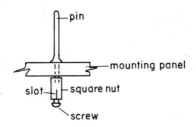

Figure A1.2. Slotted nuts and pins

In Fig. A1.3 the nodal arrangements for two types of network are illustrated: the first network is used for steady-state fixed-boundary problems; the second is suited for steady-state problems in which the position of the boundary is unknown. In Section 7.7 and in particular Fig. 7.5(b), details are given of resistance–capacitance time-variant networks.

The steady-state fixed-boundary network of Fig. A1.3(a) calls for little comment other than that the nodal spacing of 20 mm is suitable when $\frac{1}{2}$-watt resistors are used.

With the network for unknown boundaries, Fig. A1.3(b), there are three slotted nuts and three pins for each mesh intersection point. If the mesh point is within the field then a link consisting of three interconnecting sockets joins all the pins together. On the other hand, if the mesh point is outside the boundary all the pins are disconnected. An examination of Fig. A1.3(b) shows that certain of the groups of pins are connected whilst others are left disconnected. A broken line is drawn showing the approximate position of a boundary corresponding to the connections.

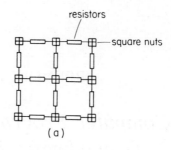

(a)

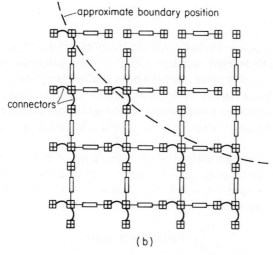

(b)

Figure A1.3. (a) Network for fixed-boundary problems.
(b) Network for unknown boundaries

APPENDIX 2

Digital Computer Program for Regional Groundwater Flow

The program discussed below is intended to form the basis for groundwater flow analysis of single-layered aquifers using a finite-difference method. Many straightforward aquifer problems have been studied using this program, though for more complicated situations (such as variable saturated depth or the change between the confined and unconfined states) additional statements or sections need to be added. In its basic form the program has been run successfully on a minicomputer.

This appendix is supported by a list of the meanings of the important variables, a listing of the computer program, a flow chart, and typical data. Reference should also be made to Chapter 8, which describes the basis of finite-difference digital solutions of groundwater flow problems.

Particular Aspects

Certain aspects of the program are discussed below.

Initial conditions

In this program the initial conditions are obtained as a steady-state solution with the inflows and outflows taking average values. The sections of the program concerned with initial conditions are indicated by IFIRST = −100. The actual steady state-values are calculated using the normal finite-difference equations but with the storage coefficient set to a low value.

After achieving the initial steady-state conditions, the storage coefficients are reset to their correct values and the calculation is cycled for a number of years with the inflows and outflows taking a typical monthly distribution. Thereafter the calculation follows the years of interest.

As an alternative, initial values of the groundwater potential can be read directly into the array HOLD( ).

Known flows

The known inflows and outflows are expressed in annual blocks of monthly values. For unconfined aquifers, monthly averages are usually adequate but for

298

confined aquifers it is often necessary to use averages over smaller periods. Each block is divided into a number of separate items, for example recharge, abstraction, flow to rivers, leakage, etc. This data is stored in the array QFC( ). Each item is distributed between a number of nodes. The distribution of the first item is controlled by the array RCHG( ) which contains the fractional recharge at each node. Other items are identified by nodal positions IW( ), JW( ), and fraction QFLOW( ).

The blocks of data are usually followed consecutively but it is always possible to investigate the effect of taking annual blocks in a different order. In this way a number of dry years can be grouped together.

Time Step

The times at which the calculations are performed are read into the array TDAY( ). The calculation is also performed on the last day of the month.

Calculation of boundary flows

It is important to calculate the flows at fixed head boundaries; the program listing shows the position within the program where this is carried out. The magnitude of the flows depends on the head differences and the transmissivities.

Solution routines

In the listing of the program the SOR routine is used. However the explicit and the ADE routines of Figs. 8.3 and 8.12 can be substituted.

Output

The output resulting from the programme listing is a minimum but it does indicate the various output options that are feasible. A vast amount of information is obtained during each simulation; the selection and output of the significant information is of great importance.

Digital Computer Program

The digital computer program of Fig. A2.1 is designed to run on a 32K minicomputer. The array dimensions can be increased for larger computers. A flow chart is included as Fig. A2.2.

```
C   REGIONAL GROUNDWATER FLOW
        DIMENSION X(16),Y(13), TX(16,13),TY(16,13),H(16,13),HFIX(16,13),
      1 RCHG(16,13),HOLD(16,13),S(16,13),A(16,13),B(16,13),C(16,13),
      2 D(16,13),RS(16,13),NFLOW(5),IW(5,20),JW(5,20),QFLOW(5,20),
      3 NDAY(20,12),QFC(5,20,12),QAV(5),TDAY(15)
C
```

```
C      TOP LEFT HAND CORNER NUMBERED (2,2)
C      M IS NUMBER OF MESH INTERVALS IN HORIZONTAL DIRECTION
C      N IS NUMBER OF MESH INTERVALS IN VERTICAL DIRECTION
C      INPUT NUMBER OF MESH INTERVALS
       READ(5,100)M,N
100    FORMAT(2I5)
C
C      INPUT OVERALL AQUIFER PARAMETERS
       READ(5,110)TRANX,TRANY,STOR,HSTART,RECH
110    FORMAT(2F10.1,F10.5,2F10.1)
       READ(5,120) OFAC,ERROR
120    FORMAT(F10.6,F15.11)
       WRITE(6,130)
130    FORMAT(4X,5HXMESH,2X,5HYMESH,2X,7HTRANS.X,2X,7HTRANS.Y,3X,
      1 27H STORAGE INITIAL.H RECHARGE    ,3X,6HFACTOR,4X,5HERROR)
       WRITE(6,140) M,N,TRANX,TRANY,STOR,HSTART,RECH,OFAC,ERROR
140    FORMAT(1X,I7,2X,I5,3X,6F9.4,F16.10)
       WRITE(6,160)
160    FORMAT(1H0)
C
C      NUMBERING OF BOUNDARY AND FICTITIOUS NODES
       MIN=M+1
       NIN=N+1
       MBOUND=M+2
       NBOUND=N+2
       MFICT=M+3
       NFICT=N+3
C
C      SET OVERALL VALUES IN ARRAYS
       DO 122 I=1,MFICT
       DO 122 J=1,NFICT
       TX(I,J)=TRANX
       TY(I,J)=TRANY
       S(I,J)=STOR
       HOLD(I,J)=HSTART
       RCHG(I,J)=RECH
       HFIX(I,J)=-999999.0
       H(I,J)=0.0
       RS(I,J)=0.0
       A(I,J)=0.0
       B(I,J)=0.0
       C(I,J)=0.0
122    D(I,J)=0.0
C
C      INPUT MESH POSITIONS
       READ(5,170)(X(I),I=1,MFICT)
       WRITE(6,180) (X(I),I=1,MFICT)
       READ(5,170)(Y(J),J=1,NFICT)
       WRITE(6,180) (Y(J),J=1,NFICT)
  170 FORMAT(F10.2)
180    FORMAT(1X,12F10.2)
C
C      INPUT PARAMETERS THAT ARE NON-STANDARD     I=1,J=1 FOR LAST CARD
C      TX(I,J) IS TO RIGHT,TY(I,J) IS BELOW I,J
200    READ (5,210)I,J,TX(I,J),TY(I,J),S(I,J),HOLD(I,J),RCHG(I,J)
210    FORMAT(2I5,2F10.2,F10.6,F10.2,F10.6)
       IF((I.EQ. 1).AND.(J.EQ. 1)) GO TO 220
       GO TO 200
220    CONTINUE
C
C      INPUT FIXED HEADS
230    READ(5,240) I,J,HFIXA
       IF((I.EQ. 1).AND.(J.EQ. 1)) GO TO 250
```

```
240     FORMAT(2I5,F10.5)
        H(I,J)=HFIXA
        HFIX(I,J)=HFIXA
        HOLD(I,J)=HFIXA
        GO TO 230
250     CONTINUE
C
C    FACTORS FOR RIVERS,WELLS ETC.    NFCS=NO OF INPUTS AND OUTPUTS
        READ(5,260) NFCS
260     FORMAT(I10)
        WRITE(6,270)
270     FORMAT(18H WELLS RIVERS ETC.)
        WRITE(6,275)
275     FORMAT(1X,8H I    J  ,8HFRACTION)
        DO 280 NF = 2,NFCS
        WRITE(6,285) NF
285     FORMAT(1X,11HINPUT GROUP,I4)
C   NFLOW(N) = NO   NODES WHERE FLOW IS DISTRIBUTED
        READ(5,260) NFLOW(NF)
        NN=NFLOW(NF)
        DO 280 L= 1,NN
C   IW(),JW() ARE LOCATION, QFLOW IS FRACTION OF FLOW
        READ(5,290)   IW(NF,L),JW(NF,L),QFLOW(NF,L)
280     WRITE(6,290) IW(NF,L),JW(NF,L),QFLOW(NF,L)
290     FORMAT(2I4,F8.5)
C
C    COEFFICIENTS FOR FINITE DIFFERENCE EQUATIONS
        DO 500 I= 2,MBOUND
        DO 500 J= 2,NBOUND
        A(I,J)=2.0*TX(I,J)/((X(I+1)-X(I-1))*(X(I+1)-X(I)))
        C(I,J)=2.0*TX(I-1,J)/((X(I+1)-X(I-1))*(X(I)-X(I-1)))
        B(I,J)=2.0*TY(I,J-1)/((Y(J+1)-Y(J-1))*(Y(J)-Y(J-1)))
500     D(I,J)=2.0*TY(I,J)/((Y(J+1)-Y(J-1))*(Y(J+1)-Y(J)))
C
C    INPUT FLOW BOUNDARIES
        DO 510 INODE=1,1000
        READ (5,520)I,J,AA,BB,CC,DD,SS
520     FORMAT(2I5,5F7.2)
        IF((I.EQ. 1).AND.(J.EQ. 1)) GO TO 540
        A(I,J)=AA*A(I,J)
        B(I,J)=BB*B(I,J)
        C(I,J)=CC*C(I,J)
        D(I,J)=DD*D(I,J)
C    -99.9 OUTSIDE NO FLOW BOUNDARY
        IF(AA.LE.0.000001) HFIX(I+1,J)=-99.9
        IF(BB.LE.0.000001) HFIX(I,J-1)=-99.9
        IF(CC.LT.0.000001) HFIX(I-1,J)=-99.9
        IF(DD.LE.0.000001) HFIX(I,J+1) =-99.9
        IF(AA.LE.0.000001) C(I+1,J) = 0.0
        IF(BB.LE.0.000001) D(I,J-1) = 0.0
        IF(CC.LE.0.000001) A(I-1,J) = 0.0
        IF(DD.LE.0.000001) B(I,J+1) = 0.0
510     S(I,J)=SS*S(I,J)
C
C   SET INITIAL HEADS
540     DO 530 I=1,MFICT
        DO 530 J=1,NFICT
        H(I,J)=HOLD(I,J)
530     IF(HFIX(I,J).EQ.-99.9) H(I,J)=-99.9
C
C   PRINT OUT INITIAL CONDITIONS
        CALL PRIN(TX,1,2,MBOUND,2,NBOUND,TIME)
        CALL PRIN(TY,2,2,MBOUND,2,NBOUND,TIME)
```

```
        CALL PRIN(RCHG,5,2,MBOUND,2,NBOUND,TIME)
        CALL PRIN(S  ,3,2,MBOUND,2,NBOUND,TIME)
        CALL PRIN(HFIX,6,1,MFICT,1,NFICT,TIME)
C
C   NBLOCK IS NO. OF YEARLY BLOCKS OF FLOW DATA
        READ(5,550) NBLOCK
550     FORMAT(I10)
        DO 580 IBLOCK = 1, NBLOCK
        WRITE(6,560) IBLOCK
560     FORMAT(10X,10HBLOCK NO.=,I3)
        DO 580 IMONTH = 1,12
        DO 765 NF = 1,NFCS
765     QFC(NF,IBLOCK,IMONTH) = 0.0
C   NDAY()= NO. DAYS IN MONTH, QFC() = FLOWS IN ML/D
        READ(5,570) NDAY(IBLOCK,IMONTH),(QFC(NF,IBLOCK,IMONTH),NF=1,NFCS)
        WRITE(6,570)NDAY(IBLOCK,IMONTH),(QFC(NF,IBLOCK,IMONTH),NF=1,NFCS)
570     FORMAT(I5,10F7.1)
C
C   CONVERT INPUT VALUES OF ML/D TO M**3/D
        DO 580 NF=1,NFCS
580     QFC(NF,IBLOCK,IMONTH) = QFC(NF,IBLOCK,IMONTH)*1000.0
C
C   CALC AV OF FIRST BLOCK FOR STEADY STATE
        DO 590 NF = 1,NFCS
590     QAV(NF) = 0.0
        DO 600 NF = 1,NFCS
        DO 600 IMONTH = 1,12
600     QAV(NF) = QFC(NF,1,IMONTH) + QAV(NF)
        DO 610 NF = 1,NFCS
610     QAV(NF) = QAV(NF)/12.0
C
C   SET IFIRST NEGATIVE FOR INITIAL STEADY STATE
        IFIRST= -100
C
C   INPUT OF TIMES IN DAYS WHEN CALCN IS PERFORMED
        READ(5,100) KDAY
        DO 620 K = 1,KDAY
620     READ(5,630) TDAY(K)
630     FORMAT(F15.5)
9000    WRITE(6,160)
C
C   TIME INCREASED  :  CHANGE IN YEAR
        DO 700 IYEAR=1,100
        IF(IFIRST.LT.0) GO TO 750
        READ(5,710) IBLOCK
710     FORMAT(I5)
C   IF IBLOCK -VE CALCN. STOPS
        IF(IBLOCK.LT.0) GO TO 8000
        WRITE(6,720) IBLOCK,IYEAR
720     FORMAT(10X,10HBLOCK NO.=,I3,10X,9HYEAR NO.=,I3)
C
C   CHANGE IN MONTH
        DO 730 IMONTH=1,12
C
C   COMBINE ALL FLOWS PER NODE IN RS(I,J); UNITS METRE**3/DAY
C   SPECIAL CALCULATION FOR INITIAL HEADS
        WRITE(6,740)IMONTH,NDAY(IBLOCK,IMONTH),(QFC(NF,IBLOCK,IMONTH),
       1 NF = 1,NFCS)
740     FORMAT(1X,7H MONTH=   ,I4,10H NO. DAYS= ,I4,8H FLOWS=   ,5F12.1)
750     DO 800 I=2,MBOUND
        DO 800 J=2,NBOUND
        IF(IFIRST.LT.0) GO TO 810
        RS(I,J)=RCHG(I,J)*QFC(1,IBLOCK,IMONTH)
        GO TO 800
```

```
810     RS(I,J)=RCHG(I,J)*QAV(1)
800     CONTINUE
        DO 820 N=2,NFCS
        NN=NFLOW(N)
        DO 820 II=1,NN
        I=IW(N,II)
        J=JW(N,II)
        IF(IFIRST.LT.0) GO TO 830
        RS(I,J)=RS(I,J)+(QFC(N,IBLOCK,IMONTH)*QFLOW(N,II))
        GO TO 820
830     RS(I,J)=RS(I,J)+QAV(N)*QFLOW(N,II)
820     CONTINUE
C
C   DIVIDE NODAL FLOW BY AREA TO GIVE METRE/DAY
        DO 840 I=2,MBOUND
        DO 840 J=2,NBOUND
840     RS(I,J)=4.0*RS(I,J)/((X(I+1)-X(I-1))*(Y(J+1)-Y(J-1)))
C
C
C   INCREASE TIME ; CALCULATE DELT
        LDAY=KDAY+1
        DO 900 IDAY=1,LDAY
        IF(IDAY.NE.1) GO TO 910
        DELT=TDAY(1)
        DAYT=TDAY(1)
        GO TO 930
910     IF(IDAY.EQ.LDAY) GO TO 920
        DELT=TDAY(IDAY)-TDAY(IDAY-1)
        DAYT=TDAY(IDAY)
        GO TO 930
920     DAYT=FLOAT(NDAY(IBLOCK,IMONTH))
        DELT=DAYT-TDAY(KDAY)
930     IF(DELT.LE.0.001) GO TO 900
        SFAC=1.0
C
C   START OF S.O.R. CALCULATION
C   MULTIPLYING FACTOR FOR STORAGE
        IF(IFIRST.LT.0) SFAC=0.00000001
C
C   MULTIPLIER AND PREVIOUS TIME STEP FACTORS; USE ARRAYS TX(),TY()
        RDELT=1.0/DELT
        DO 940 I=2,MBOUND
        DO 940 J=2,NBOUND
        HOLD(I,J)=H(I,J)
        TX(I,J)=(SFAC*S(I,J)*RDELT+A(I,J)+B(I,J)+C(I,J)+D(I,J))
940     TY(I,J)=SFAC*S(I,J)*H(I,J)*RDELT
C
C   ITERATION LOOP; MAX NO OF ITERATIONS 300
        DO 950 ICYCLE=1,300
        IND=0
        DO 960 I=2,MBOUND
        DO 960 J=2,NBOUND
        HOLD(I,J)=H(I,J)
        IF(HFIX(I,J).GE.-10000.0) GO TO 970
        AB=A(I,J)*H(I+1,J)+B(I,J)*H(I,J-1)+C(I,J)*H(I-1,J)+D(I,J)*H(I,J+1)
       1+TY(I,J)+RS(I,J)
        IF(ABS(AB-TX(I,J)*HOLD(I,J)).LT.ERROR)GO TO 980
        IND=100
980     H(I,J)=(1.0-OFAC)*HOLD(I,J)+OFAC*AB/TX(I,J)
        GO TO 960
970     H(I,J)=HFIX(I,J)
960     CONTINUE
        IF(IFIRST.LT.0) GO TO 950
        IF(ICYCLE.LT.2) GO TO 950
```

```
            IF(IND.EQ.0) GO TO 990
   950      CONTINUE
            IF(IFIRST.GT.0) GO TO 1000
   C
   C    OUTPUT SECTION FOR INITIAL STEADY HEADS
            IFIRST=100
            WRITE(6,1010)
   1010     FORMAT(1X,27HINITIAL STEADY STATE HEADS )
            CALL PRIN(H,7,2,MBOUND,2,NBOUND,TIME)
            GO TO 9000
   1000     WRITE(6,1020)
   1020     FORMAT(45H CONVERGENCE NOT ACHIEVED IN 300 ITERATIONS   )
   C    END OF SOR ROUTINE
   C
   990      CONTINUE
   C    SECTION FOR CALCULATING FLOW INSERTED HERE
            FLOW=0.0
   C
            WRITE(6,1040)ICYCLE,DAYT,H(2,3),H(3,3),H(4,4),H(5,5)  ,
           1 H(6,7),H(7,7),H(8,8),H(9,9),H(10,10),H(7,5),FLOW
   1040     FORMAT(1X, I5,F10.2,10F9.3,E9.2)
   900      CONTINUE
   730      WRITE(6,160)
            WRITE(6,160)
   C
   C    FULL PRINT OUT AT END OF EACH YEAR
            CALL PRIN(H,7,2,MBOUND,2,NBOUND,DAYT)
   700      WRITE(6,160)
            WRITE(6,160)
   8000     STOP
            END

        SUBROUTINE PRIN(FUNC,NO,IBEG,IEND,JBEG,JEND,TIME)
        DIMENSION FUNC(16,13)
   100 FORMAT(10X,29HTRANSMISSIVITY IN X DIRECTION)
   101 FORMAT(10X,29HTRANSMISSIVITY IN Y DIRECTION)
   102 FORMAT(10X,15HSTORAGE FACTORS)
   103 FORMAT(10X,22HINITIAL VALUES OF HEAD)
   104 FORMAT(10X,15HRECHARGE VALUES)
   105 FORMAT(10X,11HFIXED HEADS)
   106 FORMAT(10X,17HVALUES OF HEAD AT,F6.2,4HDAYS)
   107 FORMAT(1X,1P14E9.2)
   108 FORMAT(1H0)
   110 FORMAT(1X,14(I4,4X))
   111 FORMAT(1X,I3,1X,19F6.2)
   112 FORMAT(3X,19I6)
   115 FORMAT(1X,29H-1.00E+06 SIGNIFIES FREE HEAD,5X,34H-9.99E 1   IS N
           1 OUTSIDE BOUMDARY)
   C
   C
        IF(NO.EQ.7) GO TO 6
        IF(NO.NE.1) GO TO 1
        WRITE(6,100)
        GO TO 7
      1 IF(NO.NE.2) GO TO 2
        WRITE(6,101)
        GO TO 7
      2 IF(NO.NE.3) GO TO 3
        WRITE(6,102)
        GO TO 7
      3 IF(NO.NE.4) GO TO 4
        WRITE(6,103)
        GO TO 7
```

```
  4 IF(NO.NE.5) GO TO 5
    WRITE(6,104)
    GO TO 7
  5 IF(NO.NE.6) GO TO 6
    WRITE(6,105)
    WRITE(6,115)
    GO TO 7
  6 WRITE(6,106)TIME
    WRITE(6,112)(I,I=IBEG,IEND)
    DO 11 J=JBEG,JEND
 11 WRITE(6,111) J,(FUNC(I,J),I=IBEG,IEND)
    GO TO 10
  7 WRITE(6,110)(I,I=IBEG,IEND)
    DO 8 J=JBEG,JEND
  8 WRITE(6,107)(FUNC(I,J),I=IBEG,IEND)
 10 WRITE(6,108)
    RETURN
    END
```

Figure A2.1. Program for regional groundwater flow

Meaning of Important Variables in Computer Program

Units are given in the metre–day system, though any set of consistent units are acceptable.

$A(I,J), B(I,J), C(I,J), D(I,J)$ – coefficients of finite-difference equations (d^{-1})

$H(I,J)$ – head at node I,J (m)

$HFIX(I,J)$ – $\begin{cases} HFIX = -1.0 \times 10^{6}, \text{ head is not fixed} \\ HFIX = -99.9, \text{ node is outside boundary} \\ HFIX > -99.9, \text{ normal fixed head} \end{cases}$

$HOLD(I,J)$ – initial head; head at previous time step or previous iteration (m)

$IW(NF,L)$ – mesh number of node in x-direction

$JW(NF,L)$ – mesh number of node in y-direction

$QFLOW(NF,L)$ – fraction of flow at this node

$NFLOW(N)$ – number of nodes amongst which a flow is distributed

$NDAY(NF,IMONTH)$ – number of days in month in block (d)

$QFC(NF,IBLOCK,IMONTH)$ – quantity of flow (negative for outflow) (Ml/d or m^{3}/d)

$QAV(NF)$ – average of first years flow (m^{3}/d)

$RCHG(I,J)$ – fraction of recharge at node (I,J)

$RS(I,J)$ – total inflow at node (m/d)

$S(I,J)$ – storage coefficient

$TX(I,J)$ – transmissivity in x-direction between nodes I and $I + 1$ (m^{2}/d); later used in SOR calculation

$TY(I,J)$ – transmissivity in y-direction between nodes J and $J + 1$ (m^{2}/d); later used in SOR calculation

$TDAY(K)$ – time at which calculation is performed (d)

$X(J)$ – mesh position in x-direction (m)

$Y(J)$ – mesh position in y-direction (m)

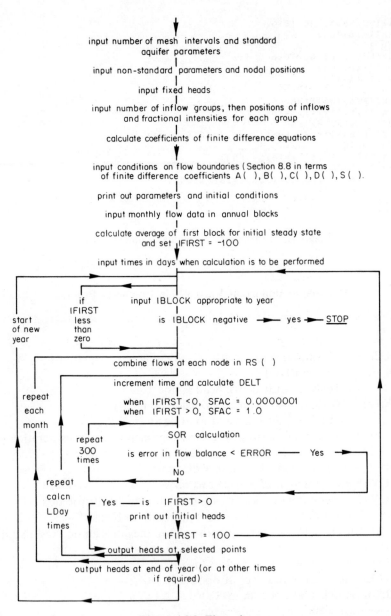

Figure A2.2. Flow chart

AB – several terms of finite-difference equation (m/d)
DAYT – time at which calculation is made (d)
DELT – time increment (d)
ERROR – maximum error in satisfying continuity equation at any one node (m/d)
FLOW – flow across a boundary, calculated by a small section of program (m³/d)
HFIXA – fixed heads (m)

HSTART – initial heads (m)
IBLOCK – counter for blocks
IMONTH – counter for months
IDAY – counter for time steps
ICYCLE – counter for iterations of SOR
IND – if IND = 100, error criterion is not satisfied
INODE – counter for flow boundary nodes
IYEAR – counter for year
IFIRST – when negative the calculation is for the average year
I – mesh number in x-direction
J – mesh number in y-direction
KDAY – number of days when calculation is performed apart from last day of month
L – counter indicating node number
LDAY – total number of time steps in month
MIN,NIN,MBOUND,NBOUND,MFICT,NFICT – nodes near to boundary
M – number of mesh intervals in x-direction
N – number of mesh intervals in y-direction
NBLOCK – total number of blocks
NFCS – number of different types of flow
NF – counter indicating type of flow (maximum value is NFCS)
OFAC – over-relaxation factor
RECH – uniform recharge fraction
STOR – uniform storage coefficient
SFAC – multiplying factor for storage coefficient
TRANX – uniform transmissivity in x-direction (m^2/d)
TRANY – uniform transmissivity in y-direction (m^2/d)

Data

This data represents the simplified problem of Fig. 6.3 but with the aquifer rotated clockwise through 90°.

INPUT

(a) M, N (2I5)–Number of mesh intervals in x- and y-directions

∧ ∧ ∧ ∧6∧ ∧ ∧10

(b) TRANX, TRANY, STOR, HSTART, RECH (2F10.0, F10.5, 2F10.1)– These are overall values of transmissivity in x- and y-directions, storage coefficient, initial heads, and recharge. Recharge is expressed as a fraction of the total recharge

∧ ∧ ∧ ∧ ∧900.0∧ ∧ ∧ ∧ ∧900.0∧ ∧ ∧0.00800∧ ∧ ∧ ∧ ∧ ∧ ∧0.0∧ ∧ ∧ ∧ ∧ ∧ ∧0.0

(c) OFAC, ERROR (F10.6, F15.11) – Over-relaxation factor and error criterion start with OFAC = 1.5, Error = 10^{-5} × average recharge (m/d)

∧ ∧1.500000∧ ∧0.00000001

(d) X(I), I = 1,MFICT – Mesh positions in x-direction including fictitious. One per card Format F10.2

```
∧∧ −1000.0
       0.0
    1000.0
    2000.0
    3000.0
    4000.0
    5000.0
    6000.0
    7000.0
```

(e) Y(I), I = 1,NFICT – As (d) for y-direction

```
∧∧ −1000.0
       0.0
    1000.0
    2000.0
       .
       .
       .
    9000.0
   10000.0
   11000.0
```

(f) I,J,TX(I,J), TY(I,J), S(I,J), HOLD(I,J) RCHG(I,J) (2I5, 2F10.2, F10.6, F10.2, F10.6) – This allows individual values of these parameters to be inserted to overwrite the values read in initially. Termination of this section by I = 1, J = 1

```
∧∧∧∧2∧∧∧∧2∧∧∧∧900.0∧∧∧∧900.0∧∧∧∧0.008000∧∧∧∧∧∧0.00∧∧0.083333
      3      2     900.0     900.0     0.008000           0.00   0.166667
      4      2     900.0     900.0     0.008000           0.00   0.166667
      5      2     900.0     900.0     0.008000           0.00   0.166667
      6      2     900.0     900.0     0.008000           0.00   0.166667
      7      2     900.0     900.0     0.008000           0.00   0.166667
      8      2     900.0     900.0     0.008000           0.00   0.083333
      1      1     900.0     900.0     0.008000           0.00   0.083333
```

(g) I,J,HFIXA   (2I5, F10.5) – Input of fixed heads, terminated by 1,1

```
∧∧∧∧2∧∧∧12∧∧∧0.00
      3    12     0.00
      4    12     0.00
      5    12     0.00
      6    12     0.00
      7    12     0.00
      8    12     0.00
      1     1     0.00
```

(h) NFCS (I10) – This gives the number of types of flows (e.g. abstraction, river, rapid recharge). The first of these is reserved for recharge.

∧ ∧ ∧ ∧ ∧ ∧ ∧ ∧ ∧2

(i) NFLOW(N) (I10)
IW(N,I), JW(N,I), QFLOW(N,I) (2I4,F8.5)
    NFLOW is the number of nodal points at which input N is distributed. IW( ), JW( ) are the locations (remember that 2,2 is top left-hand corner) and QFLOW( ) is the fraction at that node.

∧ ∧ ∧ ∧ ∧ ∧ ∧ ∧ ∧2
    4   6   0.25
    6   10   0.75

(j) I,J,AA,BB,CC,DD,SS (2I5, 5F7.2) – Local values of transmissivity and storage coefficient. This gives a means of applying no flow boundaries. It is terminated by I = 1, J = 1.

∧ ∧ ∧ ∧2∧ ∧ ∧12∧ ∧ ∧0.50∧ ∧ ∧0.50∧ ∧ ∧0.00∧ ∧ ∧0.00∧ ∧ ∧0.25
    2   11   1.00   0.50   0.00   0.50   0.50
          .
          .
          .

∧ ∧ ∧ ∧2∧ ∧ ∧ ∧3∧ ∧ ∧1.00∧ ∧ ∧0.50∧ ∧ ∧0.00∧ ∧ ∧0.50∧ ∧ ∧0.50
    2   2   0.50   0.00   0.00   0.50   0.25
    3   2   0.50   0.00   0.50   1.00   0.50
          .
          .
          .

∧ ∧ ∧ ∧7∧ ∧ ∧ ∧2∧ ∧ ∧0.50∧ ∧ ∧0.00∧ ∧ ∧0.50∧ ∧ ∧1.00∧ ∧ ∧0.50
    8   2   0.00   0.00   0.50   0.50   0.25
    8   3   0.00   0.50   1.00   0.50   0.50
          .
          .
          .

∧ ∧ ∧ ∧8∧ ∧ ∧11∧ ∧ ∧0.00∧ ∧ ∧0.50∧ ∧ ∧1.00∧ ∧ ∧0.50∧ ∧ ∧0.50
    8   12   0.00   0.50   0.50   0.00   0.25
    1   1   0.00   0.50   0.50   0.00   0.25

(k) NBLOCK (I10) – The number of yearly block of data

∧ ∧ ∧ ∧ ∧ ∧ ∧ ∧ ∧2

L

(l)   NDAY, QFC( )   (I5, 10F7.1) – There are NFC values of QFC. NDAY is
the number of days in a month. All data is in Ml/d.

∧∧∧31∧∧∧∧∧0.0∧∧∧ −6.0		
28	0.0	−6.3
31	0.0	−6.5
30	0.0	−6.2
31	2.0	−6.1
30	12.0	−5.9
31	15.0	−5.7
31	20.0	−5.5
30	15.0	−5.5
31	10.0	−5.8
30	4.0	−6.0
31	0.0	−6.5
∧∧∧31∧∧∧∧∧0.0∧∧∧ −6.5		
28	0.0	−6.7
31	0.0	−7.1
30	0.0	−7.8
31	0.0	−8.0
30	0.0	−8.2
31	16.0	−8.2
31	20.0	−7.9
30	14.0	−8.2
31	9.0	−8.3
30	0.0	−8.6
31	0.0	−8.4

(m)   KDAY   (I5) – Number of time steps per month apart from end of month
TDAY( )   (F15.5) – Days at which these steps occur

∧∧∧∧2
∧∧∧∧∧∧∧∧10.0
∧∧∧∧∧∧∧∧20.0

(n)   IBLOCK   (I5) – Order in which blocks of data are used; terminated by a
negative value.

∧∧∧∧1
     1
     1
     1
     1
     2
     2
     1
∧∧∧ −1

APPENDIX 3

Digital Computer Program for Confined Seepage

A simple digital computer program for confined seepage is included as Fig. A3.1.

```
C     CONFINED SEEPAGE
      DIMENSION X(30),Z(20),PX(30,20),PZ(30,20),H(30,20),HFIX(30,20),
     1 HOLD(30,20),A(30,20),B(30,20),C(30,20),D(30,20),CX(30,20)
C     VALUES OF FLOWS ETC. ARE FOR A UNIT THICKNESS OF DAM
C     TOP LEFT HAND CORNER NUMBERED (2,2)
C     M IS NUMBER OF MESH INTERVALS IN HORIZONTAL DIRECTION
C     N IS NUMBER OF MESH INTERVALS IN VERTICAL DIRECTION
C     THE FLOW IS CALCULATED BETWEEN SECTIONS ISECT AND ISECT+1
C     IPRINT EQUALS 1 FOR DETAILED PRINT OUT
C
C     INPUT NUMBER OF MESH INTERVALS
      READ(5,100)M,N,ISECT,IPRINT
100   FORMAT(4I5)
C     INPUT OVERALL AQUIFER PARAMETERS AND OVER RELAXATION FACTOR
C     OVER RELAXATION FACTOR IS USUALLY  1.6
C     ERROR IS 0.0001Q,  Q=MAXIMUM FLOW/UNIT VOLUME
C     Q=AREA*K*I, WHERE I=MAXIMUM HEAD DIFFERENCE/SHORTEST FLOW PATH
C     AREA=DELZ (FOR UNIT THICKNESS)
      READ(5,110)PERMX,PERMZ,OFAC,ERROR
110   FORMAT(4F10.4)
      WRITE(6,120)
120   FORMAT(4X,12HXMESH   ZMESH,5X,31HPERMEABILITY-X    PERMEABILITY-Z,
     1 3X,17HRELAXATION FACTOR,5X,5HERROR)
      WRITE(6,130) M,N,PERMX,PERMZ,OFAC ,ERROR
130   FORMAT(1X,2I7,2F17.3,10X,F7.4,10X,F8.6)
      WRITE(6,160)
160   FORMAT(1H0)
C
C     NUMBERING OF BOUNDARY AND FICTITIOUS NODES
      MIN=M+1
      NIN=N+1
      MBOUND=M+2
      NBOUND=N+2
      MFICT=M+3
      NFICT=N+3
```

```
C                           H(I,J-1)
C                              *
C
C                           PZ(I,J-1)
C
      H(I-1,J)              H(I,J)              H(I+1,J)
```

311

```
C
C                    *  PX(I-1,J)   *    PX(I,J)     *
C
C                         PZ(I,J)
C
C
C
C                              *
C                         H(I,J+1)
C
C
C    SET OVERALL VALUES IN ARRAYS
C    PX AND PZ ARE PERMEABILITY IN X AND Z DIRECTION RESPECTIVELY
C    HOLD IS OLD VALUE OF HEAD (FROM PREVIOUS ITERATION)
C    HFIX IS FIXED HEAD,FREE HEAD IS PRINTED OUT AS -1.00E+06
C    H IS HEAD AT NODE (I,J)
C
C    A,B,C AND D ARE THE COEFFICIENTS OF FINITE DIFFERENCE EQUATION
      DO 150 I=1,MFICT
      DO 150 J=1,NFICT
      PX(I,J)=PERMX
      PZ(I,J)=PERMZ
      HOLD(I,J)=0.0
      HFIX(I,J)=-999999.0
      H(I,J)=0.0
      A(I,J)=0.0
      B(I,J)=0.0
      C(I,J)=0.0
150   D(I,J)=0.0
C
C    INPUT MESH POSITIONS STARTING AND ENDING WITH FICTITIOUS NODES
      READ(5,170)(X(I),I=1,MFICT)
      READ(5,170)(Z(J),J=1,NFICT)
      WRITE(6,175)
      WRITE(6,180) (X(I),I=1,MFICT)
      WRITE(6,160)
      WRITE(6,185)
      WRITE(6,180) (Z(J),J=1,NFICT)
      WRITE(6,160)
170   FORMAT(F10.2)
175   FORMAT(1X,29HMESH POSITIONS IN X DIRECTION)
180   FORMAT(1X,12F10.2)
185   FORMAT(1X,29HMESH POSITIONS IN Z DIRECTION)
C
C    OVER WRITE NON-STANDARD PARAMETERS
C    INPUT PARAMETERS THAT ARE NON-STANDARD    I=1, J=1 FOR LAST CARD
C    PX(I,J) IS TO RIGHT,PZ(I,J) IS BELOW NODE (I,J)
200   READ (5,210)I,J,PX(I,J),PZ(I,J)
210   FORMAT(2I5,2F10.2)
      IF((I.EQ. 1).AND.(J.EQ. 1)) GO TO 220
      GO TO 200
220   CONTINUE
C
C    INPUT FIXED HEADS        I=1, J=1 FOR LAST CARD
230   READ(5,240) I,J,HFIXA
      IF((I.EQ. 1).AND.(J.EQ. 1)) GO TO 250
240   FORMAT(2I5,F10.5)
      H(I,J)=HFIXA
      HFIX(I,J)=HFIXA
      HOLD(I,J)=HFIXA
      GO TO 230
250   CONTINUE
C
C    COEFFICIENTS FOR FINITE DIFFERENCE EQUATIONS
      DO 500 I= 2,MBOUND
```

```
      DO 500 J= 2,NBOUND
      A(I,J)=2.0*PX(I,J)/((X(I+1)-X(I-1))*(X(I+1)-X(I)))
      C(I,J)=2.0*PX(I-1,J)/((X(I+1)-X(I-1))*(X(I)-X(I-1)))
      B(I,J)=2.0*PZ(I,J-1)/((Z(J+1)-Z(J-1))*(Z(J)-Z(J-1)))
500   D(I,J)=2.0*PZ(I,J)/((Z(J+1)-Z(J-1))*(Z(J+1)-Z(J)))
C
C     INPUT IMPERMEABLE BOUNDARIES   (SEE SECTION 8.8)
C     IF IMPERMEABLE, SET APPROPRIATE COEFFICIENT TO ZERO
C     IF LYING ALONG A BOUNDARY, SET COEFFICIENT TO 0.5
C                                        *
C                                       BB
C                                  * CC * AA *
C                                       DD
C                                        *
      DO 510 INODE=1,1000
      READ (5,520)I,J,AA,BB,CC,DD
520   FORMAT(2I5,4F7.2)
      IF((I.EQ. 1).AND.(J.EQ. 1)) GO TO 540
      A(I,J)=AA*A(I,J)
      B(I,J)=BB*B(I,J)
      C(I,J)=CC*C(I,J)
      D(I,J)=DD*D(I,J)
C     FOR NODES OUTSIDE BOUNDARY SET HFIX=-99.9
      IF(AA.LE.0.000001) HFIX(I+1,J)=-99.9
      IF(BB.LE.0.000001) HFIX(I,J-1)=-99.9
      IF(CC.LE.0.000001) HFIX(I-1,J)=-99.9
      IF(DD.LE.0.000001) HFIX(I,J+1)=-99.9
      IF(AA.LE.0.000001) C(I+1,J) = 0.0
      IF(BB.LE.0.000001) D(I,J-1) = 0.0
      IF(CC.LE.0.000001) A(I-1,J) = 0.0
      IF(DD.LE.0.000001) B(I,J+1) = 0.0
510   CONTINUE
C
C     SET INITIAL HEADS IN PROGRAMME
540   DO 530 I=1,MFICT
      DO 530 J=1,NFICT
530   IF(HFIX(I,J).EQ.-99.9) H(I,J)=-99.9
C     PRINT OUT PERMEABILITIES AND FIXED HEADS
      CALL PRIN(PX,1,2,MBOUND,2,NBOUND)
      CALL PRIN(PZ,2,2,MBOUND,2,NBOUND)
      CALL PRIN(HFIX,3,1,MFICT,1,NFICT)
      IF(IPRINT .EQ. 1) WRITE(6,550)
550   FORMAT(1H0,1X,9HITERATION,10X,28HPOTENTIALS AT SELECTED NODES)
C
C     START OF S.O.R. CALCULATION
C     MULTIPLIER CX(I,J) AND PREVIOUS TIME STEP POTENTIALS HOLD(I,J)
      DO 940 I=2,MBOUND
      DO 940 J=2,NBOUND
      HOLD(I,J)=H(I,J)
940   CX(I,J) = A(I,J) + B(I,J) + C(I,J) + D(I,J)
C
C     ITERATION LOOP; MAX NO OF ITERATIONS 300
      DO 950 ICYCLE=1,300
      IND=0
      DO 960 I=2,MBOUND
      DO 960 J=2,NBOUND
      HOLD(I,J)=H(I,J)
      IF(HFIX(I,J).GE.-10000.0) GO TO 970
      AB=A(I,J)*H(I+1,J)+B(I,J)*H(I,J-1)+C(I,J)*H(I-1,J)+D(I,J)*H(I,J+1)
      IF(ABS(AB-CX(I,J)*HOLD(I,J)).LT.ERROR)GO TO 980
      IND=100
980   H(I,J)=(1.0-OFAC)*HOLD(I,J)+OFAC*AB/CX(I,J)
      GO TO 960
```

```
970     H(I,J)=HFIX(I,J)
960     CONTINUE
C
C   OPTIONAL PRINT OUT, OTHER NODAL POSITIONS COULD BE CHOSEN
        IF(IPRINT .NE. 1) GO TO 950
        WRITE(6,1040)ICYCLE,H(2,3),H(3,3),H(4,4),H(5,5),H(6,7),H(7,7),
     1 H(8,6),H(9,5),H(10,4),H(7,5)
1040    FORMAT(1X, I5,11F9.3)
        IF(IND.EQ.0) GO TO 990
950     CONTINUE
990     CONTINUE
C    END OF SOR ROUTINE
C
C    OUTPUT SECTION FOR GROUNDWATER POTENTIALS
        WRITE(6,160)
        WRITE(6,160)
        CALL PRIN(H,4,2,MBOUND,2,NBOUND)
C
C    SECTION FOR CALCULATING FLOW INSERTED HERE
        FLOW=0.0
        I = ISECT
        DELX=X(I+1)-X(I)
        DO 1020 J=2,NBOUND
        DELZ=0.5*(Z(J+1)-Z(J-1))
1020    FLOW=FLOW+((H(I,J)-H(I+1,J))*A(I,J)*DELX*DELZ)
        WRITE(6,1030)ISECT,FLOW
1030    FORMAT(1X,33HFLOW TO RIGHT ACROSS SECTION I = ,I3,8H EQUALS
     1 F12.4,9H M**3/DAY)
        WRITE(6,160)
        STOP
        END

     SUBROUTINE PRIN(FUNC,NO,IBEG,IEND,JBEG,JEND)
     DIMENSION FUNC(25,15)
100  FORMAT(10X,27HPERMEABILITY IN X DIRECTION)
101  FORMAT(10X,27HPERMEABILITY IN Z DIRECTION)
102  FORMAT(10X,11HFIXED HEADS)
103  FORMAT(1X,29H-1.00E+06 SIGNIFIES FREE HEAD,5X,34H-9.99E 1  IS NODE
    1 OUTSIDE BOUNDARY)
104  FORMAT(1X,10I10)
105  FORMAT(1X,I3,1X,10F10.4)
106  FORMAT(1X,I6,12I9)
107  FORMAT(1X,I2,1X,1P13E9.2)
108  FORMAT(1H0)
109  FORMAT(1X,22HGROUNDWATER POTENTIALS)
     IF(NO .EQ. 4) GO TO 3
     IF(NO.NE.1) GO TO 1
     WRITE(6,100)
     GO TO 8
  1  IF(NO.NE.2) GO TO 2
     WRITE(6,101)
     GO TO 8
  2  IF(NO.NE.3) GO TO 3
     WRITE(6,102)
     WRITE(6,103)
     GO TO 8
  3  WRITE(6,109)
     IPBEG = IBEG
  5  IPEND=IPBEG+9
     IF(IPEND .GT. IEND) IPEND=IEND
     WRITE(6,104)(I,I=IPBEG,IPEND)
     DO 4 J=JBEG,JEND
  4  WRITE(6,105) J,(FUNC(I,J),I=IPBEG,IPEND)
     WRITE(6,108)
     IPBEG=IPEND+1
```

```
      IF(IPEND .LT. IEND) GO TO 5
      GO TO 13
 8    IPBEG = IBEG
 7    IPEND=IPBEG+11
      IF(IPEND .GT. IEND) IPEND=IEND
      WRITE(6,106)(I,I=IPBEG,IPEND)
      DO 9 J=JBEG,JEND
 9    WRITE(6,107) J,(FUNC(I,J),I=IPBEG,IPEND)
      WRITE(6,108)
      IPBEG=IPEND+1
      IF(IPEND .LT. IEND) GO TO 7
13    WRITE(6,108)
      RETURN
      END

20      5    10     1
9.6          9.6          1.6      0.0008
   -2.0
    0.0
    2.0
    4.0
    6.0
    8.0
    10.0
    12.0
    14.0
    16.0
    18.0
    20.0
    22.0
    24.0
    26.0
    28.0
    30.0
    32.0
    34.0
    36.0
    38.0
    40.0
    42.0
   -2.0
    0.0
    2.0
    4.0
    6.0
    8.0
    10.0
    12.0
 1      1    1000.0    1000.0
 2      2    5.0
 3      2    5.0
 4      2    5.0
 5      2    5.0
 6      2    5.0
12      2    0.0
13      2    0.0
14      2    0.0
15      2    0.0
16      2    0.0
17      2    0.0
18      2    0.0
19      2    0.0
20      2    0.0
21      2    0.0
22      2    0.0
```

```
 1    1   0.0
 7    2   0.5    0.0    0.5    1.0
 8    2   0.5    0.0    0.5    1.0
 9    2   0.5    0.0    0.5    1.0
10    2   0.5    0.0    0.5    1.0
11    2   0.5    0.0    0.5    1.0
 2    2   0.5    0.0    0.0    0.5
 2    3   1.0    0.5    0.0    0.5
 2    4   1.0    0.5    0.0    0.5
 2    5   1.0    0.5    0.0    0.5
 2    6   1.0    0.5    0.0    0.5
 2    7   0.5    0.5    0.0    0.0
 3    7   0.5    1.0    0.5    0.0
 4    7   0.5    1.0    0.5    0.0
 5    7   0.5    1.0    0.5    0.0
 6    7   0.5    1.0    0.5    0.0
 7    7   0.5    1.0    0.5    0.0
 8    7   0.5    1.0    0.5    0.0
 9    7   0.5    1.0    0.5    0.0
10    7   0.5    1.0    0.5    0.0
11    7   0.5    1.0    0.5    0.0
12    7   0.5    1.0    0.5    0.0
13    7   0.5    1.0    0.5    0.0
14    7   0.5    1.0    0.5    0.0
15    7   0.5    1.0    0.5    0.0
16    7   0.5    1.0    0.5    0.0
17    7   0.5    1.0    0.5    0.0
18    7   0.5    1.0    0.5    0.0
19    7   0.5    1.0    0.5    0.0
20    7   0.5    1.0    0.5    0.0
21    7   0.5    1.0    0.5    0.0
22    7   0.0    0.5    0.5    0.0
22    6   0.0    0.5    1.0    0.5
22    5   0.0    0.5    1.0    0.5
22    4   0.0    0.5    1.0    0.5
22    3   0.0    0.5    1.0    0.5
22    2   0.0    0.0    0.5    0.5
 1    1   0.0    0.0    0.5    0.5
C
C
C
      END
```

Figure A3.1. Program for confined seepage

APPENDIX 4

Digital Computer Program for Unconfined Seepage

A simple digital computer program for unconfined seepage is included as Fig. A4.1.

```
C     UNCONFINED SEEPAGE
      DIMENSION X(25),Z(15),PX(25,15),PZ(25,15),H(25,15),HFIX(25,15),
     1 HOLD(25,15),A(25,15),B(25,15),C(25,15),D(25,15),CX(25,15)
      DIMENSION AR(25,15),BR(25,15),CR(25,15),DR(25,15),HRFIX(25,15)
      DIMENSION HF(25),JN(25),JM(25),IBOUND(25)  ,JNP(25)
C
C     GENERAL SCHEME :  A SERIES OF SOLUTIONS ARE OBTAINED WITH
C     IMPROVED APPROXIMATIONS TO FREE SURFACE POSITION
C     VALUES OF FLOWS ETC. ARE FOR A UNIT THICKNESS OF DAM
C     TOP LEFT HAND CORNER NUMBERED (2,2)
C     M IS NUMBER OF MESH INTERVALS IN HORIZONTAL DIRECTION
C     N IS NUMBER OF MESH INTERVALS IN VERTICAL DIRECTION
C     THE FLOW IS CALCULATED BETWEEN SECTIONS ISECT AND ISECT+1
C     IPRINT EQUALS 1 FOR DETAILED PRINT OUT
C
C     INPUT NUMBER OF MESH INTERVALS
      READ(5,100)M,N,ISECT,IPRINT
100   FORMAT(4I5)
C     INPUT OVERALL AQUIFER PARAMETERS AND OVER RELAXATION FACTOR
C     OVER RELAXATION FACTOR IS USUALLY  1.6
C     ERROR IS 0.0001Q,   Q=MAXIMUM FLOW/UNIT VOLUME
C     Q=AREA*K*I, WHERE I=MAXIMUM HEAD DIFFERENCE/SHORTEST FLOW PATH
C     AREA=DELZ (FOR UNIT THICKNESS)
      READ(5,110)PERMX,PERMZ,OFAC,ERROR
110   FORMAT(4F10.4)
      WRITE(6,120)
120   FORMAT(4X,12HXMESH  ZMESH,5X,31HPERMEABILITY-X   PERMEABILITY-Z,
     1 3X,17HRELAXATION FACTOR,5X,5HERROR)
      WRITE(6,130) M,N,PERMX,PERMZ,OFAC  ,ERROR
130   FORMAT(1X,2I7,2F17.3,10X,F7.4,10X,F8.6)
      WRITE(6,160)
160   FORMAT(1H0)
C
C
C                              H(I,J-1)
C                                 *
C
C                              PZ(I,J-1)
C
C            H(I-1,J)            H(I,J)              H(I+1,J)
C
C              * PX(I-1,J)    *    PX(I,J)      *
```

317

```
C
C                              PZ(I,J)
C
C
C                                 *
C                              H(I,J+1)
C
C
C     NUMBERING OF BOUNDARY AND FICTITIOUS NODES
      MIN=M+1
      NIN=N+1
      MBOUND=M+2
      NBOUND=N+2
      MFICT=M+3
      NFICT=N+3
      ITRIAL = 1
C
C     SET OVERALL VALUES IN ARRAYS
C     PX AND PZ ARE PERMEABILITY IN X AND Z DIRECTION RESPECTIVELY
C     HOLD IS OLD VALUE OF HEAD (FROM PREVIOUS ITERATION)
C     HFIX IS FIXED HEAD,FREE HEAD IS PRINTED OUT AS -1.00E+06
C     H IS HEAD AT NODE (I,J)
C
C     A,B,C AND D ARE THE COEFFICIENTS OF FINITE DIFFERENCE EQUATION
      DO 150 I=1,MFICT
      JNP(I) = 0
      DO 150 J=1,NFICT
      PX(I,J)=PERMX
      PZ(I,J)=PERMZ
      HOLD(I,J)=0.0
      HFIX(I,J)=-999999.0
      H(I,J)=0.0
      A(I,J)=0.0
      B(I,J)=0.0
      C(I,J)=0.0
150   D(I,J)=0.0
C
C     INPUT MESH POSITIONS STARTING AND ENDING WITH FICTITIOUS NODES
      READ(5,170)(X(I),I=1,MFICT)
      READ(5,170)(Z(J),J=1,NFICT)
      WRITE(6,175)
      WRITE(6,180) (X(I),I=1,MFICT)
      WRITE(6,160)
      WRITE(6,185)
      WRITE(6,180) (Z(J),J=1,NFICT)
      WRITE(6,160)
      ZM(2)=Z(2)
      DO 190 J=3,NBOUND
190   ZM(J)=0.5*(Z(J-1)+Z(J))
      ZM(NFICT)=Z(NBOUND)
  170 FORMAT(F10.2)
  175 FORMAT(1X,29HMESH POSITIONS IN X DIRECTION)
  180 FORMAT(1X,12F10.2)
  185 FORMAT(1X,29HMESH POSITIONS IN Z DIRECTION)
C
C     OVER WRITE NON-STANDARD PARAMETERS
C     INPUT PARAMETERS THAT ARE NON-STANDARD    I=1, J=1 FOR LAST CARD
C     PX(I,J) IS TO RIGHT,PZ(I,J) IS BELOW NODE (I,J)
200   READ (5,210)I,J,PX(I,J),PZ(I,J)
210   FORMAT(2I5,2F10.2)
      IF((I.EQ. 1).AND.(J.EQ. 1)) GO TO 220
      GO TO 200
220   CONTINUE
```

```
C
C     INPUT FIXED HEADS          I=1, J=1 FOR LAST CARD
230   READ(5,240) I,J,HFIXA
      IF((I.EQ. 1).AND.(J.EQ. 1)) GO TO 250
240   FORMAT(2I5,F10.5)
      H(I,J)=HFIXA
      HFIX(I,J)=HFIXA
      HOLD(I,J)=HFIXA
      GO TO 230
250   CONTINUE
C
C     COEFFICIENTS FOR FINITE DIFFERENCE EQUATIONS
      DO 500 I= 2,MBOUND
      DO 500 J= 2,NBOUND
      A(I,J)=2.0*PX(I,J)/((X(I+1)-X(I-1))*(X(I+1)-X(I)))
      C(I,J)=2.0*PX(I-1,J)/((X(I+1)-X(I-1))*(X(I)-X(I-1)))
      B(I,J)=2.0*PZ(I,J-1)/((Z(J+1)-Z(J-1))*(Z(J)-Z(J-1)))
500   D(I,J)=2.0*PZ(I,J)/((Z(J+1)-Z(J-1))*(Z(J+1)-Z(J)))
C
C     INPUT IMPERMEABLE BOUNDARIES
C     IF IMPERMEABLE, SET APPROPRIATE COEFFICIENT TO ZERO
C     IF LYING ALONG A BOUNDARY, SET COEFFICIENT TO 0.5
C                                  *
C                                  BB
C                               * CC * AA *
C                                  DD
C                                  *
      DO 510 INODE=1,1000
      READ (5,520)I,J,AA,BB,CC,DD
520   FORMAT(2I5,4F7.2)
      IF((I.EQ. 1).AND.(J.EQ. 1)) GO TO 540
      A(I,J)=AA*A(I,J)
      B(I,J)=BB*B(I,J)
      C(I,J)=CC*C(I,J)
      D(I,J)=DD*D(I,J)
C     FOR NODES OUTSIDE BOUNDARY SET HFIX=-99.9
      IF(AA.LE.0.000001) HFIX(I+1,J)=-99.9
      IF(BB.LE.0.000001) HFIX(I,J-1)=-99.9
      IF(CC.LE.0.000001) HFIX(I-1,J)=-99.9
      IF(DD.LE.0.000001) HFIX(I,J+1)=-99.9
      IF(AA.LE.0.000001) C(I+1,J) = 0.0
      IF(BB.LE.0.000001) D(I,J-1) = 0.0
      IF(CC.LE.0.000001) A(I-1,J) = 0.0
      IF(DD.LE.0.000001) B(I,J+1) = 0.0
510   CONTINUE
C     SET INITIAL HEADS IN PROGRAMME
540   DO 530 I=1,MFICT
      DO 530 J=1,NFICT
530   IF(HFIX(I,J).EQ.-99.9) H(I,J)=-99.9
C
C     STORE ALL INITIAL INFORMATION IN RESERVE ARRAYS
      DO 2300 I=1,MFICT
      DO 2300 J=1,NFICT
      AR(I,J)=A(I,J)
      BR(I,J)=B(I,J)
      CR(I,J)=C(I,J)
      DR(I,J)=D(I,J)
2300  HRFIX(I,J)=HFIX(I,J)
C
C     PRINT OUT PERMEABILITIES AND FIXED HEADS
      CALL PRIN(PX,1,2,MBOUND,2,NBOUND)
      CALL PRIN(PZ,2,2,MBOUND,2,NBOUND)
      CALL PRIN(HFIX,3,1,MFICT,1,NFICT)
```

```
C
2500    CONTINUE
        IF(IPRINT .EQ. 1) WRITE(6,550)
550     FORMAT(1H0,1X,9HITERATION,10X,28HPOTENTIALS AT SELECTED NODES)
C
C   START OF S.O.R. CALCULATION
C   MULTIPLIER CX(I,J) AND PREVIOUS TIME STEP POTENTIALS HOLD(I,J)
        DO 940 I=2,MBOUND
        DO 940 J=2,NBOUND
        HOLD(I,J)=H(I,J)
940     CX(I,J) = A(I,J) + B(I,J) + C(I,J) + D(I,J)
C
C   ITERATION LOOP; MAX NO OF ITERATIONS 300
        DO 950 ICYCLE=1,300
        IND=0
        DO 960 I=2,MBOUND
        DO 960 J=2,NBOUND
        HOLD(I,J)=H(I,J)
        IF(HFIX(I,J).GE.-10000.0) GO TO 970
        AB=A(I,J)*H(I+1,J)+B(I,J)*H(I,J-1)+C(I,J)*H(I-1,J)+D(I,J)*H(I,J+1
        IF(ABS(AB-CX(I,J)*HOLD(I,J)).LT.ERROR)GO TO 980
        IND=100
980     H(I,J)=(1.0-OFAC)*HOLD(I,J)+OFAC*AB/CX(I,J)
        GO TO 960
970     H(I,J)=HFIX(I,J)
960     CONTINUE
C
C   OPTIONAL PRINT OUT, OTHER NODAL POSITIONS COULD BE CHOSEN
        IF(IPRINT .NE. 1) GO TO 950
        WRITE(6,1040)ICYCLE,H(2,3),H(3,3),H(4,4),H(5,5),H(6,7),H(7,7),
       1 H(8,6),H(9,5),H(10,4),H(7,5)
1040    FORMAT(1X, I5,11F9.3)

        IF(IND.EQ.0) GO TO 990
950     CONTINUE
990     WRITE(6,160)
C   END OF SOR ROUTINE
C
C   OUTPUT SECTION FOR GROUNDWATER POTENTIALS
        WRITE(6,1050) ITRIAL
1050    FORMAT(11H TRIAL NO= , I5)
        ITRIAL = ITRIAL + 1
        CALL PRIN(H,4,2,MBOUND,2,NBOUND)
C
C   SECTION FOR CALCULATING FLOW INSERTED HERE
        FLOW=0.0
        I = ISECT
        DELX=0.5*(X(I+1)-X(I-1))
        DO 1020 J=2,NBOUND
        DELZ=(Z(J-1)-Z(J+1))/2.0
1020    FLOW=FLOW+((H(I,J)-H(I+1,J))*A(I,J)*DELX*DELZ)
        WRITE(6,1030)ISECT,FLOW
1030    FORMAT(1X,33HFLOW TO RIGHT ACROSS SECTION I = ,I3,8H EQUALS ,
       1 F12.4,9H M**3/DAY)
        WRITE(6,160)
C
C   FIND HF(I) WHERE GROUNDWATER POTENTIAL EQUALS HEIGHT ABOVE DATUM
        DO 2000 I=3, MIN
        JK=1
2010    J=13-JK
        IF(B(I,J).LT.0.0000001)GO TO 2030
        IF(H(I,J).GT.Z(J) .AND. H(I,J-1).LT.Z(J-1))GO TO 2020
        JK=JK+1
        GO TO 2010
```

```
2020   ZZ=Z(J-1)-Z(J)
       DZ=(Z(J)-H(I,J))/(H(I,J-1)-H(I,J)-ZZ)
       HF(I)=Z(J)+DZ*ZZ
       GO TO 2000
2030   ZZ=Z(J-1)-Z(J)
       DZ=(Z(J)-H(I,J))/(H(I,J)-H(I,J+1)-ZZ)
       HF(I)=Z(J)+DZ*ZZ
2000   CONTINUE
       HF(MBOUND)=2.0*HF(MIN)-HF(MIN-1)
C
C    FIND JN(I), THE NEAREST NODE TO HF(I)
       DO 2100 I=3,MBOUND
       IBOUND(I)=-2
       DO 2100 J=2,NBOUND
       IF(HF(I).LT.ZM(J) .AND. HF(I).GT.ZM(J+1))JN(I)=J
2100   CONTINUE
C
C    IF BOUNDARY POSITION HAS NOT CHANGED, STOP CALCULATION
       ITEST=100
       DO 2140 I=3,MBOUND
2140   IF(JN(I) .NE. JNP(I)) ITEST=0
       IF(ITEST .NE. 100) GO TO 2180
       WRITE(6,2190)
2190   FORMAT(48H SUCCESSFUL CONVERGENCE TO FREE SURFACE POSITION)
       GO TO 9999
2180   DO 2170 I=3,MBOUND
2170   JNP(I)=JN(I)
C
C    LEVEL BOUND IBOUND(I)=1, STEP BOUND IBOUND(I)=2
C             LEVEL BOUNDARY                    STEP BOUNDARY
C
C          *      *      *                      *      *
C
C                                                      *      *
C
C             IBOUND(I) = 1                     IBOUND(I) = 2
C
       DO 2110 I=3,MIN
       IF(JN(I).EQ.JN(I+1))IBOUND(I)=1
       IF(JN(I).EQ.JN(I+1)-1)IBOUND(I)=2
       IF(IBOUND(I).EQ.-2)WRITE(6,2130)
       IF(IBOUND(I) .EQ. -2) STOP
2110   CONTINUE
       IF(JN(MBOUND).GT.JN(MIN)+1)WRITE(6,2130)
2130   FORMAT(43H WATER LEVEL FALLS MORE THAN ONE MESH SPACE/
      1 28H SOLUTION BECOMES INACCURATE)
C
C    OPTIONAL OUTPUT TO EXAMINE FORM OF BOUNDARY
       IF(IPRINT .NE. 1) GO TO 2390
       DO 2150 I=3,MBOUND
2150   WRITE(6,2160)I,HF(I),JN(I),IBOUND(I)
2160   FORMAT(1X,I10,F10.3,2I5)
       WRITE(6,160)
C
C    RESET PARAMETERS AS FOR FIRST STEP
2390   DO 2400 I=1,MFICT
       DO 2400 J=1,NFICT
       H(I,J) = 0.0
       A(I,J)=AR(I,J)
       B(I,J)=BR(I,J)
       C(I,J)=CR(I,J)
       D(I,J)=DR(I,J)
2400   HFIX(I,J) = HRFIX(I,J)
```

```
C
C     SET UP NEW BOUNDARY CONDITIONS
      DO 2410 I=3,MIN
      JJ=JN(I)
      IF(IBOUND(I) .EQ. 2) GO TO 2420
      IF(IBOUND(I) .NE. 1) GO TO 2410
C     LEVEL BOUNDARY
      B(I,JJ) = 0.0
      D(I,JJ-1)=0.0
      GO TO 2410
C     STEP BOUNDARY
2420  A(I,JJ) = 0.0
      B(I,JJ) = 0.0
      D(I,JJ-1)=0.0
      C(I+1,JJ)=0.0
2410  CONTINUE
C
C     SET NODES ABOVE TOP OF SEEPAGE FACE TO ZERO
      DO 2430 J=2,NIN
2430  IF(J .LT. JN(MBOUND)) HFIX(MBOUND,J) = 0.0
      GO TO 2500
C
9999  STOP
      END
      SUBROUTINE PRIN(FUNC,NO,IBEG,IEND,JBEG,JEND)
      DIMENSION FUNC(25,15)
100   FORMAT(10X,27HPERMEABILITY IN X DIRECTION)
101   FORMAT(10X,27HPERMEABILITY IN Z DIRECTION)
102   FORMAT(10X,11HFIXED HEADS)
103   FORMAT(1X,29H-1.00E+06 SIGNIFIES FREE HEAD,5X,34H-9.99E 1  IS NODE
     1 OUTSIDE BOUNDARY)
104   FORMAT(1X,10I10)
105   FORMAT(1X,I3,1X,10F10.4)
106   FORMAT(1X,I6,12I9)
107   FORMAT(1X,I2,1X,1P13E9.2)
108   FORMAT(1H0)
109   FORMAT(1X,22HGROUNDWATER POTENTIALS)
      IF(NO .EQ. 4) GO TO 3
      IF(NO.NE.1) GO TO 1
      WRITE(6,100)
      GO TO 8
  1   IF(NO.NE.2) GO TO 2
      WRITE(6,101)
      GO TO 8
  2   IF(NO.NE.3) GO TO 3
      WRITE(6,102)
      WRITE(6,103)
      GO TO 8
  3   WRITE(6,109)
      IPBEG = IBEG
  5   IPEND=IPBEG+9
      IF(IPEND .GT. IEND) IPEND=IEND
      WRITE(6,104)(I,I=IPBEG,IPEND)
      DO 4 J=JBEG,JEND
  4   WRITE(6,105) J,(FUNC(I,J),I=IPBEG,IPEND)
      WRITE(6,108)
      IPBEG=IPEND+1
      IF(IPEND .LT. IEND) GO TO 5
      GO TO 13
  8   IPBEG = IBEG
  7   IPEND=IPBEG+11
      IF(IPEND .GT. IEND) IPEND=IEND
      WRITE(6,106)(I,I=IPBEG,IPEND)
      DO 9 J=JBEG,JEND
```

```
 9 WRITE(6,107) J,(FUNC(I,J),I=IPBEG,IPEND)
   WRITE(6,108)
   IPBEG=IPEND+1
   IF(IPEND .LT. IEND) GO TO 7
13 WRITE(6,108)
   RETURN
   END
10    10    2    2
7.2        7.2         1.6       0.0007
  -2.0
   0.0
   2.0
   4.0
   6.0
   8.0
  10.0
  12.0
  14.0
  16.0
  18.0
  20.0
  22.0
  22.0
  20.0
  18.0
  16.0
  14.0
  12.0
  10.0
   8.0
   6.0
   4.0
   2.0
   0.0
  -2.0
 1     1    1000.0    1000.0
 2     2    20.0
 2     3    20.0
 2     4    20.0
 2     5    20.0
 2     6    20.0
 2     7    20.0
 2     8    20.0
 2     9    20.0
 2    10    20.0
 2    11    20.0
 2    12    20.0
12    12    4.0
12    11    4.0
12    10    4.0
12     9    6.0
12     8    8.0
12     7    10.0
12     6    12.0
12     5    14.0
12     4    16.0
12     3    18.0
12     2    20.0
 1     1    20.0
 2     2    0.5       0.0       0.0       0.5
 3     2    0.5       0.0       0.5       1.0
 4     2    0.5       0.0       0.5       1.0
 5     2    0.5       0.0       0.5       1.0
 6     2    0.5       0.0       0.5       1.0
```

```
 7    2   0.5    0.0    0.5    1.0
 8    2   0.5    0.0    0.5    1.0
 9    2   0.5    0.0    0.5    1.0
10    2   0.5    0.0    0.5    1.0
11    2   0.5    0.0    0.5    1.0
12    2   0.0    0.0    0.5    0.5
 2   12   0.5    0.5    0.0    0.0
 3   12   0.5    1.0    0.5    0.0
 4   12   0.5    1.0    0.5    0.0
 5   12   0.5    1.0    0.5    0.0
 6   12   0.5    1.0    0.5    0.0
 7   12   0.5    1.0    0.5    0.0
 8   12   0.5    1.0    0.5    0.0
 9   12   0.5    1.0    0.5    0.0
10   12   0.5    1.0    0.5    0.0
11   12   0.5    1.0    0.5    0.0
12   12   0.0    0.5    0.5    0.0
 1    1   0.0    0.5    0.5    0.0
C
C
C
     END
```

Figure A4.1. Program for unconfined seepage

APPENDIX 5

Conversion Table to Metre–day System

Convert to / System of units	Length m	Area m²	Volume m³	Time d	Discharge rate m³/d	Permeability m/d	Transmissivity m²/d
cm–sec	cm 1.0×10^{-2}	cm² 1.0×10^{-4}	cm³ 1.0×10^{-6}	sec 1.157×10^{-5}	cm³/sec 8.64×10^{-2}	cm/sec 8.64×10^{2}	cm²/sec 8.64
litre–sec	m 1.0	m² 1.0	l 1.0×10^{-3}	sec 1.157×10^{-5}	l/sec 86.40	m/d 1.0	m²/d 1.0
ft–day	ft 0.3048	ft² 9.290×10^{-2}	ft³ 2.827×10^{-2}	day 1.0	ft³/d 2.827×10^{-2}	ft/d 0.3048	ft²/d 9.290×10^{-2}
Imp gal–day–ft	ft 0.3048	ft² 9.290×10^{-2}	Imp gal 4.546×10^{-3}	day 1.0	Imp gal/d 4.546×10^{-3}	Imp gal/d-ft² 4.893×10^{-2}	Imp gal/d-ft 1.491×10^{-2}
US gal–day–ft	ft 0.3048	ft² 9.290×10^{-2}	US gal 3.785×10^{-3}	day 1.0	US gal/d 3.785×10^{-3}	US gal/d-ft² 4.075×10^{-2}	US gal/d-ft 1.242×10^{-2}
US gal–min–ft	ft 0.3048	ft² 9.290×10^{-2}	US gal 3.785×10^{-3}	min 6.994×10^{-4}	US gal/min 5.450	US gal/min-ft² 58.67	US gal/min-ft 17.88

References

Allen, D. N. de G. (1954) *Relaxation methods in Engineering and Science.* McGraw-Hill, New York, 257 pp.

Aravin, V. I., and Numerov, S. N. (1965) *Theory of Fluid Flow in Undeformable Porous Media.* Israel Program for Scientific Translation, 511 pp.

Bear, J. (1972) *Dynamics of Fluids in Porous Media.* Elsevier, New York.

Birtles, A. B., and Reeves, M. J. (1977) 'Computer modelling of regional groundwater systems in the confined-unconfined flow regime'. *J. Hydrol.*, **34**, 97–127.

Boulton, N. S. (1951) 'The flow pattern near a gravity well in a uniform water bearing medium'. *J. Inst. Civ. Eng. (London)*, **36**, 534–50.

Boulton, N. S. (1963) 'Analysis of data from non-equilibrium pumping tests allowing for delayed yield from storage'. *Proc. Inst. Civ. Eng. (London)*, **26**, 469–82.

Bouwer, H. (1962) 'Analysing groundwater mounds by resistance network'. *J. Irrig. Drain. Div., Am. Soc. Civil Eng.*, **88**, IR3, 15–36.

Bouwer, H. (1978) *Groundwater Hydrology.* McGraw-Hill, 480 pp.

Bouwer, H., and Rice, R. C. (1968) 'A salt penetration technique for seepage measurement'. *J. Sanit. Eng. Div., Am Soc. Civil Eng.*, **96**, 59–74.

Briggs, J. E., and Dixon, T. N. (1968) 'Some practical considerations in the numerical solution of two-dimensional reservoir problems'. *Soc. Petroleum Eng. J.*, **8**, 185–94.

Browzin, B. F. (1961) 'Nonsteady-state flow in homogeneous earth dams after rapid drawdown'. *Proc. 5th Int. Conf. Soil Mechs. and Foundation Eng.*, Paris, Vol. II, 551–4.

Bruin, J., and Hudson, H. E. (1955) 'Selected methods for pumping test analysis'. *Illinois State Water Survey*, Report of Investigation, No. 25.

Brustkern, R. L., and Morel-Seytoux, H. J. (1975) 'Description of water and air movement during infiltration'. *J. Hydrol.*, **24**, 21–35.

Burgess, D. B., and Smith, E. J. (1978) 'The effects of over-abstraction on a groundwater system: the case of the southern Lincolnshire Limestone'. *Man's impact on the hydrological cycle*, Memoirs of Institute of British Geographers.

Campbell, M. D., and Lehr, J. H. (1973) *Water Well Technology.* McGraw-Hill, 681 pp.

Carslaw, H. S., and Jaeger, J. C. (1959) *Conduction of heat in solids*, 2nd ed. Clarendon Press, Oxford.

Casagrande, A. (1940) 'Seepage through dams'. *Boston Soc. Civ. Eng.*, 295–337.

Cedergren, H. R. (1977) *Seepage, drainage and flow nets*, 2nd ed Wiley, New York.

Charni, I. A. (1951) 'A rigorous derivation of Dupuit's formula for unconfined seepage with seepage surface'. *Dokl. Akad. Nauk. U.S.S.R.*, **79**, No. 6.

Cheng, R. T., and Li, C. T. (1973) 'On the solution of transient free surface flow problems in porous media by the finite element method'. *J. Hydrol.*, **20**, 49–63.

Childs, E. C. (1969) *The physical basis of soil water phenomena.* Wiley, London, 493 pp.

Collis-George, N. (1977) 'Infiltration equations for simple soil systems'. *Water Resourc. Res.*, **13**, 395–403.

Connorton, B. J., and Hanson, C. A. (1978) 'Regional modelling—analogue and digital approaches'. *Thames Groundwater Scheme, Inst. Civ. Eng. (London)*, 61–76.

Connorton, B. J., and Reed, R. N. (1978) 'A numerical model for the prediction of long-term well yield in an unconfined chalk aquifer'. *Quart. J. Eng. Geol.*, **11**, 127–30.

Cooper, H. H., and Jacob, C. E. (1946) 'A generalized graphical method for evaluating formation constants and summarizing well field history'. *Trans. Am. Geoph. Un.*, **27**, 526–34.

Darcy, H. (1856) 'Les fontaines publiques de la ville de Dijon'. *Dalmont, Paris*.

Davis, J. M. (1975) 'Two-dimensional groundwater flow'. *Water Research Centre (Medmenham, UK)* TR5, 64 pp.

De Brine, B. D. (1970) 'Electrolytic model study for collector wells under river beds'. *Water Resour. Res.*, **6**, 971–8.

De Marsily, G., Ledoux, E., Levassor, A., Poitrinal, D., and Salem, A. (1978) 'Modelling of large multilayered aquifer systems: theory and applications'. *J. Hydrol.*, **36**, 1–34.

Desai, C. S. (1973) 'Approximate solution for unconfined seepage'. *J. Irrig. Drain. Div., Am. Soc. Civil Eng.*, **99**, IR1, 71–87.

De Wiest, R. J. M. (1965) *Geohydrology*. Wiley, New York.

De Wiest, R. J. M. (Ed.) (1969) *Flow through Porous Media*. Academic Press, New York and London.

Domenico, P. A. (1972) *Concepts and Models in Groundwater Hydrology*. McGraw-Hill, 405 pp.

Downing, R. A., and Williams, B. P. J. (1969) *The Groundwater Hydrology of the Lincolnshire Limestone*. Water Resources Board.

Ehlig, C., and Halepaska, J. C. (1976) 'A numerical study of confined–unconfined aquifers including effects of delayed yield and leakage'. *Water Resour. Res.*, **12**, 1175–83.

Forchheimer, P. (1930) *Hydraulik*, 3rd ed. Teubner, Leipzig, Berlin.

Forsythe, G. E., and Wasow, W. R. (1960) *Finite-difference Methods for Partial Differential Equations*. Wiley, New York, 444 pp.

Fox, I. A., and Rushton, K. R. (1976) 'Rapid recharge in a limestone aquifer'. *Ground Water*, **14**, 21–7.

Gilboa, Y., Mero, F., and Mariano, I. B. (1976) 'The Botucatu aquifer of South America, model of an untapped continental aquifer'. *J. Hydrol.*, **29**, 165–79.

Grindley, J. (1967) 'The estimation of soil moisture deficits'. *The Met. Mag.*, **96**, no. 1137, 97–108.

Gupta, S. K., and Tanji, K. K. (1976) 'A three-dimensional Galerkin finite-element solution of flow through multiaquifers in Sutter Basin, California'. *Water Resour. Res.*, **12**, 155–62.

Hantush, M. S. (1966) 'Analysis of data from pumping tests in anisotropic aquifers'. *J. Geophys. Res.*, **71**, 421–6.

Harr, M. E. (1962) *Groundwater and Seepage*. McGraw-Hill, New York.

Headworth, H. G. (1970) 'The selection of root constants for the calculation of actual evaporation and infiltration for chalk catchments'. *J. Inst. Water Eng. (London)*, **24**, 431–46.

Hefez, E., Shamir, U., and Bear, J. (1975) 'Forecasting water levels in aquifers by numerical and semi-hybrid methods'. *Water Resour. Res.*, **11**, 988–92.

Herbert, R. (1965) 'An analogue study of a ground-water lowering scheme'. Tech. Paper 41, *Water Research Association, Medmenham, UK*.

Herbert, R. (1966) 'Analogue computers to study ground water flow'. Ph.D. thesis, University of Birmingham, England.

Herbert, R. (1968a) 'Seepage under sheet piles'. *Civil Eng. Public Works Review (London)*, 977–80.

Herbert, R. (1968b) 'Analysing pumping tests by resistance network analogue'. *Ground Water*, **6**, 12–18.

Herbert, R. (1970) 'Modelling partially penetrating rivers on aquifer models'. *Ground Water*, **8**, 29–36.

Herbert, R., and Rushton, K. R. (1966) 'Ground-water flow studies by resistance networks'. *Géotechnique*, **16**, 53–75.

Howard, K. W. F., and Lloyd, J. W. (1979) 'The sensitivity of parameters in the Penman evaporation equations and direct recharge balance'. *Journal of Hydrology*, **41**.

Huisman, L. (1972) *Groundwater Recovery*. Macmillan, London, 336 pp.

Jacob, C. E. (1950) 'Flow of ground water'. *Proc. Hydraul. Conf. 4th Iowa Inst. Hydraul. Res.*, **1949**, 321–86, edited by H. Rouse. Wiley, New York.

Karplus, W. J. (1958) *Analog Simulation*. McGraw-Hill, New York.

Kirkham, D. (1964) 'Exact theory for the shape of the free water surface above a well in a semiconfined aquifer'. *J. Geophys. Res.*, **69**, 2537–49.

Kitching, R. (1975) 'A mathematical model of the Akrotiri Plio-Pleistocene gravel aquifer, Cyprus'. *Inst. Geol. Sci.*, Report No. 75/2, HMSO London, 31 pp.

Kitching, R., Rushton, K. R., and Wilkinson, W. B. (1975) 'Groundwater yield estimation from models'. *Engineering Hydrology Today*, Inst. Civil Eng. (London), 101–12.

Kitching, R., Shearer, T. R., and Shedlock, S. L. (1977) 'Recharge to bunter sandstone determined from lysimeters'. *J. Hydrol.* **33**, 217–32.

Konikow, L. F. (1974) 'Modelling mass transport in a shallow aquifer'. *Trans. Am. Geoph. Un.*, **55**, 256.

Kruseman, G. P., and de Ridder, N. A. (1970) 'Analysis and evaluation of pumping test data'. *Intern. Inst. Land Reclam. and Improvement*, Wageningen, the Netherlands, Bull. 11, 200 pp.

Lamb, H. (1932) *Hydrodynamics*, 6th ed. Cambridge University Press, 533–5.

Liggett, J. A., 1977, 'Location of free surface in porous media'. *J. Hydraul. Div.*, *Am. Soc. Civil Eng.*, **103**, HY4, 353–65.

Lloyd, J. W., Rushton, K. R., and Taylor, H. R. (1978) 'Saline groundwater studies in the chalk of Northern Lincolnshire'. *Fifth Salt Water Intrusion Meeting, Intern. Hydrological Programme*, Medmenham, UK, 88–97.

Luzier, J. E., and Skrivan, J. A. (1975) 'Digital-simulation and projection of water-level declines in Basalt aquifers of the Odessa-Lind area, East-Central Washington'. *U.S. Geol. Survey*, Water Supply Paper 2036, 48 pp.

Macneal, R. H. (1953) 'An asymmetrical finite difference network'. *Quart. Appl. Math.*, **11**, 295–310.

McWhorter, D. B., and Sunada, D. K. (1977) *Ground-water Hydrology and Hydraulics*. Water Resources Publications, Fort Collins, 290 pp.

Muskat, M. (1937) *The Flow of Homogeneous Fluids through Porous Media*. McGraw-Hill, New York.

Nassif, S. H., and Wilson, E. M. (1975) 'The influence of slope and rain intensity on run-off and infiltration'. *Hydro. Sci. Bull.*, **20**, 4, 539–53.

Neuman, S. P. (1975) 'Role of subjective value judgement in parameter identification'. *Modelling and Simulation of Water Resources Systems*, North-Holland Publishing Co., 59–82.

Neuman, S. P., and Witherspoon, P. A. (1971) 'Analysis of nonsteady flow with a free surface using the finite element method'. *Water Resour. Res.*, **7**, 611–23.

Nutbrown, D. A., Downing, R. A., and Monkhouse, R. A. (1975) 'The use of a digital model in the management of the chalk aquifer in the South Downs'. *J. Hydrol.*, **27**, 127–42.

Oakes, D. B., and Pontin, J. M. A. (1976) 'Mathematical modelling of a chalk aquifer'. *Water Research Centre (Medmenham, UK)* TR24, 37 pp.

Oakes, D. B., and Skinner, A. C. (1975) 'The Lancashire conjunctive use scheme groundwater model' *Water Research Centre (Medmenham, UK)* TR12, 36 pp.

Palmer, P. J., and Redshaw, S. C. (1957) 'An electrical resistance analogue with a graded mesh'. *J. Sci. Instruments*, **34**, 407–9.

Papadopulos, I. S., and Cooper, H. H. (1967) 'Drawdown in a well of large diameter'. *Water Resour. Res.*, **3**, 241–4.

Peaceman, D. W., and Rachford, H. H. (1955) 'The numerical solution of parabolic and elliptic differential equations'. *J. Soc. Indust. Appl. Math.* **3**, 28–41.

Penman, H. L. (1949) 'The dependence of transpiration on weather and soil conditions'. *J. Soil. Sci.*, **1**, 74–89.

Pinder, G. F., and Bredehoeft, J. D. (1968) 'Application of digital computers for aquifer evaluation'. *Water Resour. Res.* **4**, 1069–93.

Pinder, G. F., and Gray, W. G. (1977) *Finite-element Simulation in Surface and Subsurface Hydrology*. Academic Press, New York, 295 pp.

Polubarinova-Kochina, P. Y. (1962) *Theory of Ground Water Movement*. Princeton University Press, 613 pp.

Prickett, T. A. (1965) 'Type-curve solution to aquifer test under water table conditions'. *Ground Water*, **3**, 5–14.

Prickett, T. A. (1967) 'Designing pumped well characteristics into electrical analogue models'. *Ground Water*, **5**, 38–46.

Prickett, T. A. (1975) 'Modelling techniques for groundwater evaluation'. *Advances in Hydroscience*, Vol. 10, Academic Press, New York, pp. 1–143.

Prickett, T. A., and Lonnquist, C. G. (1971) 'Selected digital computer techniques for groundwater resources evaluation'. *Illinois State Water Survey*, Bull. 55.

Redshaw, S. C. (1948) 'An electrical potential analyser'. *Proc. Inst. Mech. Eng. (London)*, **159**, 55–62.

Redshaw, S. C. (1959) 'An electrical analogue for the torsion of compound bars'. *2nd International Analogue Computation Meeting*, Strasbourg, Presses Academiques Européenes, 328–32.

Redshaw, S. C. (1967) 'The representation of a free surface on a pure resistance analogue computer'. *5th International Analogue Computation Meeting*, Lausanne, 585–93.

Reeves, M. J., Birtles, A. B., Courchee, R., and Aldrick, R. J. (1974) 'Groundwater resources of the Vale of York'. *Water Resources Board (Reading UK.)*, 90 pp.

Remson, I., Hornberger, G. M., and Molz, F. J. (1971) *Numerical Methods in Subsurface Hydrology*. Wiley-Interscience, New York.

Richardson, L. F. (1911) 'The approximate arithmetical solution by finite differences with an application to stresses in masonry dams'. *Phil. Trans. Roy. Soc. A.*, **210**, 307–57.

Rushton, K. R. (1964a) 'The representation of singularities in field problems on an analogue computer'. *Proc. 4th Int. Analogue Computation Meeting*, Brighton, 521–3.

Rushton, K. R. (1964b) 'Electrical analogue solution for the deformation of skew slabs'. Aeronautical Quarterly, **15**, 169–80.

Rushton, K. R. (1973) 'Discrete time steps in digital computer analysis of aquifers containing pumped wells'. *J. Hydrol.*, **18**, 1–19.

Rushton, K. R. (1974a) 'Critical analysis of the alternating direction implicit method of aquifer analysis'. *J. Hydrol.*, **21**, 153–72.

Rushton, K. R. (1974b) 'Aquifer analysis using backward difference methods'. *J. Hydrol.*, **22**, 253–69.

Rushton, K. R. (1975) 'Aquifer analysis of the Lincolnshire Limestone using mathematical models'. *J. Inst. Water Eng. (London)*, **29**, 373–89.

Rushton, K. R. (1978) 'Estimating transmissivity and storage coefficient from abstraction well data'. *Ground Water*, **16**, 81–5.

Rushton, K. R., and Ash, J. E. (1974) 'Groundwater modelling using interactive analogue and digital computers'. *Ground Water*, **12**, 296–300.

Rushton, K. R., and Ashe, J. E. (1975) 'Modelling of aquifer behaviour for long time periods using interactive analogue–digital computers'. *Int. Conf. of Mathematical Models for Environmental Problems, Southampton*, 115–27.

Rushton, K. R., and Bannister, R. G. (1970) 'Aquifer simulation on slow time resistance-capacitance networks'. *Ground Water*, **8**, 15–24.

Rushton, K. R., and Booth, S. J. (1976) 'Pumping test analysis using a discrete time–discrete space numerical method'. *J. Hydrol.*, **28**, 13–27.

Rushton, K. R., and Chan, Y. K. (1976) 'Pumping test analysis when parameters vary with depth'. *Ground Water*, **14**, 82–7.

Rushton, K. R., and Herbert, R. (1966) 'Groundwater flow studies by resistance network'. *Géotechnique*, **16**, 264–7.

Rushton, K. R., and Herbert, R. (1970) 'Resistance network for three-dimensional groundwater problems with examples of deep well dewatering'. *Proc. Instn. Civ. Eng. (London)*, **45**, 471–90.

Rushton, K. R., and Tomlinson, L. M. (1975) 'Numerical analysis of confined–unconfined aquifers'. *J. Hydrol.*, **25**, 259–74.

Rushton, K. R., and Tomlinson, L. M. (1977) 'Permissible mesh spacing in aquifer problems solved by finite differences'. *J. Hydrol.*, **34**, 63–76.

Rushton, K. R., and Turner, A. (1974) 'Numerical analysis of pumping from confined–unconfined aquifers'. *Water Resour. Bull.*, **10**, 1255–69.

Rushton, K. R. and Ward, C. (1979) 'The estimation of groundwater recharge'. *Journal of Hydrology*, **41**.

Rushton, K. R., and Wedderburn, L. A. (1971) 'Aquifers changing between the confined and unconfined state'. *Ground Water*, **9**, 30–9.

Rushton, K. R., and Wedderburn, L. A. (1973) 'Starting conditions for aquifer simulations'. *Ground Water*, **11**, 37–42.

Scott, R. F. (1963) *Principles of soil mechanics.* Addison-Wesley, 550 pp.

Skibitzke, H. E. (1960) 'Electronic computers as an aid to the analysis of hydrologic problems'. *Int. Assn. Sci. Hydrol.*, Publ. 52, 347–58.

Smith, G. D. (1965) *Numerical Solution of Partial Differential Equations.* Oxford University Press.

Smith, D. B., Wearn, P. L., Richards, H. J., and Rowe, P. C. (1970) 'Water movement in the unsaturated zone of high and low permeability strata by measuring natural tritium'. *Symposium on the use of isotopes in hydrology, I.A.E.A.*, Vienna, 73–87.

Sokolnikoff, I. S., and Sokolnikoff, E. S. (1941) *Higher Mathematics for Engineers and Physicists.* McGraw-Hill, New York.

Southwell, R. V. (1940) *Relaxation Methods in Engineering Science.* Clarendon Press, Oxford.

Southwell, R. V. (1946) *Relaxation Methods in Theoretical Physics.* Clarendon Press, Oxford, 248 pp.

Stallman, R. W. (1959) 'Analog Model Analyser—for steady state flow, operating instructions'. *U.S. Geol. Survey, Washington.*

Streltsova, T. D. (1974) 'Method of additional seepage resistances—theory and application': *J. Hydraul. Div., Am. Soc. Civil Eng.*, **100**, HY8, 1119–31.

Sukhija, B. S., and Shah, C. R. (1976) 'Conformity of groundwater recharge rate by tritium method and mathematical modelling'. *J. Hydrol.*, **30**, 167–78.

Tanaka, H. H., Hansen, A. J., and Skrivan, J. A. (1974) 'Digital-model study of groundwater hydrology, Columbia Basin Irrigation Project Area'. *State of Washington Water Supply Bulletin No. 40*, 60 pp.

Taylor, G. S., and Luthin, J. N. (1969) 'Computer methods for transient analysis of water table aquifers': *Water Resour. Res.*, **5**, 144–52.

Thames Conservancy (1967, 1968, 1970) 'Thames Conservancy river flow augmentation scheme'. *Water and Water Eng.*, March 1967, 100–102; April 1968, 141–44; August 1970, 332–6.

Thames Conservancy (1971) 'Report on the Lambourn Valley pilot scheme 1967–1969', *Thames Conservancy (Reading, UK)* 172 pp.

Thames Conservancy, 1972, 'Thames Conservancy groundwater scheme'. *Water and Water Eng.*, Dec. 1972, 452–3.

Theis, C. V. (1935) 'The relation between the lowering of the piezometric surface and the rate and duration of discharge of a well using groundwater storage'. *Trans. Am. Geoph. Un.*, **16**, 519–24.

Thomas, R. G. (1973) 'Groundwater models'. *Irrigation and Drainage Paper 21*, Food and Agricultural Organization, UN, Rome, 192 pp.

Todd, D. K. (1959) *Ground Water Hydrology*. Wiley, New York.

Todsen, M. (1971) 'On the solution of transient free-surface flow problems in porous media by finite difference methods'. *J. Hydrol.*, **12**, 177–210.

Tomlinson, L. M., and Rushton, K. R. (1975) 'The alternating direction explicit method for analysing groundwater flow'. *J. Hydrol.*, **27**, 267–74.

Trescott, P. C., Pinder, G. F., and Larson, S. P. (1976) 'Techniques of water-resource investigations of the United States Geological Survey'. *U.S. Geol. Survey*, Book 7, Chapter C1, 116 pp.

UNESCO, 1975, *Ground-water Studies*. The UNESCO Press, Paris.

Vemuri, V., and Karplus, W. J. (1969) 'Identification of non-linear parameters of groundwater basins by hybrid computation'. *Water Resour. Res.*, **5**, 172–85.

Volker, R. E. (1975) 'Solutions for unconfined non-Darcy seepage'. *J. Irrig. Drain Div.*, *Am. Soc. Civ. Engrs.*, **101**, 53–65.

Volker, R. E., and Guvanasen, V. (1977) 'Calibration of a model of a coastal aquifer in Northern Australia'. *Proc. Hydrol. Symp.*, *Inst. Eng. (Australia)*, Brisbane.

Walton, W. C. (1970) *Groundwater Resources Evaluation*. McGraw-Hill, New York.

Water Resources Board (1973) 'Artificial recharge of the London Basin. II, Electrical analogue model studies; III, Economic and engineering desk studies'. *Water Resources Board (Reading, U.K.)*

Young, R. A., and Bredehoeft, J. D. (1972) 'Digital computer simulation for solving management problems of conjunctive groundwater and surface water systems'. *Water Resour. Res.*, **8**, 533–56.

Zienkiewicz, O. C. (1977) *The Finite Element Method*, 3rd ed. McGraw-Hill, UK, 787 pp.

Index